2 0 0 7
STATE OF THE WORLD
Our Urban Future

Other Norton/Worldwatch Books

2 0 0 7
STATE OF THE WORLD

Our Urban Future

A Worldwatch Institute Report on
Progress Toward a Sustainable Society

Molly O'Meara Sheehan, *Project Director*

Zoë Chafe   Danielle Nierenberg
Christopher Flavin   Janice Perlman
Brian Halweil   Mark Roseland
Kristen Hughes   David Satterthwaite
Jeff Kenworthy   Janet Sawin
Kai Lee   Lena Soots
Lisa Mastny   Peter Stair
Gordon McGranahan   Carolyn Stephens
Peter Newman

Linda Starke, *Editor*

W·W·NORTON & COMPANY

NEW YORK   LONDON

The text of this book is composed in Galliard, with the display set in Gill Sans. Book design by Elizabeth Doherty;
cover design and composition by Lyle Rosbotham; manufacturing by Victor Graphics.

First Edition
ISBN 978-0-393-06557-2
ISBN 978-0-393-32923-0 (pbk)

W.W. Norton & Company, Inc., 500 Fifth Avenue, New York, N.Y. 10110
www.wwnorton.com

W.W. Norton & Company Ltd., Castle House, 75/76 Wells Street, London W1T 3QT

1 2 3 4 5 6 7 8 9 0

Acknowledgments

The Worldwatch Institute could not assemble a book as ambitious as *State of the World 2007: Our Urban Future* without an amazing global network. Over the past year, as we have sought to understand the disparate realities of our rapidly urbanizing world, we have relied on guidance and insights from every continent.

We owe much to the tremendous support and leadership of our Board of Directors. This group consists of Chairman Øystein Dahle, Vice Chair and Treasurer Tom Crain, Secretary Larry Minear, President Christopher Flavin, Geeta B. Aiyer, Adam Albright, L. Russell Bennett, Cathy Crain, James Dehlsen, Robert Friese, Lynne Gallagher, Satu Hassi, Jerre Hitz, Nancy Hitz, John McBride, Akio Morishima, Izaak van Melle, Wren Wirth, and Emeritus members Abderrahman Khene and Andrew E. Rice. In 2006, the Worldwatch Board of Directors named Worldwatch's conference room for Andy Rice for his years of thoughtful leadership.

State of the World would not exist were it not for the generous financial contributions of our many supporters. More than 3,500 Friends of Worldwatch fund nearly one third of the Institute's operating budget.

This *State of the World* report is part of a larger Worldwatch project analyzing the historic transition to a world in which most people live in urban areas. We greatly appreciate the funds provided for this venture by the United Nations Population Fund and the Winslow Foundation.

In addition, Worldwatch's research program is backed by a roster of organizations. We thank the following for their generous support over the last year: Blue Moon Fund, Chicago Community Trust, Energy Future Coalition and Better World Fund, Ford Foundation, Goldman Environmental Prize, W. K. Kellogg Foundation, Marianists of the USA, Noble Venture Gift Fund of the Community Foundation Serving Boulder County, Natural Resources Defense Council, Prentice Foundation, V. Kann Rasmussen Foundation, Rockefeller Brothers Fund, Shared Earth Foundation, Shenandoah Foundation, Wallace Genetic Foundation, Wallace Global Fund, The Johanette Wallerstein Institute, and the Governments of Germany and Norway.

We are also indebted to our international network of publishing partners, who bring State of the World to a global audience. They provide advice, translation, outreach, and distribution assistance. We give special thanks to Univeridade da Mata Atlântica in Brazil; Global Environmental Institute in China; Oy Yliopistokustannus University Press in Finland; Germanwatch, Heinrich Böll Foundation, and Westfälisches Dampfboot in Germany; Evonymos Ecological Library in Greece; Earth Day Foundation in Hungary; Winrock International in India; World Wide Fund for Nature and Edizioni Ambiente in

Italy; Worldwatch Japan; Worldwatch Norden at IVL Swedish Environmental Research Institute in Scandinavia; Center of Theoretical Analysis of Environmental Problems and International Independent University of Environmental and Political Sciences in Russia; Korean Federation for Environmental Movement in South Korea; Politika Newspapers in Serbia; Centre UNESCO de Catalunya for the Catalan version and Fundacion Hogar del Empleado and Editorial Icaria for the Castilian version in Spain; L'État De La Planète in Switzerland, which also connects us to France and French-speaking Canada; Taiwanwatch; Turkiye Erozyonla Mucadele, Agaclandima ve Dogal Varliklari Koruma Vakfi (TEMA) in Turkey; and Earthscan/James & James in the United Kingdom.

Worldwatch's longest publishing relationship is with W. W. Norton & Company in New York. Thanks to their team—especially Amy Cherry, Leo Wiegman, Nancy Palmquist, Lydia Fitzpatrick, and Anna Oler—*State of the World*, *Vital Signs*, and other Worldwatch books make it into bookstores and classrooms across the United States.

Authors of this year's *State of the World* benefited from a distinguished international panel of reviewers who took time from hectic schedules to read draft chapters. For their penetrating comments, we thank Donald Aitken, Madhavi Malalgoda Ariyabandu, Xuemei Bai, John Byrne, Anne Carlin, Olufunke Cofie, Jason Corburn, Rob de Jong, Pay Drechsel, Sandro Galea, Peter V. Hall, Walter Hook, Paul Kerz, Peter Kimm, Günter Langergraber, Kai Lee, Michael Levenston, Jan Lundqvist, Sean Markey, Eric Martinot, Barjor Mehta, Michaela Oldfield, Mark Pelling, Kami Pothukuchi, Susan Roaf, Tom Roper, David Satterthwaite, Alan Silberman, Jac Smit, Eva Sternfeld, John Twigg, Timeyin Uwejamomere, Christine Wamsler, and several anonymous reviewers.

Contributors to *State of the World* also appreciate the information and guidance given by prominent and knowledgeable individuals from around the world. We were fortunate to receive help from Sarika Agrawal, Eva Anisko, Eduardo Athayde, Carine Barbier, Sheridan Bartlett, Timothy Beatley, Marc Berthold, Susan Blaustein, Sarah Brachle, Jeb Bruggman, Yves Cabannes, Majora Carter, Kiran Chhokar, Toshiko Chiba, Billy Cobbett, Penny Cuff, Glenn D'Alessio, Carlton Eley, Gordon Feller, Greg Franta, Dan Goodman, Rajat Gupta, William Holmberg, Tim Honey, Dan Hoornweg, Me'An Ignacio, Tetsunari Iida, Min Jin, Dan Kammen, Gavin Killip, Bowdin King, Mike Kossey, Benoit Lambert, Frannie Léautier, Peter Marcotullio, Dale Medearis, Richard Munson, Nguyen Le Quang, Yoshi Nojima, Soki Oda, David Painter, Scott Paul, Richard Perez, Blair Ruble, Mona Serageldin, Jutta Schmieder, Parin Shah, Jacob Songsore, Freyr Sverrisson, Kaarin Taipale, Carmelle J. Terborgh, Thi Le Thi Minh, Ibrahim Togola, John Tomlinson, Masami Toyofuku, John Waugh, Marc Weiss, Elizabeth Westrate, Jorge Wilheim, Angelika Wirtz, and Kurt Yeager.

We are particularly grateful for overall project support volunteered by Jill Greaney, a former lawyer and student of urban planning with keen analytical skills and a passion for the urban environment. In addition to carefully reviewing the first draft of the manuscript, Jill scoured major newspapers and the Internet to find relevant articles, drafted text on urban development, and translated documents from French.

For this edition of *State of the World*, we enlisted a record number of gifted scholars and leading thinkers on urban issues from outside the Institute. Kai Lee of Williams College in Massachusetts wrote and revised our introductory chapter with alacrity and also helped focus the wide-ranging discussions of other chapters. David Satterthwaite of the International Institute for Environment and Development in London gave welcome guidance to the project as a whole, and with his colleague Gordon McGranahan contributed the chapter on water and sanitation. Peter Newman of Murdoch University in Australia endured late-night conference calls that bridged a 12-hour time difference to discuss the chapter on transportation he prepared with colleague Jeff Kenworthy. Kristen Hughes at the University of Delaware wrote part of the chapter on energy. Carolyn Stephens, who teaches at the London School of Hygiene & Tropical Medicine and the Federal University of Paraná in Brazil, contributed the chapter on public health. Her coauthor Peter Stair, a former MAP Fellow at Worldwatch, is now at the University of California in Berkeley. Mark Roseland of Simon Fraser University in Vancouver, with the assistance of Lena Soots, contributed the chapter on local economies and helped organize a discussion at the World Urban Forum in Vancouver in June 2006, where *State of the World* authors received feedback on their outlines from an international audience. Janice Perlman, founder and president of the Mega-Cities Project in New York, took time away from writing a book on Rio de Janeiro to prepare the chapter on urban poverty.

We thank the wonderful group of academics, journalists, and urbanists who contributed the two-page stories on individual cities that appear between each chapter. Charlie Benjamin of Williams College collaborated with colleagues in Mali, Aly Bocoum of the Near East Foundation and Aly Bacha Konaté of Réseau GDRN5, to write the Timbuktu piece. The Lagos story was contributed by Ayodeji Olukoju of the University of Lagos. Rob Crauderueff of Sustainable South Bronx wrote about Loja, Ecuador, where he once lived. Thomas Winnebah of Njala University in Sierra Leone and Olufunke Cofie of the International Water Management Institute in Accra, Ghana, put together the article on Freetown. Dana Cuff of the University of California at Los Angeles wrote about her hometown. Xuemei Bai of the Commonwealth Scientific and Industrial Research Organization in Australia prepared the story about Rizhao, China. Tom Roper, a former Minister in the Victoria Government in Australia, wrote the Melbourne piece. Ivana Kildsgaard of IVL Swedish Environmental Research Institute gave us the Malmö story. Biko Nagara of Stanford University wrote about Jakarta, the city of his birth. Kalpana Sharma, Deputy Editor of *The Hindu*, corresponded from Mumbai. Architect Eva Staňková of the Vaňkovka Civic Association sent us the story about Brno. Dana Firas, a Jordanian author on sustainable development, wrote about Petra. And Rasna Warah, a freelance writer in Kenya and editor of UN-HABITAT's *State of the World's Cities 2006/07*, sent us the story from Nairobi.

During the summer of 2006, the *State of the World* team was fortified by a crew of exceptionally talented research assistants and interns. Kai Lee, author of Chapter 1, recruited Paaven Thaker to become one of Worldwatch's youngest and most enthusiastic interns. His assistant at Williams College, Fathimath Musthaq, found additional data for the first chapter. Biko Nagara helped research Chapter 2 and commissioned the Brno story. Dana Artz lent some of her prodigious energy to Chapter 3. Monideepa Talukdar, working from her university in Louisiana, aided the Australia-based authors of Chapter 4. Hanna Vovk tracked down information for Chapter 5, as did Stephanie Kung, who joined Worldwatch as a Stanford MAP Fellow in 2006. Angela Choe ferreted out books and data for Chapter 6. Corey Tazzara used his intimate knowledge of Georgetown University

Library to bolster Chapter 7 and the Year in Review timeline. Working from Canada, Candace Bonfield helped with Chapter 8. Chapter 9 owes much to the dedication and speedy work of Kenro Kawarazaki, whose thorough research also turned up useful information for Chapters 1 and 4. Matt Friese jumpstarted research for the timeline, and Mark Friese assisted with our Campus Greening Initiative.

In the fall of 2006, more interns enlisted to fine-tune the electronic version of this book as well as the companion website, www.worldwatch.org/urban. Patrick Cyrus Gilman and Semiha Caliskan took time out from their studies to locate satellite images of many of the cities featured in this book on Google Earth and other online sources. Neelam Singh, who aided us with the previous edition, returned to help us produce web content.

Buoyed by these marvelous funders, advisers, volunteers, and colleagues, the Worldwatch staff brings dedication to *State of the World*. The Institute would not be able to function without Director of Finance and Administration Barbara Fallin, who has kept the office running smoothly for nearly 18 years. Like Barbara, Joseph Gravely joined the Institute in 1989 and quickly became integral to its daily operation, taking charge of mail and publication fulfillment. After many years of keeping information flowing between Worldwatch and the world, Joseph retired in December 2006. Among the other notable changes on the Worldwatch staff this year, Librarian Lori Brown left to work full-time on her organic farm after 13 years of unearthing much of the data used in the *State of the World* series.

Early discussions of this book were enriched by the participation of many staff members, including Vice President Georgia Sullivan, who has inspired the Institute with her vigorous leadership. Patricia Shyne, Direc-

tor of Publications and Marketing, moved swiftly to put authors in touch with Worldwatch's international partners. Tom Prugh, editor of *World Watch* magazine, weighed in on Chapters 4 and 5. Amid many other activities, Darcey Rakestraw, Communications Manager, helped organize a roundtable discussion with UN-HABITAT Executive Director Anna Tibaijuka. Research Associate Erik Assadourian not only secured the International Student House in Washington, D.C., for meetings that included discussion of this book, he also played a role in recruiting interns and linking the Institute to international partners.

Our development team, which maintains Worldwatch's ties to its supporters, also actively shaped this book. We welcomed the creativity of Mary Redfern, Manager of Foundation Relations. Courtney Berner, Friends of Worldwatch Program Manager, researched the achievements of local governments led by forward-thinking mayors. We were also boosted by Laura Parr, Development Assistant and Assistant to Worldwatch's President, and Drew Wilkins, Administrative Assistant and jack of all trades. Before leaving in the summer of 2006, Director of Development John Holman helped build the Institute's base of support.

New staff members lifted morale. Alana Herro joined us as the Staff Writer for e2 (eye on the earth), our news service launched in 2006. Just as this book was going to press, we welcomed Ling Li as Worldwatch's new China Fellow and Ali Jost as interim Communications Manager.

This edition of *State of the World* owes much to Research Director Gary Gardner. Although Gary's recent book, *Inspiring Progress*, kept him from writing a chapter in this volume—a first in his 12 years at the Institute—he provided careful reviews of many chapters, Spanish translation when needed, and even some last-minute research assistance.

Other researchers pitched in to improve the book. Hilary French, Senior Advisor for Programs, shared thoughts on the project, reviewed several chapters, and forged partnerships with reviewers and other experts. Senior Researcher Michael Renner's probing questions strengthened Chapters 4, 6, 8, and 9. Yingling Liu, China Program Manager, commented on chapters and connected us to Chinese sources of information, as did China Fellow Zijung Li, who joined the World Bank when she completed her fellowship in September 2006. Suzanne Hunt, Biofuels Program Manager, sharpened authors' arguments with her incisive comments on Chapters 4 and 5. Before leaving in the fall for the International Resources Group, Biofuels Project Assistant Lauren Sorkin shared her contacts in European cities. Lisa Mastny took time away from editing *World Watch* Magazine and other projects to compile the timeline in this book.

Beyond contributing research and writing to Chapter 7, Research Assistant Peter Stair bolstered the entire book by recruiting interns, organizing discussions with urban experts, and working with Web Manager Steve Conklin to create an in-house Web site for authors to share information. Peter, who joined Worldwatch as a MAP Fellow in 2005, is now pursuing his interests in urban planning and public health in Berkeley.

As always, we are indebted to independent editor Linda Starke, who cleaned up the rough drafts of our far-flung correspondents with breathtaking speed. Since 1983,

Linda has devoted her autumn days—and nights and weekends—to *State of the World*. Matching Linda's pace was Worldwatch's Art Director Lyle Rosbotham, who rapidly turned the manuscript into eye-catching page proofs.

For many years, Worldwatch was aided by Magnar Norderhaug who founded and led our Scandinavian affiliate Worldwatch Norden. He crafted Worldwatch research into op-eds, letters to the editor, and articles, skillfully linking Institute research to the issues of the day, and helping to make Worldwatch Norden an authoritative voice on environmental sustainability in Scandinavia. Magnar passed away this year after a long illness. We will miss him greatly.

The circle of life teaches us that loss is not the last word. Last year, we noted the arrival of Finnían Freyson Sawin, whom we thank for the joy he has brought us during visits to the office and the sacrifices he endured so his mother could contribute to this book. We now welcome Amel Rakestraw Benhamouda, born in September 2006 to Darcey Rakestraw and Atef Benhamouda. Her tiny face reminds us of our hopes for a healthy, peaceful, and equitable future.

Molly O'Meara Sheehan
Project Director

Worldwatch Institute
1776 Massachusetts Ave., N.W.
Washington, DC 20036
worldwatch@worldwatch.org
www.worldwatch.org

Contents

List of Boxes, Tables, and Figures

Boxes

Units of measure throughout this book are metric unless common usage dictates otherwise.

Foreword

Anna Tibaijuka
Executive Director, UN-HABITAT

When I first came to UN-HABITAT with a background in agricultural economics and international trade negotiations, I brought my own set of professional and personal prejudices. Like many other development theorists, I felt that although urban development was important, rural development was the first priority. Like many people of my generation in Africa and around the world, I thought of urban areas as a necessary evil. Though they were economic centers, cities led to overcrowding, pollution, and, inevitably, slums.

I had given little thought to the possibilities, even less to the problems and process of urbanization. However, in the years since I became Executive Director of UN-HABITAT I have traveled far and wide. I have experienced firsthand the appalling results of rapid chaotic urbanization.

In city after city, I have been stranded in traffic jams; I have visited men in hospitals suffering from preventable diseases caused by industrial pollution; I have seen slum dwellers living in conditions that do not bear describing and met young women who were raped on their way to the closest public toilet shared by over 500 people; I have walked through flattened terrain that once housed whole communities destroyed by floods and other natural disasters.

Whereas in 1950 New York and Tokyo were the only cities with more than 10 million people, today there are 20 megacities, most of which are in the developing world. As cities sprawl, turning into unmanageable megalopolises, their expanding footprint can be seen from space. These hotbeds of pollution are a major contributor to climate change.

Though urbanization has stabilized in the Americas and Europe, with about 75 percent of the population living in urban areas, Africa and Asia are in for major demographic shifts. Only about 35 percent of their populations are urban, but it is predicted that this figure will jump to 50 percent by 2030. The result is already there for all to see: chaotic cities surrounded by slums and squatter settlements.

Of the 3 billion urban dwellers today, it is estimated that 1 billion are slum dwellers. What is worse, if we continue with business as usual that figure is set to double by 2030.

If ever there was a time to act, it is now. Though cities are important engines of growth and provide economies of scale in the provision of services, most of them are environmentally unsustainable. In addition, in this age of increasing insecurity, with more than 50 percent of their residents living in slums without adequate shelter or basic services, many cities are rapidly becoming socially unsustainable.

The U.N. General Assembly first explicitly cited its concern at the "deplorable world housing situation" in 1969, and it declared human settlements a priority for the twenty-fifth anniversary of the United Nations in 1971. The next year, the first U.N. conference on the human environment, in Stockholm, marked a conceptual shift from global environmental degradation to its causes—largely urbanization and the impact of human settlements.

In 1977, the Secretary-General of the first U.N. Human Settlements Conference (Habitat I), Enrique Peñalosa, asked "whether urban growth would continue to be a spontaneous chaotic process or be planned to meet the needs of the community." Yet the urban agenda never received the full attention it deserved. For decades now, donors have given priority to rural development. The Human Settlements Foundation, established at the same time as UN-HABITAT to fund slum upgrading, was never financed. Perhaps this was because in 1977, only one third of the world lived in urban areas.

Today, urbanization is being taken increasingly seriously. In 1996, at Habitat II, 171 countries signed the Habitat Agenda, a comprehensive guide to inclusive and participatory urban development. In 2000, concerned about the number of people who were being marginalized by the rapidly globalizing economy, world leaders committed themselves to the Millennium Development Goals. Many of these address the living conditions of the urban poor, in particular Targets 9 and 11 within Goal 7 on environmental sustainability. In 2001, the General Assembly passed a resolution that promoted UN-HABITAT from a center into a full-fledged U.N. program and called on UN-HABITAT to establish the World Urban Forum as a think tank on all things urban.

With more than 10,000 delegates, the third session of the World Urban Forum, in Vancouver in 2006, proved that people are increasingly concerned about the future of human settlements. Ministers and mayors,

industrialists and slum dwellers, all recognized that their combined efforts are required to overcome the urban crisis.

As we struggle to change our cities, authors and journalists are ever more critical. Charles Dickens, Emile Zola, Jacob Riis, and Edward Mayhew were instrumental in improving the urban policies of their day. Today, researchers and authors of reports like this *State of the World 2007* help sensitize the larger public to the major issues of our time.

Surprisingly, there was no commonly agreed-upon definition of slums until 2003, when the United Nations published *Global Report on Human Settlements: The Challenge of Slums*. Where there was a lack of information about urban indicators, there is now a network of Global Urban Observatories. The World Bank, with UN-HABITAT, has established the Cities Alliance that coordinates donor activity in urban areas, particularly in slum upgrading. The United Nations has also launched major campaigns to promote security of tenure and better urban governance.

The political machinery is finally beginning to recognize urbanization. In 2006, the United States Senate held it first hearing on African urbanization, while the British Parliament held its first debate on urbanization in developing countries. United Cities and Local Governments, founded in 2004, has become a legitimate partner in the international arena.

These kinds of international, regional, and local political institutions help create legitimacy for change; more important, they provide a locus for interventions. If our campaigns of advocacy and awareness do not translate into action, we will have failed.

There are signs of hope. There are more and more best practices showing what measures can be taken to improve housing conditions for the urban poor while enforcing environmental laws. Many cities in Southeast and South Asia, in particular, are beginning to reduce the share of their people living in urban poverty. Though all Habitat Agenda partners have contributed to this improvement, it has been spearheaded by central governments and local authorities. Their political will has spurred increased investment in making cities and towns sustainable.

As an African, living in the world's fastest urbanizing continent, I am aware that we need to persuade everyone—from presidents to ordinary policymakers—of the urgency of urban issues. The Commission for Africa, of which I was a member, highlighted urbanization as the second greatest challenge confronting the continent after HIV/AIDS. As we move into the urban age, we have to change how we see the world, how we describe it, and how we act in it.

Fortunately, the leaders of Africa have taken note. At the Maputo Summit in 2003, the African Heads of State adopted Decision 29 reiterating their commitment to sustainable urbanization, an agenda that was subsequently encouraged by Joaquim Chissano during his term as President of the African Union. In Nigeria, concerned about the country's urban problems, President Olusegun Obasanjo personally set up the Ministry of Housing and Urban Development. In his inaugural address in 2006, President Jakaya Kikwete of Tanzania emphasized the need for well-managed cities as a basis for national development. To coordinate urban issues at the regional level, African ministers recently established the African Ministerial Conference on Housing and Development. At the same time, AFRICITIES has been at the forefront of organizing local authorities on the continent.

This is just the beginning. As I walk through the slums of Africa, I find it hard to witness children suffering under what can only be described as an urban penalty. I am astonished at how women manage to raise their families under such appalling circumstances, without water or a decent toilet. The promise of independence has given way to the harsh realities of urban living mainly because too many of us were ill prepared for our urban future. Many cities are confronting not only the problems of urban poverty, but the very worst of environmental pollution. From Banda Aceh to New Orleans, whole communities are being wiped out through no fault of the innocent victims.

We will, all of us, bear the responsibility of a world gone wrong. If we continue as usual, a disastrous future beckons: whole cities swamped by slums, whole societies destroyed by climate change.

Working at UN-HABITAT and with other agencies worldwide, I hope that together we can correct the past failures of urban planning. I hope that the work of organizations like the Worldwatch Institute will motivate more people to take up the cause of environmentally and socially sustainable cities. We are warned, it cannot be business as usual.

Foreword

The Honorable Jaime Lerner
Former Governor of Paraná, Brazil, and former Mayor of Curitiba

The twentieth century was, *par excellence*, the century of urbanization. Around the world the supremacy of rural populations over urban ones was reversed and cities experienced an accelerated growth, often beyond the desirable. They have been through unthinkable transformations, which left a fantastic array of challenges and possibilities as a legacy.

If the last century was the century of urbanization, the twenty-first will be the century of cities. It is in the cities that decisive battles for the quality of life will be fought, and their outcomes will have a defining effect on the planet's environment and on human relations.

Therefore, what can we expect from an urban planet? What will the cities of the future be like? There are those who portray an urban world in apocalyptic colors, who depict cities as hopeless places where a person cannot breathe, move, or live properly due to excess population and automobiles. I, however, do not share these views. My professional experience has taught me that *cities are not problems, they are solutions*. So I can face an urban world only with optimism.

My strongest hope resides in the speed of transformation. For instance, the demographic projections based on the high birth rates of 20–30 years ago have not been confirmed, allowing us a more encouraging view on the growth of cities for the next years and decades. Renewable energy sources, less-polluting automobiles, new forms of public transportation, and communication technologies that reduce the need for travel are all pushing away the chaos that was predicted for large urban centers. The evolution of technology and its democratization are presenting new perspectives for cities of all sizes and shapes.

In terms of physical configuration, the cities of the future will not differ significantly from the ones of yesterday and today. What will differentiate the good city will be its capacity for reconciling its residents with nature. Socially just and environmentally sound cities—that is the quest!

By having to deal directly with economic and environmental issues, this quest will foster an increasingly positive synergy between cities, regions, and countries. As a consequence, it will motivate new planetary pacts focused on human development.

Still, a certain sense of urgency is vital to positively transform our cities. The idea that action should only be taken after having all the answers and all the resources is a sure recipe for paralysis. The lack of resources cannot be an excuse not to act. The planning of a city is a process that allows for corrections, always. It is supremely arrogant to believe that planning can be done only after figuring out every possible variable.

To innovate is to start! Hence, it is necessary to begin the process. Imagine the ideal, but do what is possible today. Solutions for 20, 30 years ahead are pointless, because by then the problems will probably be different. Therefore we need urban policies that can generate change beginning now, that will not need decades to show results. The present belongs to us and it is our responsibility to open paths.

In the roots of a big transformation there is a small transformation. Start creating from simple elements, easy to be implemented, and those will be the embryos of a more complex system in the future. Although we are living a phase of our history when events happen at a galloping pace, and information travels in the blink of an eye, the decisions regarding urban problems are postponed due to a systematic lack of synchrony with the speed of the events.

The world demands increasingly fast solutions, and it is the local level that can provide the quickest replies. But it is necessary to plan to make it happen. Plan for the people and not for centralized and centralizing bureaucratic structures.

Those responsible for managing this urban world must have their eyes on the future, but their feet firmly on the ground in the present. Those who only focus on the daily needs of people will jeopardize the future of their city. On the other hand, those who think only about the future, disregarding the daily demands, will lose the essential support of their constituents and will not accomplish anything.

It is necessary then not to lose track of the essence of things; to discern within the amazing variety of today's available information what is fundamental and what is important, the strategic from the daily demands. A clear perspective on future objectives is the best guide for present action—that is, to bind the present with a future idea.

There are three crucial issues that need to be addressed: mobility, sustainability, and identity.

For mobility, the future is on the surface. Entire generations cannot be sacrificed waiting for a subway line while in less than two years complete networks of surface transportation can be set up. In Curitiba, starting in 1974 we gave priority to public buses carrying 25,000 passengers a day in exclusive lanes on a north-south axis. Today, the network carries 2 million passengers throughout the metro area with a single fare.

The key to mobility is the combination and integration of all systems: subway, bus, taxi, cars, and bikes. But these systems cannot compete in the same space. People will select the most convenient combination according to their own needs and travel with a "mobility card." Operators of each transportation mode will be partners in the system.

Regarding sustainability, the main idea is to focus on what we know instead of what we don't know. And, above all, to transfer this knowledge to the children, who will then teach their parents. Curitiba's Garbage That Is Not Garbage Program encouraged separation of recyclable waste in households; children learned about the program at school and helped mobilize their parents.

Simple things from the day-by-day routine of cities can be decoded for children: for instance, how each person can help by reducing the use of the automobile, living closer to work or bringing the work closer to home, giving multiple functions during the 24 hours of the day to urban infrastructure, saving the maximum and wasting the minimum.

Sustainability is an equation between what is saved and what is wasted. Therefore, if sustainability=saving/wasting, when wasting is "zero," sustainability tends to infinity. Waste is the most abundant source of energy.

A sustainable city cannot afford the luxury of leaving districts and streets with good infrastructure and services vacant. Its downtown area cannot remain idle during great portions of the day. It is necessary to fill it up with the functions that are missing. The "24 hours city" and multiple-use equipment are essential for sustainability.

Finally, identity. Identity is a major factor in the quality of life; it represents the synthesis of the relationship between the individual and his or her city. Identity, self-esteem, a feeling of belonging—all of them are closely connected to the points of reference that people have about their own city.

Rivers, for instance, are important references. Instead of hiding them from view or burying them in concrete, cities should establish riverbanks as valuable territories. By respecting the natural drainage characteristics, cities can make sure the preserved areas provide necessary episodic flooding relief channels and are still used most of the time for recreation in an economic and environmentally friendly way. Parks can work within a similar logic, providing areas that people can relate to and interact with.

Historic districts are also major reference points, closely related to each city since its inception. But these areas often suffer a process of devaluation and degradation. Finding ways to keep these districts alive by connecting identity elements, recycling outdated uses, and hosting a mix of functions is vital. In Curitiba, a deactivated gunpowder storage facility was transformed into one of the city's most cherished theaters—Teatro do Paiol.

A city is a collective dream. To build this dream is vital. Without it, there will not be the essential involvement of its inhabitants. Therefore, those responsible for the destinies of the city need to draw scenarios clearly—scenarios that are desired by the majority, capable of motivating the efforts of an entire generation.

A city is a structure of change even more than it is a model of planning, an instrument of economic policies, a nucleus of social polarization. The soul of a city—the strength that makes it breathe, exist, and progress—resides in each one of its residents.

Cities are the refuge of solidarity. They can be the safeguards of the inhumane consequences of the globalization process. They can defend us from extraterritoriality and the lack of identity.

On the other hand, the fiercest wars are happening in cities, in their marginalized peripheries, in the clash between wealthy enclaves and deprived ghettos. The heaviest environmental burdens are being generated there too, due to our lack of empathy for present and future generations. And this is exactly why it is in our cities that we can make the most progress toward a more peaceful and balanced planet, so we can look at an urban world with optimism instead of fear.

Jaime Lerner

Preface

Christopher Flavin
President, Worldwatch Institute

Sometime in 2008, the world will cross an invisible but momentous milestone: the point at which more than half the people on the planet—roughly 3.2 billion human beings—live in cities. The combined impact of a growing population and an unprecedented wave of migration from the countryside means that over 50 million people—equivalent to the population of France—are now added to the world's cities and suburbs each year. More than at any time in history, the future of humanity, our economy, and the planet that supports us will be determined in the world's cities.

Urban centers are hubs simultaneously of breathtaking artistic innovation and some of the world's most abject and disgraceful poverty. They are the dynamos of the world economy but also the breeding grounds for alienation, religious extremism, and other sources of local and global insecurity. Cities are now both pioneers of groundbreaking environmental policies and the direct or indirect source of most of the world's resource destruction and pollution.

This modern "tale of two cities," to borrow the title of Charles Dickens' famously grim book about nineteenth-century London, is something that every policymaker and citizen needs to understand. The battles against our greatest global problems, from unemployment and HIV infections to water shortages, terrorism, and climate change, will be largely won—or lost—in the world's cities.

Although our species existed for over 100,000 years before the first small cities were built between the Tigris and Euphrates Rivers around 4000 BC, the growing dominance of cities is one of the most dramatic changes we have experienced and one for which we are poorly equipped. As recently as the early twentieth century, the vast majority of the world's people lived in the countryside and practiced subsistence farming. Even today, the electoral systems of many predominantly urban countries—Japan is a good example—give disproportionate political influence to rural citizens. And the international development community often neglects cities when allocating its aid.

In 1950, only New York and Tokyo had populations of more than 10 million. Today there are 20 of these so-called megacities, the bulk of them in Asia and Latin America. But most of the growth in the decades ahead will come in smaller cities. By 2015, demographers project there will be 59 cities with populations between 1 million and 5 million in Africa, 65 such cities in Latin America and the Caribbean, and 253 in Asia. As early as 2030, four out of five of the world's urban residents will be in what we now call the "developing" world.

The demographic and political impacts of this transformation will test us. In China, for example, millions of people are moving to cities each year, and while that nation has done better than most in meeting the needs of new urban residents, the social strains are showing. And Africa, the least urban continent today, is the area that is urbanizing the fastest—a trend that will undoubtedly put additional social, economic, and political pressure on this already stressed part of the world.

The great majority of the population growth in the new urban centers of Africa and Asia is in the unplanned and underserved settlements commonly known as slums. Over one quarter of urban residents in the developing world—more than half a billion people—lack clean water and sanitation, and 1.6 million die each year as a result. The face of twenty-first century cities is often that of a small, malnourished child living in a vast slum in a city such as Abidjan, Kolkata, or Mexico City, not far from the newly built opera houses, gleaming office buildings, and automobile-choked highways that are now common even in poor countries.

This child frequently lacks electricity, clean water, or even a nearby toilet. While air quality has improved markedly in many European and American cities in recent years, it has become far worse in most cities in the developing world; China alone has 16 of the world's most polluted cities. For that child in the slum, pollution-related sickness and violence are daily threats, while education and health care are a distant hope.

Our ability to meet the needs of the urban poor is one of the greatest humanitarian challenges of this century. It is also going to shape key global developments—from the security of those who live in nearby luxury apartments to the stability of Arctic ice sheets near the planet's poles. It is particularly ironic that the battle to save the world's remaining healthy ecosystems will be won or lost not in the tropical forests or coral reefs that are threatened but on the streets of the most unnatural landscapes on the planet.

At stake is the ability of those ecosystems to provide the food, fiber, fresh water, and climate stability that all cities depend on. Nearly two thirds of these "ecosystem services" have already been degraded, according to the latest scientific estimates. Our challenge is to avoid the fate of the great Mayan cities that lie in ruins in the jungles of southern Mexico and Guatemala—cities that were abandoned not just because of forces at work within their borders but because of the collapse of the surrounding agricultural lands and water resources after centuries of overexploitation.

The task of saving the world's modern cities might seem equally hopeless—except that it is already happening. This book documents the problems facing the world's cities, but also a remarkable array of promising advances that have begun to mushroom over the past few years. Particularly striking is the self-reliance being demonstrated by both rich and poor communities that have stepped in to fill gaps left by governments. Even necessities such as food and energy are increasingly being produced by urban pioneers inside city limits.

In Accra, at least 1,000 urban farmers grow food in backyard plots, in empty lots, along roadsides, and in abandoned dumps, fertilizing their crops with "greywater" from kitchens and bathrooms. In Barcelona, over half the new and refurbished buildings now have solar hot water. In Karachi, the urban poor have organized themselves to provide sewer services by having the inhabitants take responsibility for planning, building, and managing the local piping system. In Bogotá, many residents move easily around on the spiffy new bus rapid transit system. On an island in the Yangtze River near Shanghai, a new ecological city is being built from scratch. And in Johannesburg, cooperative businesses have been formed to sell eco-friendly construction materials while creating hundreds of new jobs for city residents.

As these examples suggest, *State of the World 2007* covers a topically and geographically diverse urban landscape as we explore the many ways in which cities are key to both human progress and ecological sustainability. My colleague Molly O'Meara Sheehan, who directed this year's *State of the World* project, has assembled an inspired team of Institute researchers and outside experts to write this volume. It includes in-depth discussions of many of the challenges facing today's cities as well as exciting stories about the innovators who are finding new ways to address these problems, often in the poorest corners of the developing world. The short "Cityscape" stories that appear between the chapters were prepared by people who know firsthand what is happening in these cities.

We are particularly pleased that two of the world's great leaders on urban issues—both from the global South—have written eloquent Forewords to *State of the World 2007*. Anna Tibaijuka, Executive Director of UN-HABITAT, the U.N. body devoted to the well-being of human settlements, has brought the plight of urban slum dwellers to the attention of world leaders. As a woman who grew up in rural Tanzania and studied agricultural economics at university, Anna Tibaijuka provides the perspective of a person who has professionally and personally straddled the rural-urban divide.

Jaime Lerner, former mayor of Curitiba in Brazil and former governor of Paraná, who developed the bus rapid transit system that inspired Bogotá's system and is now being replicated in cities such as Los Angeles and Beijing, wrote our second Foreword. In contrast to those who portray today's cities as hopeless and apocalyptic places, Jaime Lerner views cities as exciting laboratories of change. That sense of optimism is central to the future of cities—and the world itself.

Christopher Flavin

State of the World: A Year in Review

Compiled by Lisa Mastny

This timeline covers some significant announcements and reports from October 2005 through September 2006. It is a mix of progress, setbacks, and missed steps around the world that are affecting environmental quality and social welfare.

Timeline events were selected to increase awareness of the connections between people and the environment. An online version of the timeline with links to Internet resources is available at www.worldwatch .org/features/timeline.

State of the World: A Year in Review

POPULATION
UN experts predict that by 2010, as many as 50 million people will be environmental refugees, fleeing the effects of worsening environmental conditions.

TOXICS
Explosion at a Chinese petrochemical plant releases 100 tons of benzene and other toxins into the Songhua River, forcing disruptions in drinking water supply.

AGRICULTURE
WTO members approve a declaration agreeing to end trade-distorting agricultural export subsidies by 2013.

ECOSYSTEMS
Mexico designates the Sierra del Carmen mountains Latin America's first "wilderness area," creating a transnational park with Big Bend National Park in Texas.

CLIMATE
Report warns that half the world's coral reefs may die within 40 years unless urgent action is taken to protect them from climate change.

BIODIVERSITY
Twelve West African countries sign pact to improve cross-border cooperation to conserve elephant populations and their habitats.

OCTOBER   NOVEMBER   DECEMBER

2005 STATE OF THE WORLD: A YEAR IN REVIEW

2 4 6 8 10 12 14 16 18 20 22 24 26 28 30   2 4 6 8 10 12 14 16 18 20 22 24 26 28 30   2 4 6 8 10 12 14 16 18 20 22 24 26 28 30

SECURITY
Rioters in Paris suburbs, mainly second-generation immigrant youths, ignite unrest across France, drawing attention to unemployment and discrimination.

FORESTS
FAO reveals that net forest loss worldwide has slowed somewhat over the past five years, to 7.3 million hectares annually.

GOVERNANCE
Seven states in the US Northeast agree to the nation's first mandatory plan for reducing greenhouse gas emissions from power plants.

NATURAL DISASTERS
Earthquake of 7.2 magnitude strikes northwest Pakistan, killing more than 73,000 people by early November.

MARINE SYSTEMS
Fijian chiefs establish marine protected areas in the Great Sea Reef, the world's third largest reef system, aiming to expand system coverage to 30 percent by 2020.

CLIMATE
Scientists project that as much as 90 percent of the near-surface permafrost area in the Arctic could disappear by 2021.

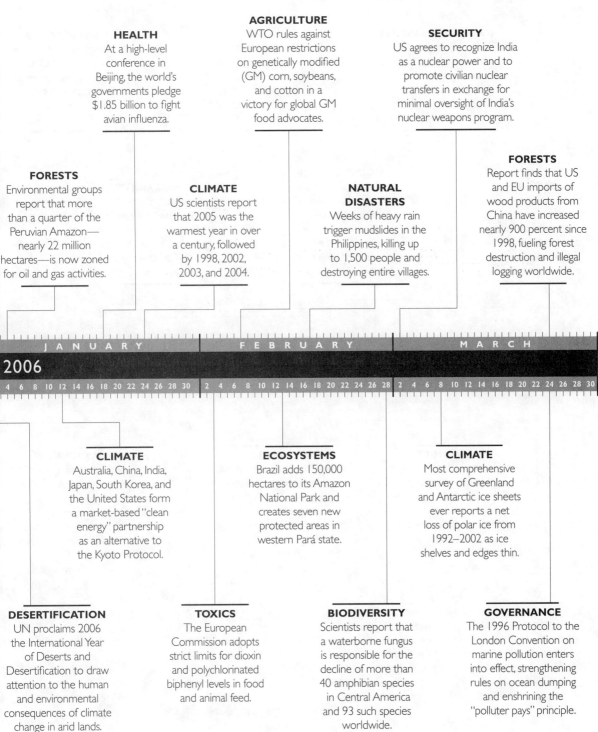

HEALTH
At a high-level conference in Beijing, the world's governments pledge $1.85 billion to fight avian influenza.

AGRICULTURE
WTO rules against European restrictions on genetically modified (GM) corn, soybeans, and cotton in a victory for global GM food advocates.

SECURITY
US agrees to recognize India as a nuclear power and to promote civilian nuclear transfers in exchange for minimal oversight of India's nuclear weapons program.

FORESTS
Environmental groups report that more than a quarter of the Peruvian Amazon—nearly 22 million hectares—is now zoned for oil and gas activities.

CLIMATE
US scientists report that 2005 was the warmest year in over a century, followed by 1998, 2002, 2003, and 2004.

NATURAL DISASTERS
Weeks of heavy rain trigger mudslides in the Philippines, killing up to 1,500 people and destroying entire villages.

FORESTS
Report finds that US and EU imports of wood products from China have increased nearly 900 percent since 1998, fueling forest destruction and illegal logging worldwide.

JANUARY   FEBRUARY   MARCH

2006

2 4 6 8 10 12 14 16 18 20 22 24 26 28 30   2 4 6 8 10 12 14 16 18 20 22 24 26 28   2 4 6 8 10 12 14 16 18 20 22 24 26 28 30

CLIMATE
Australia, China, India, Japan, South Korea, and the United States form a market-based "clean energy" partnership as an alternative to the Kyoto Protocol.

ECOSYSTEMS
Brazil adds 150,000 hectares to its Amazon National Park and creates seven new protected areas in western Pará state.

CLIMATE
Most comprehensive survey of Greenland and Antarctic ice sheets ever reports a net loss of polar ice from 1992–2002 as ice shelves and edges thin.

DESERTIFICATION
UN proclaims 2006 the International Year of Deserts and Desertification to draw attention to the human and environmental consequences of climate change in arid lands.

TOXICS
The European Commission adopts strict limits for dioxin and polychlorinated biphenyl levels in food and animal feed.

BIODIVERSITY
Scientists report that a waterborne fungus is responsible for the decline of more than 40 amphibian species in Central America and 93 such species worldwide.

GOVERNANCE
The 1996 Protocol to the London Convention on marine pollution enters into effect, strengthening rules on ocean dumping and enshrining the "polluter pays" principle.

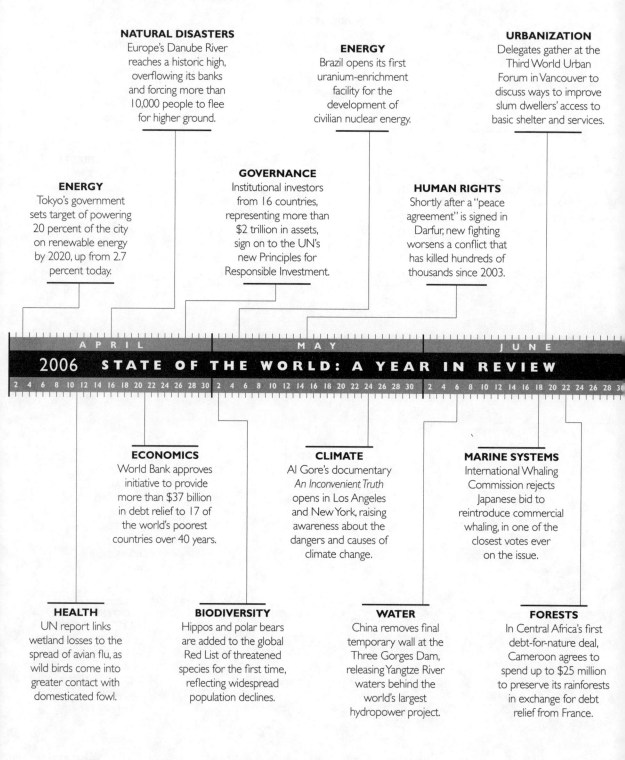

NATURAL DISASTERS
Europe's Danube River reaches a historic high, overflowing its banks and forcing more than 10,000 people to flee for higher ground.

ENERGY
Brazil opens its first uranium-enrichment facility for the development of civilian nuclear energy.

URBANIZATION
Delegates gather at the Third World Urban Forum in Vancouver to discuss ways to improve slum dwellers' access to basic shelter and services.

ENERGY
Tokyo's government sets target of powering 20 percent of the city on renewable energy by 2020, up from 2.7 percent today.

GOVERNANCE
Institutional investors from 16 countries, representing more than $2 trillion in assets, sign on to the UN's new Principles for Responsible Investment.

HUMAN RIGHTS
Shortly after a "peace agreement" is signed in Darfur, new fighting worsens a conflict that has killed hundreds of thousands since 2003.

APRIL   MAY   JUNE

2006 STATE OF THE WORLD: A YEAR IN REVIEW

2 4 6 8 10 12 14 16 18 20 22 24 26 28 30   2 4 6 8 10 12 14 16 18 20 22 24 26 28 30   2 4 6 8 10 12 14 16 18 20 22 24 26 28 30

ECONOMICS
World Bank approves initiative to provide more than $37 billion in debt relief to 17 of the world's poorest countries over 40 years.

CLIMATE
Al Gore's documentary *An Inconvenient Truth* opens in Los Angeles and New York, raising awareness about the dangers and causes of climate change.

MARINE SYSTEMS
International Whaling Commission rejects Japanese bid to reintroduce commercial whaling, in one of the closest votes ever on the issue.

HEALTH
UN report links wetland losses to the spread of avian flu, as wild birds come into greater contact with domesticated fowl.

BIODIVERSITY
Hippos and polar bears are added to the global Red List of threatened species for the first time, reflecting widespread population declines.

WATER
China removes final temporary wall at the Three Gorges Dam, releasing Yangtze River waters behind the world's largest hydropower project.

FORESTS
In Central Africa's first debt-for-nature deal, Cameroon agrees to spend up to $25 million to preserve its rainforests in exchange for debt relief from France.

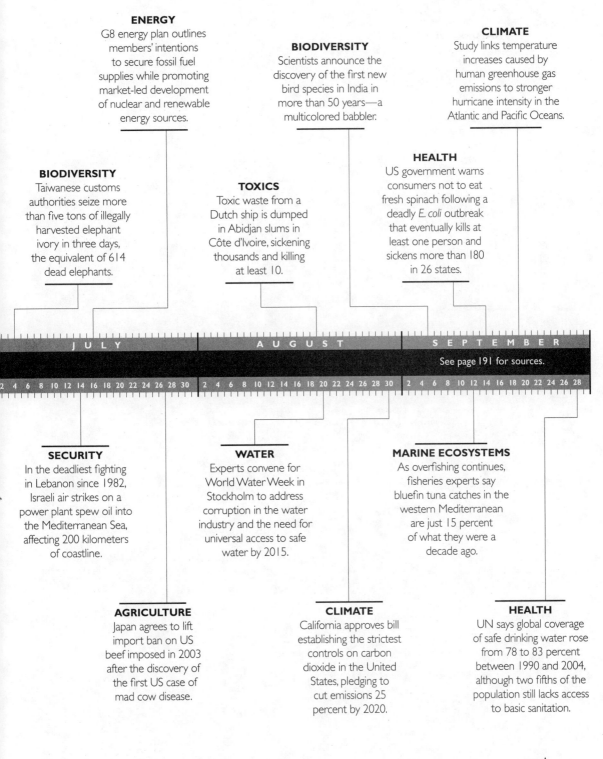

ENERGY
G8 energy plan outlines members' intentions to secure fossil fuel supplies while promoting market-led development of nuclear and renewable energy sources.

BIODIVERSITY
Scientists announce the discovery of the first new bird species in India in more than 50 years—a multicolored babbler.

CLIMATE
Study links temperature increases caused by human greenhouse gas emissions to stronger hurricane intensity in the Atlantic and Pacific Oceans.

BIODIVERSITY
Taiwanese customs authorities seize more than five tons of illegally harvested elephant ivory in three days, the equivalent of 614 dead elephants.

TOXICS
Toxic waste from a Dutch ship is dumped in Abidjan slums in Côte d'Ivoire, sickening thousands and killing at least 10.

HEALTH
US government warns consumers not to eat fresh spinach following a deadly E. coli outbreak that eventually kills at least one person and sickens more than 180 in 26 states.

JULY   AUGUST   SEPTEMBER

See page 191 for sources.

2 4 6 8 10 12 14 16 18 20 22 24 26 28 30   2 4 6 8 10 12 14 16 18 20 22 24 26 28 30   2 4 6 8 10 12 14 16 18 20 22 24 26 28

SECURITY
In the deadliest fighting in Lebanon since 1982, Israeli air strikes on a power plant spew oil into the Mediterranean Sea, affecting 200 kilometers of coastline.

WATER
Experts convene for World Water Week in Stockholm to address corruption in the water industry and the need for universal access to safe water by 2015.

MARINE ECOSYSTEMS
As overfishing continues, fisheries experts say bluefin tuna catches in the western Mediterranean are just 15 percent of what they were a decade ago.

AGRICULTURE
Japan agrees to lift import ban on US beef imposed in 2003 after the discovery of the first US case of mad cow disease.

CLIMATE
California approves bill establishing the strictest controls on carbon dioxide in the United States, pledging to cut emissions 25 percent by 2020.

HEALTH
UN says global coverage of safe drinking water rose from 78 to 83 percent between 1990 and 2004, although two fifths of the population still lacks access to basic sanitation.

2 0 0 7
STATE OF THE WORLD
Our Urban Future

CHAPTER I

An Urbanizing World

Kai N. Lee

In 2007, engineers and construction workers are to begin transforming rural Chongming Island, in the Yangtze River near Shanghai, into a city. Arup, the firm preparing the master plan for this new development, called Dongtan, touts it as "the world's first sustainable city." Plans call for a city of 50,000 by 2010, with the population expected to reach 500,000 by 2040. The development will cover 4,600 hectares, less than a fifth of the entire island. Windmills will dominate the skyline, and turf, vegetation, and solar panels will cover the roofs. Some 80 percent of solid waste will be recycled, while organic waste will be composted or burned to supply heat and power. The only motorized vehicles allowed on the streets will be powered by electricity or fuel cells.[1]

In theory, Dongtan will be self-sufficient in energy, food, and water, with close to zero carbon emissions from transportation. If this

is accomplished, each person living in Dongtan will exert much less pressure on nature than a New Yorker does today. Although New York City's density of settlement is similar to that envisioned for Dongtan, the American city relies on electricity that is virtually all generated by fossil fuels and nuclear fission, and its wastes are carried by truck to landfills up to 650 kilometers away. New York's recycling rate is less than 20 percent.[2]

The Dongtan eco-city project is one of the latest attempts to design an urban form that brings the needs of people into line with the needs of the environment. A century ago, Ebenezer Howard, a British reformer, advocated "garden cities," self-contained towns of roughly 30,000 people living on 1,000 acres (405 hectares) surrounded by greenbelts. Within these new towns, zoning was to separate houses and gardens from factories and farms. The first garden city, Letchworth, was

Kai N. Lee is Rosenburg Professor of Environmental Studies at Williams College in Williamstown, Massachusetts. He gratefully acknowledges the contributions of Molly O'Meara Sheehan, who also wrote parts of this chapter.

founded some 60 kilometers from London in 1902, and the idea spread to other countries, including the Netherlands and Japan. But the new towns did not create their own workplaces, as planned, and were instead absorbed into the regional growth of the cities they surrounded.[3]

Dongtan has yet to be built and tested. Like utopian projects before it, there are sure to be negative consequences. Some may be expected, such as the possible displacement of farmers now living on the island or the potential disruption of the protected wetlands that house a bird sanctuary, while others have yet to surface. Nonetheless, this project comes at a time when humanity needs new models for urban development.[4]

The Global Challenge of Urbanization

Thanks to rapid urban growth not only in China but elsewhere in Asia and Africa, some time in the coming year the population of the world will become mostly urban. By 2005, the world's urban population of 3.18 billion people constituted 49 percent of the total population of 6.46 billion. Very soon, and for the first time in the history of our species, more humans will live in urban areas than rural places.[5]

This is a significant milestone on the long road of civilization. Ten thousand years ago, humans were hunter-gatherers who moved with their food sources. With the discovery of agriculture came permanent settlements and, in time, the imperial cities of the ancient world. More than two centuries ago, improvements in agriculture in northwest Europe made it possible for a smaller fraction of the population to feed everyone. In 1740 about two thirds of the labor force in England and Wales worked in agriculture. By 1840 this had dropped to less than a quarter, even though

the English exported food throughout that century. On the heels of this increase in agricultural productivity came the invention of machines that could transform the heat of burning coal or wood into useful work. The Industrial Revolution spread from Europe to North America and then Japan, and cities grew to house and serve the new factory workers, many of whom had left farms where their labor was no longer needed. By 1900 humanity stood on the threshold of modernity: a new way of life anchored in cities that would rewrite the conditions of human life. (See Box 1–1.)[6]

In parallel, human activity has emerged as an environmental force of planetary proportions: replumbing watercourses, exterminating species, and altering the global climate. These changes have brought unprecedented material gains to our species, particularly in the high-income nations. Whether these gains can be shared with all of humanity, and whether they can be sustained, are questions that now seem increasingly urgent, as the impact of humans on the natural world can no longer be considered negligible. These are also matters that will be decided by urban inhabitants, because although human population growth may well cease in this century, cities and their environmental pressures are continuing to expand through economic growth, migration, natural increase, and the transformation of rural areas into urban settlements.[7]

U.N. projections suggest that nearly all of the world's population growth in the coming generation will be in cities in low- and medium-income nations. Asia and Africa, the most rural continents today, are expected to double their urban populations to about 3.4 billion in 2030. Already, about 1 billion urban dwellers live in "slums" or informal settlements—areas where people live without one or more of life's basic necessities: clean water,

Box 1–1. Transitions of Modernity

The cities made possible and necessary by industrialization have incubated a mutually reinforcing set of transitions that have redrawn the outlines of material existence. What we now think of as globalization is the latest phase in a set of linked transitions in population, health, economy, politics, social relations, and environment.

Families have grown smaller all over the world, led by the high-income countries and by China's surprisingly successful one-child policy. This demographic transition seems likely to result in the end of human population growth before the end of the twenty-first century. Stabilization of population has not been imposed by disease, famine, or war. With the exception of African countries hard hit by AIDS, health has improved almost everywhere over the past century, with declines in child mortality, decreases in infectious illness, and longer life spans, especially in high-income countries. This epidemiological transition had an important consequence for cities: just over a century ago, cities had so much disease that urban populations declined unless people moved into them. That urban penalty was erased by improvements in sanitation and clean water—although the poor public health conditions of slums still sicken and kill on a large scale.

Industrialization has brought an unprecedented economic transition, one still unfolding in the process of globalization. Average per capita income has grown, with interruptions, since the early nineteenth century. But wealth and the indirect benefits of prosperity have been shared unequally, even as the output of the world economy has grown enormously. Economic power has been rooted in cities; the purchases of urban people, who cannot live off the land, form the foundation of national economies. Technological changes allowed cities to become larger and to spread further; steel-framed buildings made skyscrapers possible, while faster transportation brought people to and from jobs in high-density downtowns even as they came to live in distant, low-density suburbs.

As dramatic as the economic transformations of the past two centuries has been the uneven but unmistakable rise of democracy. The fraction of the world's people ruled by democratic governments rose from about 4 percent in 1840 to about 12 percent in 1900, and crossed 50 percent around 2000. This remarkable transition reflects the end of colonialism, which added India and many other medium- and low-income countries to the list of democracies. The widening reach of competitive elections has given city dwellers a chance to demand accountability for the conditions of urban life, and governments are now considered responsible for matters ranging from education to parks to women's rights that were simply not on the agenda of feudal societies.

SOURCE: See endnote 6.

sanitation, sufficient living space, durable housing, or secure tenure, which includes freedom from forced eviction.[8]

Urbanization thus presents a global challenge of human development and human rights. The shift in where we live brings to the fore the question of how we live—the challenge of sustainable development, defined in a widely quoted form 20 years ago as meeting the needs of the present without compromising the ability of future generations to meet their own needs.[9]

Many scientists agree that the global economy is not on a path toward sustainable development. More than a decade ago, the Intergovernmental Panel on Climate Change found that the burning of fossil fuels was altering the composition and heat balance of the atmosphere; the group has since documented signs of a changing climate, from shrinking glaciers to the decline of some plant and animal populations. An international analysis of the world's ecosystems, written by more than 1,300 scientists, found that 60 percent of the services of nature—including those provided by farmlands, fisheries, and forests—are being degraded or used unsustainably. This Millennium Ecosystem Assessment warned in 2005 that "these problems, unless addressed, will substantially diminish the benefits that future generations obtain from ecosystems."[10]

This chapter reviews the state of the world's urban areas, highlighting the way in which urbanization and sustainable development are linked. At first sight, cities seem to be the problem rather than the solution: the number of people living in slums has steadily increased, and industrial pollution in rapidly growing economies fouls water and air. Yet the flow of people toward cities seems unlikely to stop or even slow, in part because life chances and economic opportunities are often better in cities, even for many of the poor.[11]

From that perspective, urbanization provides a crucial opportunity: to create living patterns harmonized with nature's rhythms as people continue to create urban habitat. Cities offer economies of scale for recycling water and materials, for instance, and for using energy efficiently. Yet today's high-income cities use resources unsustainably, and the high-consumption approach is plainly unaffordable for slum dwellers. Finding ways to create better urban settlements in all societies is central to sustainable development.

A city may be thought of as a physical and social mechanism to acquire and deliver essential natural services, such as clean water, to a concentrated human population. The physical part of this mechanism is often called infrastructure, while the social part includes markets, government, and community organizations. Cities vary tremendously, but

thinking of urban areas as linked to nature reveals an important pattern: the environmental problems of low-income cities are different in kind and in scale from the problems facing industrializing, medium-income cities. And the challenges brought by rapid industrialization in Guangzhou, China, or by poverty in Cochabamba, Bolivia, differ from those found in high-income cities like Phoenix or Turin.

Cities are tied to nature through markets and technology. Virtually all cities rely on food, fuels, and materials from elsewhere, and all cities are marketplaces. Thus, "sustainable" does not mean self-sufficient. Rather, a city moving toward sustainability improves public health and well-being, lowers its environmental impacts, increasingly recycles its materials, and uses energy with growing efficiency. Note the word "toward": it is unrealistic for a human economy to have no impacts on the natural world, but clearly it is necessary for the human economy to share its wealth more equitably and in ways that enable our species to endure on a finite planet.

Urban Areas Today

While the trend of urbanization is clear, the measurement of what is urban is not. When the United Nations projects that the world's population will become predominantly urban in 2008, it is drawing upon information provided by member nations, who define "urban" differently. More than two dozen nations do not document their definitions at all. Urban populations can be identified using at least three different ideas: the number of people living within the jurisdictional boundaries of a city; those living in areas with a high density of residential structures (urban agglomeration); and those linked by direct economic ties to a city center (metropolitan area).

These definitions yield quite different pictures of the "city." The U.S. National Research Council remarked in 2003 that "cities such as Buenos Aires, Mexico City, London, and Tokyo can correctly be said to be declining or expanding in population, depending on how their boundaries are defined." Moreover, about two dozen low-income countries have not had a census in more than a decade, and the populations attributed to them are projections. Then there is the matter of how large a settlement must be to count as urban. India, for example, would change from being mostly rural to being mostly urban if it adopted the definition of urban area used in Sweden. Despite these weaknesses, the data published by the United Nations are widely used (as they are throughout this book) for lack of better estimates.[12]

In the second half of the twentieth century, according to the United Nations, the urban population of the world increased nearly fourfold, from 732 million in 1950 to 2.8 billion in 2000 and to more than 3.2 billion in 2006. As shown in Table 1–1, growth has been rapid in Africa, Asia, and Latin America, but much slower in Europe and North America, where more than half the population already lived in urban areas by 1950. Only 40 percent of the urban population lived in low- and middle-income nations in 1950, but that fraction will reach three quarters shortly after 2010.[13]

Since 1975, more than 200 urban agglomerations in low- and medium-income nations have grown past 1 million inhabitants. Their local governments are faced with the sanitation, housing, transportation, water, energy, and health needs of more than a million constituents—a striking new challenge that arose in a single

generation. Many municipal governments trying to cope with these matters lack trained workers, the budgets to pay them, and traditions of civic governance on a mass scale.[14]

The trends of the past generation are projected to continue into the coming one. More important, as noted earlier, the overwhelming majority of net additions to the human population—88 percent of the growth from 2000 to 2030—will be urban dwellers in low- and medium-income countries. Already, Africa has 350 million urban dwellers, more than the populations of Canada and the United States combined. In terms of absolute numbers of inhabitants, urban growth is unprecedented and will continue to be so. But in percentage terms, the rate at which national populations are becoming urban lies within the historical range experienced by the high-income countries.[15]

The rapid swelling of urban populations is due to both migration into cities and natural increase of the people already there. Although policymakers tend to emphasize the role of migration, which is high compared with historical levels in the places where rapid growth is taking place, natural increase actually accounts for over half of

Table 1–1. Urban Populations by Region, 1950–2000, with Projection for 2010

Region	1950	1970	1990	2000	2010
	(million inhabitants)				
Africa	33	85	203	294	408
Asia	234	485	1,011	1,363	1,755
Latin America and the Caribbean	70	163	315	394	474
Europe	277	411	509	522	529
North America	110	171	214	249	284
Oceania	8	14	19	22	25
World	732	1,329	2,271	2,845	3,475

Note: Columns may not add up to world total due to rounding.
SOURCE: See endnote 13.

the rise in urban population.[16]

Perhaps the most visible aspect of global urbanization has been the rise of megacities, large urban agglomerations with more than 10 million inhabitants. (See Figure 1–1.) These cities only account, however, for about 9 percent of total urban population. Just over half of the world's city dwellers live in settlements with fewer than 500,000 inhabitants. (See Figure 1–2.)[17]

The rapid urbanization of the world's population is unfolding in distinctive ways in different parts of the world. Latin America, at 77 percent urban, has already gone through an urban demographic transition like those of North America and Europe, with national population growth rates declining since the 1960s. Growth in the region's megacities has slowed considerably, as the costs of congestion have made smaller urban areas more attractive. Yet thanks to the world's highest levels of economic and social inequality, Latin American cities have large slum populations that continue to grow.[18]

In Africa, where some 38 percent of the population lives in urban areas, urbanization is more recent and more rapid in proportional terms because of higher population growth rates, rural poverty due to low agricultural productivity, and wars that drive people into cities. The spatial and economic structure of African cities reflects choices made by Europeans in the colonial era, when trading centers for agricultural products and natural resources produced for international export replaced an older network of market settlements serving an agrarian population. The colonial cities were designed by Europeans with small enclaves for themselves; adjoining indigenous districts were built with little attention to water and sanitation, roads, transportation, or energy supply. The lack of infrastructure for the poor, followed by rapid urban growth, has produced large slum populations living at high levels of risk from disease and environmental hazards like flooding.[19]

Poor macroeconomic performance in sub-

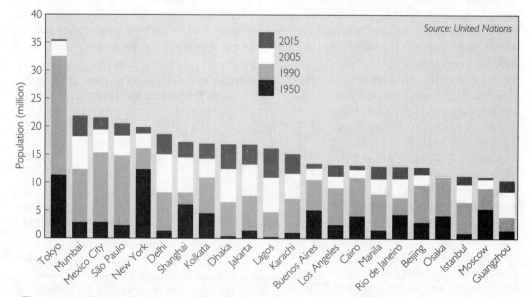

Figure 1–1. Urban Agglomerations Projected to Exceed 10 Million Population by 2015

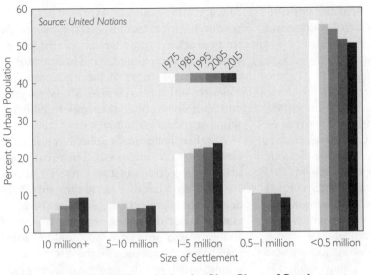

Figure 1–2. Urban Population by Size Class of Settlement

Saharan Africa since independence, nearly half a century ago, has led to urban economies dominated by informal work such as food hawking and small-scale commerce, but little industrial employment. More than three quarters of nonagricultural employment is in the informal sector, yet it accounts for only 41 percent of economic output, because nearly all its jobs are in low-wage and low-profit activities. African economies are little integrated into the global economy, and they still depend on exports of natural resources and agricultural products in order to import manufactured products—as in colonial times.[20]

Asia, the world's most populous region, is roughly 40 percent urban, with a varied urban landscape. Pacific Asia—the coastal regions from Japan to Southeast Asia—has undergone a remarkable economic transformation over the past generation as China and the newly industrializing countries of East Asia have rapidly increased incomes and levels of urbanization. China is now home to 16 of the world's 20 most polluted cities, as rapid eco-

nomic changes have pressed the ability of governments to protect and improve public health. In western China, South Asia, and interior Asia, urbanization is also rapid, but economic growth has not been so meteoric, and poverty burdens nearly a third of India's urban population. Population growth rates remain high in Bangladesh and Pakistan, although they are declining. Urban populations shrank or grew slowly in Central Asia in the severe economic and political disruptions that followed the collapse of the Soviet Union.[21]

Beyond these regional generalizations, each city has a history and a population that will lead the city in its own direction. Cities attract settlers and retain residents because they offer opportunities for employment, for meeting and being with people, for becoming someone different. As migrants to every slum will affirm, they are there because they want to be. Seizing the opportunities and taking the risks of city life, some will fail and others succeed. Often, people will do things they could not have done in rural settings, and sometimes they will push the urban community and economy in a new direction altogether, whether this involves opening up a new kind of business like an organic food market, or making new links to distant communities by sending earnings back home, or disrupting the community by committing a crime or contracting a previously unknown disease like avian flu.

The dynamism of cities makes each urban

area a place, a distinctive social and environmental setting around which loyalties and antipathies can form.

Dark Alleys

That each city is different has an important implication for policy: sensible support of or investment in an urban area requires knowledge of its relevant characteristics. Yet our understanding of cities is strikingly limited, and this constrains the ability of international institutions, governments, and nongovernmental organizations to act intelligently. One result is that there is no simple model of how to spur economic growth—not to mention sustainability—in cities. By comparison, it is an axiom of rural development that raising agricultural productivity is an ingredient of economic growth.

A basic limitation lies in the paucity of information collected on cities. Urban population is tabulated by the United Nations, but the variables that provide a picture of sustainability—human well-being, environmental conditions, and economic data—are measured mostly at the national level, with scant coverage in the cities of the developing world, where the most rapid urbanization is occurring.[22]

As noted earlier, the delineation of cities varies from country to country. Drawing a boundary between rural and urban spaces implies a sharp separation between countryside and built-up area that simply does not exist in most places. Many families depend on both urban and rural settings to make a living. Cecilia Tacoli of the International Institute for Environment and Development points to studies that estimate the share of rural households' incomes from nonfarm sources, including migrants' remittances, at 40 percent in Latin America, 60 percent in South Asia, and 30–50 percent in sub-Saharan Africa, reaching as much as 80–90 percent in Southern Africa. In addition, the changes in land use at the fuzzy edges of urban agglomerations are complicated, rapid in comparison to the reaction time of local government, and often difficult to control. Municipal governments do not often take responsibility for slums outside their boundaries.[23]

Another challenge for policy is our unclear view of population growth. The urban population projections of the United Nations are not quite what they seem: they are purely demographic projections, showing the population trajectories if parameters observed and estimated now were to unfold in future years. They do not include the effect of economic, social, or environmental factors that might alter, for instance, birth or migration rates. Although the assumptions behind them are spelled out in U.N. documents, these projections are commonly taken to be predictions incorporating the best understanding of all the forces at play. Yet, studies of demographic projections have shown that U.N. estimates have tended to overestimate urban growth in developing countries by about 19 percent for estimates made 20 years earlier. Thus the projections of rapid urbanization in sub-Saharan Africa may prove to be high in the future if the economies in those countries remain weak.[24]

A further limitation arises from the lack of data about variations within cities. Wealthy districts and slums occupy areas well known to a city's residents, but little information is available to analysts about how housing or employment conditions vary between them. Even accurate maps—essential for planning, building, and maintaining streets or sewers— are missing in many developing countries, especially for informal settlements that were occupied without authorization or recorded land transactions. This lack of knowledge compounds the difficulties arising from multiple, conflicting sets of property rights. In Ghana, as in many other former colonies, one system of property is rooted in precolonial family claims while another is inherited from the colonial legal system. Sorting these out, without maps or records of the traditional property claims, clogs courts and hinders the development of housing and businesses to accommodate rising urban populations.[25]

Geographic information systems (GIS) that use computers to assemble data from different sources and to overlay them on maps are beginning to make a substantial contribution. Building on studies of specific cities that find that slum dwellers may have worse health conditions than people in rural areas, GIS analyses of conditions in Accra in Ghana and Tijuana in Mexico, among other places, are showing how poverty is much more than a lack of money. It includes higher prevalence of disease, exposure to flooding, and other adversities.[26]

Why do these limitations of understanding matter? There is about $150 billion spent each year on physical infrastructure in developing countries. The United Nations Millennium Project estimates that meeting the Millennium Development Goal of improving the lives of 100 million slum dwellers would cost $830 billion over the next 17 years. These expenditures could help move poor people toward sustainable living and decent lives, but only if the donor agencies and governments allocating these substantial sums are able to target spending sensibly. Most development assistance has been aimed at rural poverty on the assumption that urban poverty is a transitory phenomenon for those migrating into cities. Yet with more than half of urban growth due to natural increase, it is far from clear when slum dwellers will escape or be able to improve their dwellings and neighborhoods. Development that moves people toward sustainable patterns will surely

need investments based on understanding who lives where in growing cities and how they earn their living.[27]

Development that moves people toward sustainable patterns will need investments based on understanding who lives where in growing cities and how they earn their living.

Research and learning are accordingly practical necessities. UN-HABITAT has been collecting a large set of indicators in its Global Urban Observatory, and the important 2003 study *Cities Transformed* undertook analyses of the international Demographic and Health Survey in order to illuminate urban health and social conditions. Another noteworthy effort is led by economist Stephen Sheppard and urban planner Shlomo Angel. They chose 120 cities of various sizes from all regions of the inhabited world and developed a fast-track protocol to assess a wide range of variables in each city, ranging from housing prices to air pollution to urban planning policies. The protocol is designed so that a student who is a native speaker of the language can collect information on several hundred indicators in about a week. The project includes remote-sensing analyses of the 120 cities, using satellite images from 1990 and 2000. This study, supported by the World Bank and the U.S. National Science Foundation, is building a database for analysts worldwide in order to investigate social, environmental, and economic changes over a decade for a large sample of cities.[28]

As important as research is learning from experience, translating failures and surprises into better choices going forward. This has proved to be a challenge in both development and environmental management.

Although surprises happen so often in social interventions that they should be expected, it is rare for those implementing a plan even to consider unexpected results. Merely stating goals clearly enough that failures can be identified is risky for politicians. Systematic methods for learning from policy implementation have been developed and tried out at a small scale, but uneven learning hangs like fog over the path to urban sustainability.[29]

Wealth and Environment

The environmental challenges of cities vary with their level of economic activity. To oversimplify, poor city dwellers face direct, everyday environmental problems, while the wealthiest urban residents cause environ- mental problems that they do not experience in their daily lives. A child in Soweto, South Africa, risks dying from waterborne illnesses that his distant cousin in Birmingham, England, will not be exposed to. A factory worker in Wuhan, China, may suffer from asthma triggered by air pollution, while her counterpart in Nagoya, Japan, is less likely to encounter pollutants in the air she breathes. The college student in Denver, Colorado, contributes more to global warming as he drives to the campus each day than does someone riding a bus to classes at the Universidad de los Andes in Bogotá, Colombia.

These stories of individual experiences correspond to statistical differences among low-, medium-, and high-income cities. Table 1–2 compares indicators for three cities drawn from these three categories,

Table 1–2. Sustainability Indicators for Ghana, Mexico, Singapore, Accra, and Tijuana

Indicator	Ghana	Mexico	Singapore
Population	21.2 million (2003)	104.3 million (2003)	4.37 million (2005)
Share of population urban, 2003	45.4 percent	75.5 percent	100 percent
GDP per capita (in purchasing power parity), 2003	$2,238	$9,168	$24,481
Human Development Index rank, out of 177, in 2005	138	53	25
Life expectancy at birth, 2003	56.8 years	75.1 years	78.7 years
Probability of dying before age 5 (male/female) per 1,000 population, 2001	107/100	31/25	4/3
Health expenditure per capita (in purchasing power parity), 2002	$73	$550	$1,105
Energy use (oil equivalent per capita), 2003	400 kg per year	1,564 kg per year	5,359 kg per year
	Accra	Tijuana	Singapore
Population (2005)	1.97 million	1.57 million	4.37 million
Share of population without access to "improved" sanitation	48 percent (1991–92)	17 percent (2000)	0 percent (2002)
Share of population without access to an "improved" water source	46 percent (1991–92)	29 percent (2000)	0 percent (2002)

SOURCE: See endnote 30.

showing economic and health indicators at the national level together with indicators of the cities' environmental conditions; energy use, a national statistic, is used as an indicator of carbon emissions. National statistics do not accurately represent conditions in cities, and variations within cities can also be large. The economic and health indicators are included here to show the divergence in the national contexts of low-, medium-, and high-income cities.[30]

These numbers illustrate a pattern of spatial, environmental, and economic variation. A low-income city like Accra faces direct threats to health: water contaminated with human waste, housing infested with insect and rodent pests, streets and neighborhoods that flood in the rainy season. Each person and family must cope with these environmental problems in daily life. An industrializing city like Tijuana may face additional environmental problems from polluting factories and toxins from manufacturing processes. The rapid rise of energy use during industrialization, often in inefficient foundries and furnaces, imposes a large burden of air pollution on workers and residents, with substantial public health consequences. But industrialization also generates earnings that can be invested in environmental controls and public health measures, as the data on Tijuana and Singapore indicate.[31]

With the transition to economies dominated by service industries, high-income cities competed with one another on quality of life, seeking to attract the professional talent to staff service firms such as software engineering or finance. Good environmental conditions and amenities help create the clean, interesting places that draw and keep highly mobile people in cities like Singapore. The rising economies of wealthy cities also powered increasing energy consumption and exploitation of forests, oceans, and other natural resources—with effects that were often far removed from the comfortable offices and homes of those living there.

High- and medium-income cities today are caught in the paradox of losing sight of nature just as they become more dependent on it through increasing consumption and the globalization of production. The paradox itself is a gift of markets of ever greater reach: if a coffee crop fails in Indonesia, the supply from Guatemala or Kenya will smoothly fill the cup in Rouen or Buenos Aires. A disaster for rural growers is an unnoticed blip for the urban coffee drinker. But there is another paradox of planetary-scale markets. Cities are places. Yet as cities become wealthier, their residents buy goods from around the world and invest in global companies. The widening spatial range of urban economies has frequently eroded a city's distinctiveness. This process is accelerating. Industrialization took more than a century to unfold in Europe, the United States, and Japan. The spread of industrial production to the once-poor lands of Asia has transformed economies in a few decades. And the rise of the information-intensive service economy brings change measured in years.[32]

The variations among low-, medium-, and high-income cities have been discussed in terms of a curious empirical pattern known as the environmental Kuznets curve, named for American Nobel laureate in economics Simon Kuznets. (See Figure 1–3.) Drawing together a wide array of data, analysts have framed a generalized scenario of urban environmental development: local environmental problems that pose an immediate threat, such as lack of sanitation, tend to improve with increasing wealth, while global ones such as carbon emissions worsen, slowly undermining large-scale life-support systems such as climate. And as a city industrializes, environmental problems at the scale of the city and metro-

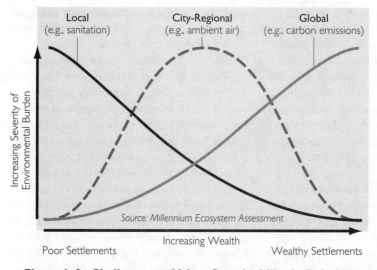

Local (e.g., sanitation)   City-Regional (e.g., ambient air)   Global (e.g., carbon emissions)

Increasing Severity of Environmental Burden

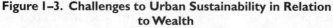

Source: Millennium Ecosystem Assessment

Increasing Wealth

Poor Settlements   Wealthy Settlements

Figure 1–3. Challenges to Urban Sustainability in Relation to Wealth

politan region first worsen, as pollution increases, and then improve, as resources for engineered controls and regulation became available. In some cases, those regulations, combined with economic changes, force polluting activities to other locations.[33]

This pattern does not mean that environmental problems automatically improve with greater wealth, as has sometimes been suggested. Yet the fact that there are different types of environmental challenges at different income levels does have significant implications for sustainable development. The idea of meeting the needs of the present has a sharply different meaning for someone living in a slum than for someone with a high-income lifestyle. Similarly, the obligation not to compromise the ability of future generations to meet their own needs has a different resonance for the poor and the rich.

Sorting cities by income alone is a drastic oversimplification, of course. A key difference is the range of inequality in different cities, something that is missed by focusing only on average income. The poorest residents of

Tijuana or Accra face more difficult environmental conditions than those at the high end of the income scale in these cities, while the exposure to health and environmental risk varies less in Singapore. Poverty does not necessarily translate into high risks to health and poor environmental conditions, however, as has been demonstrated by the success of community-level organizations within a small number of slums. (See Chapter 8.) And transportation patterns vary widely among high-income cities, with large effects on energy use, air quality, and land use. (See Chapter 4.)[34]

Despite the complex circumstances of individual cities, however, it is useful to understand overall trends. The rapidly growing cities of India, China, and other industrializing nations need to organize and pay for their environmental cleanups. In rich nations' cities, reducing consumption of fossil fuels and other finite resources and redirecting investment toward sustainably managed industries—from renewable energy to sustainably harvested wood to well-managed fisheries—are critical to managing global threats to biodiversity, climate, and renewable resources. Poor cities whose populations are also growing rapidly must deal with worsening environmental and health conditions, in many cases without comparable increases in locally generated income. They might not be able to afford high-cost, long-term solutions such as expensive drinking-water purification plants and citywide pipelines.

Nature: Still Essential to Human Well-being

All humans rely on the natural world: water comes from wells and streams; food from farms and fisheries; and wastes are returned to nature. Some ancient civilizations may have been weakened by degrading the natural resources they needed to keep their cities operating. Archaeological evidence suggests that although Sumerians had figured out by 3500 BC how to draw water from the Tigris and Euphrates Rivers to their fields to grow wheat and barley, their irrigation systems did not drain well, so salts built up in the soil and caused wheat production to cease by 1700 BC. Overuse of resources is also implicated in the fall of the Mayan cities of Central America.[35]

Today, we still need trees and wetlands to protect us from floods and storms, and we rely on nature for the raw materials of everyday life. These ecosystem services are essential to life and well-being. Securing necessary ecosystem services is a daily preoccupation of hunter-gatherer societies and a seasonal reality to farmers and fishers today. But it is barely glimpsed amid the hustle and cacophony of city life, at least until disaster strikes in the form of supply disruptions, skyrocketing prices, or a "natural" disaster for which engineering proves inadequate. As noted earlier, a city can be thought of as a mechanism to provide its inhabitants with ecosystem services; this is a large task, performed imperfectly for many urban dwellers around the world.

In the early 1990s, William Rees and Mathis Wackernagel devised the ecological footprint to measure human communities' reliance on nature. The footprint is an estimate of "how much land and water area a human population requires to produce the resources it consumes and to absorb its wastes under prevailing technology." The appealing notion of a footprint evokes the picture of a prein-dustrial city, drawing its sustenance from the farmlands around it. The ecological footprint attempts to adapt this picture to cities and nations deeply enmeshed in a global economy. Using this approach, the per capita footprint of high-income countries is eight times as large as that of low-income countries.[36]

Perhaps as important as the magnitude of humans' reliance on nature is how people take care of the ecosystems that supply services. There are pastures in the Alps and irrigation systems in Bali that have been used for centuries with no diminution of their productive capability. In many other cases, ecosystems have been overused with disastrous economic and social consequences, particularly when the social institutions to govern use were absent or ineffective. Many ocean fisheries are now suffering this fate. Or consider litter: no one owns it, and few want to pick it up—so government refuse collectors have to be hired. There is a broader pattern here: when human responsibility does not match the cycles and patterns of nature, irresponsibility is likely to flourish. This is a problem of human institutions, including markets, governments, and concepts of property. These mismatches between nature's logic and the rules and incentives that shape human behavior are called problems of the commons.[37]

In cities, the high intensity of human activities leads often to problems where individual interests are at odds with the common good. The central task of urban sustainability is effectively managing commons problems in the ecosystems that sustain cities.

During the Industrial Revolution, pollution in cities of North America and Western Europe spurred a renegotiation of the relationship between humans and the environment as more people came to live at much higher densities than in rural areas. As industrialization drew workers to cities, water from wells was supplanted by piped water. The availability of large quantities of relatively inexpensive water, in turn, spurred rapid growth in use. Between 1856 and 1882, for example, water consumption in Chicago rose from 125 to 545 liters per day per person.[38]

For poor people in low-income cities today, nature's services are both expensive and arduous to obtain, as illustrated by the cost of water. Water free of disease-causing germs is available to a small and sinking share of residents of low-income cities. An assessment of 116 cities by the World Health Organization in 2000 estimated that only 43 percent of urban dwellers in Africa had access to piped water. The fraction is declining as more people settle in urbanizing areas without water service and as the existing delivery systems falter from inadequate maintenance, corruption, and the exhaustion of their sources from growing demand. Yet people still require water for drinking, cooking, washing, and bathing.[39]

Where water is not available by pipe, the costs can be steep. (See Table 1–3.) In two informal settlements in Accra, Ghana, a bucket of water from a standpipe costs about 5¢, a price that seems modest to someone from Houston who readily pays more than $1 for 500 milliliters of bottled "spring" water. But water can command more than 10 percent of a poor family's budget. In Addis Ababa, Ethiopia, the poorest fifth of the population spends roughly a sixth of its household income on water. Moreover, water sold by private vendors in small quantities is far more expensive than water from a pipe: 37 times more expensive in Accra, nearly twice as high in a study done in East Africa. Water that is costly is used sparingly: usually, hygiene suffers and disease is more prevalent. These problems have become so widespread that they affect families that are not poor. One scientist in an Accra research institute rises before dawn each day to fetch water for his family to bathe, carrying buckets up four flights of stairs. He lives in the tony district among diplomatic compounds in his nation's capital.[40]

Infrastructure and Governance

The economies of scale possible with high-density settlement in urban areas may offer the best chance to bring decent conditions to all of the world's poor and to conserve the resources on which we all depend. Health, education, and other measures of human development are highest in countries with mostly urban populations. Our need to build at least as much urban habitat as exists today, with its attendant infrastructure and governance systems, is the central hope for sustainable development—if we can learn the lessons of the world we are

Table 1–3. Cost of 100 Liters of Water in Accra and East Africa from Different Sources

Water Source	Accra, 2006	East Africa, 1997	Users
Sachet (500 milliliters)	$8.01		General public for street drinking
30-pack (sachets)	$4.45		General public for household drinking
Bucket from kiosk	$1.87	18¢	Households relying on shared standpipe
Community shower	$1.33 (bathing only)		Informal settlement dwellers
Vendor	27¢	45¢	Mixed-income neighborhood without piped water
Water pipe	5¢	10¢	Households with pipe connection

SOURCE: See endnote 40.

now modifying apace.[41]

Human life in cities is structured by infrastructure: water and food supply, sewers, transportation and communications networks, technologies to improve air quality, and buildings to house people and production. Physical infrastructure is largely inflexible. Streets and water pipes are expensive, long-term commitments, shaping urban form for decades to come. A decision against mass transit locks in a commitment to buses and autos. If a power plant is built to burn coal, that implies a stream of greenhouse gases for two generations.

A good deal of the infrastructure in high-income nations, built in response to the health threats in industrializing cities of the nineteenth century, spawned new problems. Engineers built dams and aqueducts to boost urban water supplies, but the environment has suffered. A study of 292 large river systems in the United States, Canada, Europe, and the former Soviet Union concluded that 42 percent of the ecosystems drained by these streams were strongly affected by impoundments and diversions, putting at risk habitats for a wide range of plant and animal species in the rivers and their watersheds.[42]

Systems designed to channel waste away from people as quickly as possible have also improved human health while damaging the environment in other ways. When rainwater runs off pavement and into drains and sewers, the rivers at the end of the pipes flood more often and more severely than would be the case if plants, soil, and wetlands soaked up some of the deluge. Much as storm drains short-circuit the water cycle, urban waste disposal systems disrupt the nutrient cycle. Roughly half of the food brought into New York City is transformed into human energy; the other half is shunted to sewers or trucked to increasingly remote landfills. Yet organic waste is a valuable resource if composted into a product that invigorates agricultural soils. (See Chapter 3.)[43]

Twentieth-century transportation infrastructure has allowed urban food and energy supply lines to stretch to new lengths, as production has become ever larger and more centralized. Although large farms and power plants excel at producing more food and electricity with fewer employees, they generate pollution and require complex distribution routes, with use occurring far from the site of production. And despite gains in productivity, some 852 million people remain hungry, and roughly 1.6 billion lack electricity.[44]

Infrastructure is largely invisible—pipes are usually underground and water, sewer, electricity, and telecommunications services are widely taken for granted. But it is not inexpensive. When it is financed and maintained from the public purse, governments take on debt that usually can be repaid only if there is economic growth. In the decades ahead, access to capital to build infrastructure will play a central role in the quality of life in cities, particularly for poor people.

As the large costs of physical infrastructure suggest, the social institutions needed to build, maintain, and pay for cities' connections to nature are also complex and varied. Foremost among them is the market, which affords access to ecosystem services to those who pay. The magnitude of poverty in rapidly growing low-income cities thus poses a stark dilemma. People who cannot afford reliable access to vital ecosystem services suffer risks to health and well-being. Yet many governments are so poor and so overwhelmed by the pace of urban settlement that they cannot afford to build the infrastructure that would, in many cases, bring down the cost of water, sanitation, and other services. The problem of creating a business model that can bring affordable and clean water to residents of low-income cities has beset corporations and governments experimenting with privatization of water supply over the past 20 years. (See Chapter 2.)[45]

In short, markets are not a complete solution for provision of nature's services to urban inhabitants, especially where there are commons problems. The institutions that complement and substitute for markets range from bureaucracies to assure control of ecosystem interactions, such as public health and environmental protection agencies, to traditions such as informal harvest limitations in some fishing communities. Many of these social arrangements can be costly, requiring educated workers and durable organizations. As with physical infrastructure, sustaining formal organizations requires steady revenues and effective management.

The concepts of development and investment in infrastructure implicitly assume arrangements found in rich nations: public institutions that are largely free of corruption, economic activity that takes place in a formal economy, and per capita incomes high enough that water, food, shelter, and transportation become settled matters for most people rather than persistent crises of daily living.

In low-income cities, however, these conditions are not present for many people, particularly those living in slums and in an informal economy. Yet people cope and mostly survive. Safe drinking water, secure claim to a dwelling, protection from criminals, and much else may not come from government at all. In these circumstances, community-level organizations have sometimes been able to supply some social and public services. (See Chapter 8.) Urban initiatives such as the Orangi Pilot Project in Karachi have demonstrated that very poor people need not live without sanitation or clean water. (See Chapter 2.)[46]

Recognizing that potential and linking it

to development assistance have risen higher on the agenda of international donors over the past decade. Major efforts to decentralize government have been instituted from Mexico to Mali to Thailand with notable success. Decentralization has channeled more resources to municipal governments, and innovations such as participatory budgeting have in turn given poor communities a voice in allocating public funds. There is still a long way to go, however, in realizing the self-help potential of slum dwellers and other poor people in meeting their needs for nature's services. (See Chapter 9.)[47]

Circular Metabolism

Giving the poor a voice in solving local environmental problems would be a big step toward meeting the needs of the present, one of the two criteria of sustainable development. But for sustainability in the long term, more is needed: to move institutions and infrastructure toward forms that also protect the ability of future generations to meet their own needs. In this arena there are ideas that cut across income groups. While there have been some promising beginnings, particularly in high-income nations, here too there is much work to do.

A key conceptual step is to reconsider infrastructure. In theory, much of the waste from the water, food, fuels, and materials that course into cities could be reused or recycled. Herbert Girardet called for substituting a "circular metabolism," in which wastes are reused, for the linear metabolism of a city that simply converts resources into wastes. The notion of closing nutrient loops in a way parallel to the operation of natural ecosystems can be pursued at different scales, ranging from an individual building to the design of a metropolitan area. (See Box 1–2.)[48]

"Green architecture" is the name of an

Box 1–2. Circular Urban Metabolism in Stockholm

Stockholm's new urban ecological district Hammarby Sjöstad is the best demonstration to date of putting circular urban metabolism into practice through creative design and building in a new, dense neighborhood. From the outset, planners tried to think holistically—to understand the resources that would be required by residents and the wastes that would result and could be used productively. For instance, about 1,000 apartments are equipped with stoves that use biogas extracted from the community's wastewater. Biogas also provides fuel for buses that serve the area.

People in the neighborhood put their solid waste into a vacuum-based underground collection system, which allows efficient separation of recyclables and organic and other wastes. Combustible waste is burned and returned to the neighborhood in the form of electricity and hot water, with the latter deliv-ered through a district heating grid.

Stormwater from streets is directed into a purification and filtration system, and stormwater from buildings is guided to greenroofs and wetlands. Both streams of water are kept apart from wastewater, which is treated separately.

Carbon emissions from residents' transportation are minimized, as the neighborhood is close to central Stockholm, with a high-frequency light rail system, the Tvärbanan, and an extensive pedestrian and bicycle network. There are also 30 car-sharing cars distributed throughout the neighborhood.

While not a perfect example, Hammarby represents a new and valuable way of seeing city buildings, and it requires a degree of inter-disciplinary and intersectoral collaboration that is unusual in most cities.

—*Timothy Beatley, University of Virginia*

SOURCE: See endnote 48.

---

approach to building design that moves toward circular metabolism by using technologies that reuse water and generate electricity. (See Chapter 5.) Vegetation planted on a building's exterior captures water that would otherwise be wasted, for instance, while reducing the energy needed for cooling. The 15-story IBM headquarters in Kuala Lumpur designed by Ken Yeang is a good example of this. In New York City, photovoltaic cells embedded in the south and east facades of the Condé Nast building in Times Square, combined with two fuel cells, provide enough electricity to operate the building at night.[49]

A well-known example of a circular design on a larger scale is the eco-industrial park in Kalundborg, Denmark, where waste gases from an oil refinery are burned by a power plant, waste heat from the plant warms commercial fish ponds, and other companies use byproducts of combustion to make wallboard and concrete.[50]

The concept of green infrastructure is gaining adherents in Europe and America. This is a planning idea for a whole metropolitan region, applicable also to rural areas facing strong development pressures. By thinking at a regional level, planners identify natural areas and corridors that can sustain the ecological fabric of the area, allowing plants and wildlife to continue ecological functions such as migration and seed dispersal, even as land is converted to urban uses. The network of green spaces also provides flood control, clean air and water, and recreational services to the urban residents.[51]

Cities in low-income countries could, in theory, leapfrog directly to twenty-first-century technologies, and many ideas are being tried in industrializing economies, such as the large-scale adoption of solar energy in Rizhao, China. Already, many cities have skipped directly to wireless phone systems, and a number of projects are under way to test the viability of water recycling and decentralized renewable energy in medium-income countries. A key element of leapfrog innovation is making it possible for municipal governments to benefit politically from solving problems in new ways. One effort to do this in the rapidly urbanizing landscapes of Asia is described in Box 1–3.[52]

Which Urbanizing World?

As most humans come to call an urban environment home in this century, we are learning from the determined inhabitants of cities like Accra. In the long-established slum of Nima, a member of the municipal assembly has organized his neighbors to manage the mountain of refuse that emerges each day from Nima's numerous small businesses and markets, helping an overtaxed city deal with the garbage that clogs drains and worsens

---

Box 1–3. The Mayors' Asia-Pacific Environmental Summit

In May 2006, mayors and other local government officials from 49 cities in 17 countries around the Pacific Rim gathered in Melbourne, Australia, at the fourth Mayors' Asia-Pacific Environmental Summit (MAPES), a conference first held in 1999. Of 47 municipal leaders at the 2003 meeting who pledged to meet specific environmental goals within two years, 7 received awards for doing so in 2006. Their achievements ranged from building new composting plants for municipal wastes in Nonthaburi, Thailand, to extending water pipes to 4,000 poor households in Phnom Penh, Cambodia, and installing a new sewage-disposal system in Male, the capital of the Maldives.

Once a mayor makes a commitment, the summit organizers provide technical support and help line up funding. In 2006, each city that received a MAPES award also won a scholarship for a resident of the city to study at the Royal Institute of Technology in Stockholm in a program leading to a master's degree in urban sustainable technology. The American software publisher ESRI also offered assistance to cities that wanted to adopt geographic information systems in municipal management.

The brainchild of Jeremy Harris, former mayor of Honolulu, and Karl Hausker, an economist who worked at the U.S. Environmental Protection Agency, MAPES is funded by the Asia Development Bank and the U.S. Agency for International Development.

"We know how to build sustainable cities," Harris says, "but it is still hard to marshal the political will to do so." With an eye on Asia's rapidly growing cities, he and Hausker devised MAPES to give mayors political rewards for moving toward sustainability, together with technical resources and the chance to learn from one another.

Other attempts to transfer sustainable development projects from one city to another have shown that lasting change is difficult to achieve. Even within a city, environmental reforms can be started and stopped with the changing of administrations. In Honolulu, energy-efficient lightbulbs in City Hall and a water recycling facility seem likely to last beyond Harris's tenure. However, his successor has ripped out some street trees, arguing that Harris was fiscally irresponsible and that the city could not afford the cost of maintaining the trees. Whether MAPES, with its focus on political leaders, will be able to facilitate environmentally responsible development that is also sustainable in a governmental sense is an important question.

SOURCE: See endnote 52.

---

flooding in the rainy season.[53]

We are seeing cities with rapidly growing economies like Tijuana experiment with reforms that have streamlined municipal finance, enabling the city to support residents' self-initiated upgrading of their housing.[54]

We have seen the transformation of Singapore in only one generation from a struggling, newly independent city-state to a modern postindustrial city prospering in a global economy through financial services, manufacturing, and a major port. Singapore's slums were replaced with modern housing, most of it built by a competent, incorruptible government. With its mass transit system and compact urban design, Singapore also enjoys a high standard of living with lower energy use per person than in the United States.[55]

None of these cities is sustainable yet. Low-income cities stagger under their growing populations' unmet needs. Industrializing cities' demands for energy and materials compound the demands already placed on nature by high-income urban and suburban consumers. All cities depend on many ecosystems. There are few cases now in which that dependence would be durable over the long run, even if population and consumption were to stabilize.

The urbanizing world must coexist with the natural world if both are to endure. The extraordinary diversity of human experience and human enterprise provides ample evidence of the threat of wider irreversible damage to ecosystems but also promising paths toward a sustainable future. Urbanization, perhaps surprisingly, is leading us to rediscover nature and the ecosystem services on which all humans rely. Creating urban habitats that deliver the bounty of nature in a sustainable fashion to the inhabitants of cities in all societies is an opportunity within our reach, as well as a cardinal test of our humanity.

CITYSCAPE:
TIMBUKTU

Greening the Hinterlands

As an ancient trading center at the Sahara's edge and an early center of culture and learning, Timbuktu occupies a unique place in popular imagination. More recently, the city captured attention during the Sahelian droughts of 1968–74, when famine led to the deaths of 100,000 people and millions of livestock, and 1984–85. Many survivors from northern Mali migrated to Timbuktu, which grew from 8,000 people in the mid-1960s to 32,000 in 1998. They found work in rice fields and market gardens, in the handicrafts industry, and in tourism and small enterprise. Timbuktu remains an active commercial center for trans-Saharan trade.[1]

Since the late 1960s, increasingly arid conditions have led to the dramatic contraction of woodlands and flood pastures around the city. Drought devastated several critical species, including bourgou, an aquatic grass that provides important dry-season pasture for livestock, and the doum palm, a slow-growing tree whose trunk is widely used in construction.

Yet Timbuktu's growth has created incentives for rural people to invest in natural resource management. Along the banks of the Niger River near the city, eucalyptus woodlots and bourgou pastures have taken root on once barren and fissured land. Over the past two decades, rural people have been planting these species to supply the city with construction materials and fuel. Bourgou and eucalyptus provide significant money in a region where the average per capita income is $240 per year and 77 percent of the population lives in poverty.[2]

Aguissa Bilal Touré, mayor of Bourem-Inaly, a village of 8,700 people that is 35 kilometers from Timbuktu, has been at the cutting edge of this trend. Through careful management, he earns about $1,335 a year from his five-hectare eucalyptus forest along the Niger River. Touré explains: "Since we started planting eucalyptus, we have not had to buy wood to cook or to build our houses. With the sale of its wood, we can buy everything we need—grain, dry goods, livestock, clothing, motorbikes. It is with this income that we pay our taxes, purchase medicine and school supplies for our children, and even save money. People continue to plant rice here, because it's our tradition. But everyone here has a woodlot."[3]

Eucalyptus was first introduced by the Malian forest service in the late 1980s with projects supported by the U.N. Capital Development Fund and the International Labour Organization that mobilized community groups to plant trees along irrigated rice fields as windbreaks and for fuelwood. Development projects supported associations of individual producers by providing hand tools and shared motor pumps to irrigate saplings. One such effort, the Projet de Lutte Contre l'Ensablement, worked with 124 associations and 54 individuals before closing in 2001.[4]

A hectare of eucalyptus yields about 10.6 cubic meters per year in this region. The net value of construction wood from a hectare after five years is about $1,335 on the local market. Coppiced trees can be harvested up to three times, every five years, before they must be replanted.[5]

By the early part of this decade, according to the regional director of the Forest Service, people were planting trees so quickly that foresters could not keep track. By then the eucalyptus forest stretched along the Niger for 50 kilometers on either side of Timbuktu. Some producers have begun making charcoal, and others are negotiating with processing facilities to sell eucalyptus leaves for use in insecticides.[6]

In nearby Hondo Bomo Koyna, villagers

Charles Benjamin

Young girl washes dishes in a bourgou plain, fish trap in the background, near the town of Bourem-Inaly

plant bourgou flood pastures, which they cut to sell for fodder in Timbuktu. A single hectare of well-managed bourgou yields an average income of $800 per year. Amadou Mahamane, the village chief, explains: "Bourgou is more interesting for us than rice. After paying for fertilizer and water fees, we earn around $180 per hectare from rice. We sometimes even use our income from bourgou to pay our water fees for growing rice."[7]

Bourgou was introduced in 1985 by the French nongovernmental group Vétérinaires Sans Frontières (VSF) to rehabilitate flood pastures damaged by drought. From 1985 to 1989, VSF worked with 24 nomadic groups and 19 villages. Hondo Bomo Koyna was the first to ask for help; the number of individuals cultivating bourgou there climbed from zero to over 150, and they planted about 1,200 hectares. Villagers soon discovered demand for bourgou in Timbuktu, where people were increasingly investing in home livestock. As a result, they nearly doubled the area planted in bourgou.[8]

The woodlots and flood pastures around Timbuktu are unintended outcomes of projects intended to improve the natural resource base of rural production systems. For hundreds of people like Touré and Mahamane, a healthier natural resource base has meant increased income and reduced vulnerability to drought. The income from these products provides rural people with a few more choices. They are not forced into cities as environmental refugees. A young man in Hondo Bomo Koyna explains: "Last year, my brother who lives in Senegal called me to help with his business. I responded that I couldn't come until after the bourgou had been harvested.

In other words, the income from bourgou is more important than what I could have earned abroad."[9]

Driven by urban demand for wood and fuel, this greening of Timbuktu's hinterlands has not only reconstituted natural resources but also increased the resilience of the environment. Woodlots have stabilized the banks of the Niger River, slowing the movement of sand into the riverbed. They have slowed erosion by protecting the soil from wind and rains. And the bourgou grasslands have buffered fields from seasonal floods and provided spawning grounds for fish.

Although development organizations jump-started the process, local initiative spurred the rapid spread of bourgou and eucalyptus. The relationship between Timbuktu and the surrounding countryside offers a powerful counterpoint to commonly held beliefs that rural people in the Sahel are hopelessly trapped in a cycle of poverty and environmental degradation and that urbanization has exacerbated negative environmental trends.

—*Charles Benjamin, Williams College, Massachusetts*
—*Aly Bocoum, Near East Foundation, Mali*
—*Aly Bacha Konaté, Réseau GDRN5, Mali*

23

Ecological and Healthy City

When Dr. José Bolívar Castillo was elected mayor of Loja, Ecuador, in 1996, land use policies permitted this impoverished Andean city of 160,000 to sprawl uncontrollably, including in precarious parts of the city. Deforestation resulted in flooded rivers, while lead-fueled buses and cars polluted the air. Garbage filled the city's streets, polluted rivers, overflowed collection bins, and ultimately filled a site across the street from the world-renowned Podocarpus National Park.[1]

For Mayor Castillo, "The inspiration for an ecological city came from within Loja. I remember when I was a child, before the city became so polluted." During his eight years as mayor, Loja's municipality turned the city from an "average" Ecuadorian city, into a *ciudad ecologica y saludable*—an ecological and healthy city—by implementing policies that underscore the correlation between a healthy ecosystem, a healthy human population, and a healthy economy.[2]

Comprehensive land use planning and environmental policies that were carried out at the scale of the county limited degradation of the land, improved public health, and facilitated the municipality's management of necessary infrastructure—all while saving material and construction costs for important municipal projects, such as adding water lines to the poorest neighborhoods. Scientist Dr. Ermel Salinas explained that the water supply was drinkable "because our rivers have been cleaned up, protected and are treated to United States Standard Method requirements," preventing many illnesses caused by drinking dirty water.[3]

A well-enforced ordinance required real estate developers to leave 20 percent of their land undeveloped for public open space, resulting in many popular parks. Architect Jorge Muños Alvarado, Loja's Director of City Planning, explained that the greenery "acts as a sponge by retaining stormwater, which prevents the rivers from flooding," while Dr. Humberto Tapia, Director of Public Health, noted that exercising in parks reduces preventable illnesses such as obesity, diabetes, and heart disease, which in turn can lower the death rate from these ailments.[4]

In addition, Wilson Jaramillo, a municipal transportation planner, noted that air quality improved following what he called "a more sustainable transportation policy" by implementing a new rule requiring all cars to run on unleaded gas with catalytic converters and by running cleaner public buses throughout Loja.[5]

To handle the waste problem, the city's recycling program required residents to separate organic from inorganic trash. Residents have been very receptive to this program: 95 percent of them separate their garbage perfectly every day. Meanwhile, the municipality collected all of the city's trash at least once daily and swept streets several times a day, leaving them eerily clean.[6]

The benefits of this program were widespread. Ecologically, the city recycled all organic waste and over 50 percent of the inorganic waste generated in the city, dumping nonrecyclables and hazardous waste in a sanitary landfill. Economically, the city earned about $50,000 a year (7 percent of the program's $685,000 operating cost) from selling recycled materials, and it created more than 50 related full-time jobs citywide. And in terms of health, the cleanliness of the streets lowered the presence of rodents and vermin.[7]

Understanding why the recycling program works so well can help others bring about environmental change in their own cities. Most obviously, Lojanos joined in because the municipality fined any household or busi-

Municipality of Loja

Trash containers at the San Sebastián market

Municipality of Loja

before (left) and after the improved collection and recycling program was implemented (right)

ness that did not comply with a local regulation that required participation. Furthermore, the municipality shut off a building's water supply if its owner did not pay the fine. The system's organization also assured a high rate of participation: each collection truck met its schedule, plus or minus 10 minutes, seven days a week—and inspectors on the trucks recorded and rigorously enforced infractions. Fernando Montesinos, Director of Sanitation, insisted that the program was "an investment of Lojanos in our own city."[8]

Moreover, the municipality created incentives for participation, such as providing pipes for water lines and materials for public parks, which were constructed through communal work projects called *mingas*. Lolita Samaniego, president of La Floresta, a women's housing organization, explained that a *minga* is an obligatory event where "everyone works for everyone's benefit…women provide food and men distribute the work amongst themselves, working from sunup to sundown." These gestures strengthened residents' perception of a direct relationship between waste management, natural resources, and civic improvement. Local leaders asserted that civic awareness and cultural solidarity were values that stemmed from Loja's indigenous past.[9]

Furthermore, hiring a local workforce gave city workers a tremendous sense of ownership and pride. One evening, Marlon Cueva and other Lojano engineers spoke of the significance of having designed Loja's recycling facilities: "Designing

something I had never seen before—that's what it means to live….When I achieved my goal it satisfied me and validated me as a person and professional, raising my confidence and increasing my expectations of myself."[10]

In addition, explained Fernando Montesinos, it helped that no single "recycling tax" existed. Rather, 20 percent of each household's water bill (about 20¢ per month) funded a portion of the recycling program, while smaller percentages of taxes were taken from the highway tax and other public funds. Thus recycling became not a project unto itself, but a small part of a greater network of public works and development projects.[11]

The city has won three international prizes for its efforts: the International Awards for Livable Communities' Nations in Blooms first prize for community involvement (endorsed by the United Nations Environment Programme), its bronze prize for a global ecological city (behind the Swedish cities Norrkoping and Malmo), and Promoter of Natural Environment's City of the Americas first prize for public recreation and physical activity. If a poor city that has many needs beyond the health of its environment can learn to embrace the ecological city, then such a concept—with appropriate cultural adjustments—surely can succeed elsewhere.[12]

—*Rob Crauderueff*
Sustainable South Bronx, New York

25

CHAPTER 2

Providing Clean Water and Sanitation

David Satterthwaite and Gordon McGranahan

Few readers of this chapter spend any time or effort collecting water for their daily needs. We have water on demand, when we need it, 24 hours a day, in kitchens, bathrooms, and toilets. We do not have to fetch and carry it from a distant public tap or water tanker or from a polluted river. But hundreds of millions of urban dwellers in low- and middle-income nations still do.

A household of five needs at least 120 liters per day (24 liters per person) to meet basic needs—for drinking, food preparation, cooking and cleaning up, washing and personal hygiene, laundry, house cleaning. This means someone in the household (usually a woman or older child) has to fetch and carry 120 kilograms of water—the equivalent of six heavy suitcases—every day. This heavy load often needs to be carried over considerable distances. The water tap or the place where a water tanker delivers is usually more than 50 meters from people's homes, and often much further. In many places, there are long lines at the tap or tanker, greatly increasing the time needed to get water. The water often has to be paid for—and it costs much more per liter than water piped into our homes.[1]

Almost everyone reading this chapter has a toilet in their home and easy access to toilets in their workplace, with piped water to a basin for hand washing. Hundreds of millions of urban dwellers in low- and middle-income nations do not have this. They have no toilet at all or only communal toilets that are hard to get to or public toilets that are far away, dirty, and expensive to use—and these are often dangerous for women and children to use after dark. As a result, hundreds of millions of people have to defecate in the open or into waste materials they throw away.

David Satterthwaite is a Senior Fellow at the Human Settlements Programme and Gordon McGranahan is Director of the Human Settlements Programme at the International Institute for Environment and Development in London.

Sanitary Revolution Still Needed

Clean, convenient water supplies for drinking and bathing and convenient toilets only became routine in the last 100–150 years. The cities of the industrializing world in the nineteenth century were notorious for being far less healthy than their rural surrounds, until the sanitary revolution provided them with piped water and sewerage systems. Today, about a billion urban dwellers still need their sanitary revolution.[2]

The health costs of inadequate provision are very large. A million or more infants and children still die each year from diseases directly related to inadequate provision of water and sanitation, and hundreds of millions are debilitated by illness, pain, and discomfort. Their nutritional status is often compromised by water-related diseases (especially diarrhea and intestinal worms).[3]

It is still common for one child in 10 to die before their fifth birthday in urban areas in low-income nations, with much higher mortality rates among low-income urban dwellers. Such statistics are not surprising, considering that around half the people in African and Asian cities lack water and sanitation to a standard that is healthy and convenient. For Latin America and the Caribbean, more than a quarter lack such provision. (See Table 2–1.)[4]

How is this possible, given that in the mid-1970s governments and international agencies committed themselves to making safe water and sanitation accessible to all by 1990? It might be explained by governments and international agencies concentrating on needs in rural areas—but provision for much of the rural population is also very inadequate. It might be explained by disappointing interna-

tional funding levels and the low priority that most funding agencies give to this. But provision is also poor in many cities that have received substantial international funding. Clearly, existing approaches are not working—but what has to change?[5]

At first glance, the solutions seem simple: expand international funding for systems that have worked well in high-income nations—piped water supplies and flush toilets connected to sewers in each home. These may be water-intensive and poor at recycling. But this is hardly a reason for letting hundreds of millions of urban dwellers suffer the ill health, hardship, and indignity of distant, overused, or contaminated water supplies and unsanitary living conditions.

Flush toilets and sewers are generally the safest, most convenient, most easily maintained forms of provision in urban areas for homes, schools, workplaces, and public places. They provide public health advantages by reducing the risk of human contact with excreta and preventing groundwater contamination. Water and sewer pipes require very little space within buildings—an important advantage in most city contexts—and flush

Table 2–1. Number and Share of Urban Dwellers Lacking Adequate Provision of Water and Sanitation, by Region, 2000

Region	Lacking Water	Lacking Sanitation
	(number of people)	
Africa	100–150 million (35–50 percent)	150–180 million (50–60 percent)
Asia	500–700 million (35–50 percent)	600–800 million (45–60 percent)
Latin America and the Caribbean	80–120 million (20–30 percent)	100–150 million (25–40 percent)

SOURCE: See endnote 4. Note that these figures are "indicative estimates" because most governments do not report on the share of their population with adequate provision of water and sanitation.

toilets work better than most other forms of sanitation in multistory buildings. Their unit costs are reduced with higher densities, in some cases making them cheaper than on-site latrines. And if most or all households and workplaces have piped water supplies, sewers are needed to collect wastewater even if they do not collect toilet wastes. This system also works well in many urban centers in middle-income nations such as Brazil and Mexico, where a high proportion of the population is well served by this model.[6]

But in low-income urban settings flush toilets and sewers are rarely cost-effective, often become dysfunctional, and—where they do function—often reach only a small elite of urban residents. This conventional "high-income-nation" solution will not work in most urban centers in Africa and Asia or many in Latin America and the Caribbean because there is no local capacity to provide the needed design, implementation, management, and financing, as well as accountability to residents. The effectiveness of this model depends almost entirely on the service provider. A pit latrine may seem inadequate in most urban contexts, but it is an option that households can build and manage themselves without depending on an external agency. It also works without a regular piped water supply (unlike flush toilets). Most urban centers in low- and middle-income nations—including many with more than a million inhabitants—have no sewers at all.[7]

In most urban centers in low- and middle-income nations, many of the people lacking adequate water supplies and sanitation live in "slums" or informal settlements. More than 900 million urban dwellers live in "slums" in low- and middle-income nations, and this is where most of the deficiencies in provision are evident. Most of their housing is informal or even illegal—on land illegally occupied or subdivided and with a house not compliant with building and planning regulations. This presents difficulties for any agency as there is no official, registered "owner" to whom they can offer a service—and it is risky to extend piped supplies or sewers to illegal settlements that the government may bulldoze. In fact, it is often illegal for official providers to do so.[8]

In addition, the ways in which most cities have expanded in low-income and most middle-income nations is haphazard, with most new developments being outside of any plan or regulatory framework. Most cities have expanded with very inadequate provision of land use management (including the need for new land development, watershed protection, and flood control) and basic infrastructure. New homes or enterprises sprout up in areas unserved by water and sanitation infrastructure—and often in places that are expensive to reach. Meanwhile, all water users and wastewater generators seek the cheapest and most convenient means to meet their needs, regardless of whether it depletes or contaminates groundwater resources or pollutes water for those downstream.

The larger the city and the greater the volume of wastewater generated by households and businesses, the more pressing is the need for good wastewater management. Difficulties providing this are compounded when, as is often the case, built-up areas limit water infiltration and channel storm water flows in ways that cause or exacerbate flooding.

Measuring Deficiencies in Provision

To say that everyone needs access to water and sanitation is not enough. No one can live without drinking water and defecating, so in a sense 100 percent of the population has some form of access. But what everyone needs is adequate provision—provision that is safe, affordable, and accessible. To achieve this, we need to know who already has this and, more important, who does not.

This is difficult to measure. In most low- and middle-income nations, there is no information on who has adequate water and sanitation. While a discussion of how to measure who has adequate provision might seem to be a technical issue, actually it should be at the center of all discussions. Get the measurements wrong, and the problem is misstated, which usually means that the policies to address it are also wrong. For too long, that is what has happened.

It is easy to list the attributes of adequate provision of water and sanitation, but it is difficult to measure most of them. For water, the issue is whether each person has easy, affordable access to safe and sufficient water for all needs. Easy access implies that water supplies are close to the home, are regular, and can be obtained without waiting in a long line. Water quality is also very important, at least for the water to be used for drinking and food preparation. Of course, all of this water needs to be available at a price that even low-income groups can afford.

The attributes of adequate sanitation are also easily listed. Clean and readily available places for women, men, and children to defecate in private at home—and also at work, at school, and in public places—with good provision for anal cleaning (and managing the safe disposal of paper or other materials used) and washing. If there is no provision within each person's home, good-quality, easily accessible, well-maintained toilets should be available close by at costs that low-income people can afford. Excreta must be disposed of in ways that prevent contamination by direct contact, contaminated water, or flies.

Although these criteria seem uncontroversial, it is hard to find out who has water and toilets that meet them. Most official statistics are based on questionnaires administered to a representative sample of households. But it is impossible to know whether a household has adequate water and sanitation on the basis of a few predetermined questions, particularly where people rely on a range of water sources and technologies for different uses and at different times and seasons. Adequate provision depends not only on the technologies involved but on how they are used, in what sort of setting, and by whom. Men, women, and children have different needs, and people with impaired mobility, including many elderly, have their own special needs. Finding out whether a household's pit latrine is adequate, for example, would require a long list of questions and observations, most of which would be quite different from those relevant to flush toilets or a communal toilet.

Adequate provision depends not only on the technologies involved but on how they are used, in what sort of setting, and by whom.

Large-scale household surveys typically involve samples that underrepresent precisely the areas where water and sanitation provision is at its worst—people living in informal, low-income settlements. They are also not designed to provide useful information for local water providers; they can say what proportion of households in a nation lack piped water supplies, but they do not indicate exactly which households lack them—or which neighborhoods have the worst provision. Censuses provide the coverage that household surveys do not because they draw information from all households (though they too usually undercount the homeless and those living in informal settlements). But censuses at best happen every 10 years, and they generally include even less information on water and sanitation. It is also rare for the information they contain to made available to local governments or utilities to identify where improvements are most needed.

This creates a dilemma for those responsible for presenting international statistics on water and sanitation. They can either aspire to estimates of the population shares without adequate water and sanitation, despite the absence of any systematic empirical basis for such estimates, or they can develop indicators that can be estimated empirically but that do not correspond to adequate provision. During the Water and Sanitation Decade of the 1980s, governments were asked to provide estimates of access to "safe" water and sanitation even though most had no real basis for doing so. Starting in 2000, and setting the basis for monitoring progress toward the Millennium Development Goals (MDGs), the latter approach prevailed instead, and governments have been helped to develop estimates of the proportion of urban and rural populations with "improved provision" on the basis of household surveys wherever possible.[9]

The statistics in Table 2–1 might seem to be at odds with official U.N. statistics, which show a much smaller proportion of urban dwellers lacking "improved provision." But the Table provides estimates of access to adequate water and sanitation, while U.N. statistics provide estimates of "improved" coverage, which may or may not be adequate. As the United Nations explains, there are no data available for most nations on the proportion of people with good quality or "adequate" provision from a health perspective or with "sustainable access to safe drinking water," as called for in the MDGs. Unfortunately, many who use the official statistics are unaware that these do not measure "adequate" or "safe" provision. This often means erroneous conclusions are drawn not only about the scale of the problem but also about where the worst situations are found.[10]

The fact that U.N. statistics do not measure who has adequate provision helps explain why urban populations that show "improved" water and sanitation still have very high infant and child mortality rates. For instance, according to a U.N. report published in 2002, over three quarters of the urban population of Kenya and Tanzania had "improved" provision of sanitation in 2000—but most of these have poor-quality, poorly maintained pit latrines, often shared with many other people. Infant and child mortality rates remain high in the urban areas of both these nations.[11]

Although it is tempting to argue for more-detailed, internationally comparable household surveys, these do not provide the information needed at the local level. Action on the ground often requires detailed data about each household, each structure, each plot boundary in the poorly served areas, and the forms of provision already found there. It also requires maps that include contours and details of roads, paths, and plot boundaries. This is not the same as the information that governments and international agencies need to monitor provision. While it is important to recognize the deficiencies in the statistics commonly used to monitor progress toward the MDGs, it is more important to correct the deficiencies in the information needed to drive local action.

First Things First

Although the most animated discussions on water and sanitation generally focus on which technologies should be used, the choices of how to finance and manage provision and who takes responsibility for different aspects of it are just as important. Discussions on finance and management tend to focus on public versus private, but here too there are a range of options, often with public, private, and nonprofit organizations having complementary roles.

A key issue is whether the providers and their associates—whether government agencies, private enterprises, or nongovernmental organizations (NGOs)—are accountable to local residents. What is needed in each urban center or district is more competent local water and sanitation organizations in which those who are unserved or ill served have influence. In many of the cities where provision has improved and expanded, it was not technological innovation that drove the improvements, but financial and management innovations underpinned by responses to demands from those without adequate provision.

Thus, there are no universal truths in regard to technology, management, or financing, although there are some useful working principles. For one, what is provided has to work for low-income groups, which also means being convenient and affordable by both users and providers. This also means seeking solutions that do not undermine the financial viability of providers—whether public or private. It is nonsense to think that it is always possible to recover all costs, including all infrastructure investments, through user fees or connection charges. The poorest groups may have incomes so low that full cost recovery is impossible, making cross-subsidies necessary, through either utility tariffs or taxes. But it is good practice to seek solutions that keep down unit costs and minimize the need for subsidies. The less these are needed, the more the possibilities of expanding provision and reaching those with very limited incomes, and the less the dependence on external funding.

Seeking to reach everyone with adequate provision does not necessarily mean reaching everyone with the same form of provision. It is much better to have well-managed cheaper provision in low-income areas than no provision at all—well-managed communal water taps, for instance, rather than house connections. The lower-cost forms of provision can allow incremental improvements as more finance becomes available. For instance, the piped water system serving the communal taps needs to have the capacity to allow household connections when people can afford them. Sometimes it is cheaper to expand water provision by tapping local water resources rather than waiting for the expansion of the conventional water mains. Externally funded "solutions" that replace rather than build on existing systems are often overly expensive (as well as ill suited to local circumstances) and can discourage or even crowd out more appropriate locally funded solutions.[12]

Provision can also be improved in ways that are not classified as "water and sanitation" projects. For instance, housing loans may allow low-income households to improve provision in existing homes. There is a growing trend for micro-finance institutions to provide loans for home improvements—which often include connections to piped water and sewer systems or better provision within the home through better internal plumbing or toilets. Or mortgage finance can allow people to buy or build a new home with better provision. Programs providing those living in informal settlements with legal tenure may allow local utilities to extend piped water and sewer connections that were not allowed when the settlements were illegal. Many city government support "slum and squatter" upgrading programs, which usually include better provision of water and sanitation. Federations of "slum" and "shack" dwellers can work with the relevant agencies on the household enumerations and mapping of informal settlements that are needed for local action—and at a small fraction of the cost of professionally managed surveys.[13]

Providing Adequate Water Supplies

The best system for providing water remains a city-wide piped network supplying as many households and other buildings as possible with in-house piped connections. The cost of serving each structure with piped supplies comes down with increasing densities. Cities also provide economies of scale for water treatment and bill collection. Willingness and capacity to pay is higher in most urban areas, if a good service is offered.

But installing, managing, and, where needed, expanding a piped water system is complex. This is especially so in rapidly growing cities, where there is usually a patchwork of new developments scattered around the periphery of the built-up area to which it is difficult and expensive to extend pipes for water. In many cities and small urban centers where local capacity is lacking, the piped system serves only a small proportion of inhabitants—generally the better-off. Water supplies are also often intermittent—which commonly means that supplies in many parts of the system are contaminated (as pressure in the piped system drops, contaminants can easily seep into the pipes through cracks).

One critical issue is how to reduce costs and increase the possibility of covering them, through either user fees or some form of public finance.

However desirable it is to recommend a citywide system with piped connections for everyone, this is often beyond local capacities. A pro-poor strategy in such circumstances would focus on two aspects. First,

ensure that those who do receive piped connections pay the full cost (within a system that seeks to keep down such costs). Second, look for ways to ensure better provision for low-income groups and other groups currently unserved. This might involve connecting them to the piped network, but if the cost of house or yard connections is too high, provision could be through shared taps or communal or public taps or through supporting separate provision (for instance, local shared water tanks that draw from groundwater).

In all urban centers there is a need to review who is using poor-quality water sources—for example, unprotected shallow wells and untreated water from local rivers or lakes—and to discuss with them the best means of improvement. This means considering what arrangements with low-income neighborhoods can improve provision in ways that can be afforded by providers and inhabitants. (See Table 2–2.) Providing taps in backyards is cheaper than in-house connections. Water taps shared and managed by 5–25 households are much cheaper than yard taps, especially if billing is shared (since it is much cheaper for the water provider to collect the money from the group). Water kiosks (where water is sold by the bucket) are also a possibility—and these can be managed by a community organization. But if water costs more than low-income groups can afford, they will draw on unimproved sources for other domestic needs.[14]

Of course, all these forms of provision need a regular, sufficient, good-quality water supply. Communal and public taps need regular water supplies; it is often the intermittency of supplies that causes long lines. In all this, perhaps the most fundamental change sought is the quality of the relationship between water providers and the population.

Most households would prefer to be as far

down the steps in Table 2–2 as possible, so one critical issue is how to reduce costs and increase the possibility of covering them, through either user fees or some form of public finance. Even if the upstream infrastructure and trunks are government-funded—and there is often considerable justification for this—the costs for connection and for operation and maintenance can be burdensome for low-income households.

One innovation first developed in Brazil and now being applied in several other nations is "condominial" water supplies. Here, the cost of the public water network is much reduced by the water agency providing the water pipes to groups of households (including condominiums or cooperatives)—with the households within this group taking responsibility for installing the pipes to their homes or yards. This can cut the cost per household served so much that many low-income households can afford this instead of communal or public standpipes.[15]

There are also many examples of local ingenuity and partnerships that allowed improved provision despite limited budgets and users' capacity to pay. One example is the community taps program in San Roque Parish, Mandaue City (which is part of Metro Cebu, in the Philippines). A local federation formed by savings groups of urban poor people provides the legal framework for various community-managed development projects. The Metro Cebu Water District's Community Faucet Program gives poor communities permission to tap into the mains and get water at a low cost if they plan, finance, and manage the laying of the pipes and install the taps.[16]

Table 2–2. Ladder of Water Supply Improvement Options for Households

Types of Provision	Management
Shared unimproved water sources	Individual or communal
Shared public taps	
Cooperative taps (such as 1 per 20 households)	Mixture of formal (for instance, for supply) and community
Yard taps for each household	
Individual house supplies (including multiple-tap)	Formal organization (government or private sector)

Shift from left to right leads to or involves:
- increasing convenience (and decreasing time needed to obtain water), usually increasing unit costs; and
- increasing use of water, usually with health benefits (especially for washing, laundry, and personal hygiene).

Shift from left to right is helped by increasing size of population, more commercial and industrial demand (which helps fund infrastructure), and more households with greater capacity to pay.

Shift from management by households and communities to professional management involves increasing sophistication for technical management and higher capital costs.

SOURCE: See endnote 14.

Providing Adequate Sanitation

The ideal adequate sanitation is for everyone to have a well-functioning toilet within their home, with facilities for washing after using it. But for many low-income groups, this is not possible—it may be too expensive, or the room or rooms they live in are rented and the owner does not want to provide toilets. One defining characteristic of cities is the high cost of space, so poorer people live in overcrowded conditions in order to keep accommodation costs down. Many low-income households have only one or two square meters per person. In addition, the most "space-efficient" form of sanitation—a toilet connected to a sewer—is often not an option because there are no sewers.[17]

For the full public health benefits of sanitation, every woman, man, and child needs to be able to use a toilet that safely disposes of their feces. If even a small portion of the people in a neighborhood defecate in the open, this can put everyone's health at risk. But whatever the health and other risks posed by defecation in the open or into waste materials (known as "wrap and throw"), these remain the cheapest sanitation options. (See Table 2–3.)[18]

Eco-sanitation, which takes an ecosystem perspective, emphasizes the closure of material flow cycles, including the recycling of human excreta and water from households. Returning urine and feces to the soil makes wastewater treatment much cheaper. Most forms of eco-sanitation also use little or no water. For the system to work well, there needs to be a good, safe way to store and remove the urine and feces—and then transport the materials to farmers who want to use them. It also requires good management to control smells, keep the toilets clean, and avoid flies. Most designs use more space than toilets connected to sewers—although the dehydration model (where urine is separated from feces) reduces the space needed. The unit costs vary widely, depending on local circumstances and the type of system used.

A study in the city of Kunming in China asked stakeholders about introducing two different kinds of eco-sanitation: "no mix" toilets in which urine is separated for use as a fertilizer (with feces disposed of through a conventional flush) and dry toilets, also with urine separation but with feces kept in a chamber within the house, with ash being added and the dry wastes collected regularly. Most people surveyed recognized the need for such toilets: Kunming is by a lake that is now heavily polluted. The dry toilets are significantly cheaper and imply much less water use, but most stakeholders considered them less appropriate. This suggests the need for innovative technical and organizational solutions to make dry toilets more acceptable in urban contexts. This form of eco-sanitation is likely to find widespread application first in lower-density developments on the edges of cities, where there are fewer space constraints and no sewers to connect to and where demand for the "nutrients" is close by.[19]

A Pakistani NGO, the Orangi Pilot Project (OPP), has helped transform debates about provision of sanitation by demonstrating that it is possible to install good-quality sewers serving each house, even in low-income settlements, with full cost recovery. This NGO was formed in 1980 to support new models of providing infrastructure and services in Orangi, a large cluster of low-income, informal settlements in Karachi that now has some 1.2 million inhabitants. Since then, its work has expanded to encompass many other areas of Karachi and to support partner organizations working in other urban centers in Pakistan.[20]

Table 2–3. Different Sanitation Options and Costs

Type of Provision of Sanitation	Cost per Household	Benefits and Drawbacks
	(dollars)	
A flush toilet connected to a sewer or septic tank within each home plus piped water to the home for personal hygiene	400–1500	Costs per person rise a lot if provision is made for sewage treatment using conventional treatment plants with high levels of treatment.
Condominial sewers (the Orangi Pilot Project model of "component sharing")	40–300	With high densities and strong community organization and input, unit costs per household can compete with pit latrines.
An "improved" latrine or pour-flush toilet linked to a latrine within each home	40–260	No need for sewers. Improved latrines control smells better than conventional pit ones and limit or prevent insect access to excreta. Difficult to find space for this in most urban contexts; not suitable for multistory buildings. Children often frightened of using them (dark, large pit).
Eco-sanitation	90–350+	In most models, no need for sewers. Many models with provision for urine diversion, which has advantages for nutrient recycling and on-site decomposition but usually adds significantly to unit costs.
Basic latrine	10–50	No need for sewers. If well-managed, can be as healthy as more expensive options but difficult to find space for in most urban contexts; not suitable for multistory buildings.
Access to a public or communal toilet/latrine (assuming 50 persons per toilet seat)	12–40	Effectiveness depends on how close it is to users, how safe to use at night, how well maintained, and how affordable by poorest groups.
Possibility of open defecation or defecation into waste material	none	Obvious problems both for those who defecate and for others in the community.

SOURCE: See endnote 18.

At the core of OPP is the concept of "component-sharing," whereby the inhabitants of each street or lane take responsibility for planning, financing, overseeing the construction of, and managing the "internal" pipes—in the case of sanitation, the lane sewer to which each household's toilet connects—that then connect to a government-provided "external" sewer or to a natural drain. As the inhabitants of each lane work together to install and manage the pipes, advised by OPP or another local organization, they cut unit costs dramatically—typically to a fifth of what they would have been charged by the official water and sanitation agency. This brings unit costs down to the point where low-income households can afford to pay and so allows full cost recovery. Thus, each lane organization offers the local sanitation utility a partner who can manage the most time-consuming aspect of improved sanitation—the work at each household and lane.

The intention of the locally supported initiatives in Pakistan is to form partnerships with local governments. OPP has also supported the mapping of all informal settlements in Karachi, with the help of local youth teams, to strengthen inhabitants' claims to stay on the land and to provide the basis for installing or improving water, sanitation, and other infrastructure. For people living in an informal settlement, negotiations with any government agency are enormously strengthened if they have accurate and detailed maps showing existing plot boundaries, roads, pipes, and drains that also allow detailed discussions of what is needed to improve conditions and where this has to be built.[21]

After more than two decades of work—and often opposition from government as well as external consultants who told them they had the wrong approach—OPP has not only supported or catalyzed good-quality sanitation and drainage for hundreds of thousands of people in Karachi. It has changed the way that city governments design and plan sanitation provision. It succeeded in changing a sanitation master plan for the whole of Karachi from one that was expensive and poorly designed (and unlikely to work well) to one that was far more effective, cheaper, and better for low-income groups and that avoided the need for a large foreign loan. OPP is helping to identify priorities for work all over the city—influencing what government investments are made but without relying on funding from government to do so in order to safeguard its independence.

This example raises an interesting issue— the role of local civil society organizations in developing the best possible mix between good-quality and convenient provision, affordability, and local management. The annual budget of this local NGO is tiny compared with the value of the pro-poor sanitation and other activities it has supported—and also compared with the money it has saved the government in unnecessarily expensive and poorly designed infrastructure systems. The lessons from OPP lie less in what was built and more in how it was done: the work done directly with urban poor organizations, building their capacities; the demonstration of new ways of designing, building, managing, and financing water and sanitation infrastructure; the very detailed documentation and mapping of needs; the credibility established with local governments; and, finally, the quality of the advice provided to communities and to government.

Communal Provision of Water or Sanitation

Communal or public provision of water through public taps or water kiosks has long been seen as a "solution" for poorer groups. But the same does not hold for sanitation, although this has been challenged recently by some local innovations (with some receiving support from the international NGO WaterAid). The communal approach works better for water, as people can store water in their homes and can avoid lines or the need to fetch water at night. Communal or public provision of sanitation is never ideal—but in many places it is the best compromise between better provision, affordability, and the potential for local management. It is cheaper than household provision and often much easier to provide in existing high-density settlements.[22]

Efforts to provide public toilet blocks in India provide one example of this working well. (See Box 2–1.) As with OPP in Pakistan, this was initiated by local NGOs and community-based organizations but not as autonomous provision. The project improved sanitation for hundreds of thousands of people and showed how to reduce the gap between the cost of achieving better provision and what can be afforded by low-income groups. The cities and communities where this program was implemented have some of the country's

---

Box 2–1. Toilet Blocks in India Designed and Managed by the Community

During the late 1980s and early 1990s, two community organizations and a local NGO designed, built, and managed public toilet blocks because provision was poor to nonexistent in their neighborhoods. The alliance they formed was between the National Slum Dwellers Federation, Mahila Milan (a network of savings groups formed by women "slum" and pavement dwellers), and the Mumbai-based NGO SPARC—the Society for Promotion of Area Resource Centers.

The construction of each toilet block was usually preceded by a community-managed survey to document the inadequacies in provision. Local savings groups from the "slum" helped design, implement, and manage the toilet blocks. These blocks sought to avoid the deficiencies in the siting, design, and management of existing public toilets. The community toilets gave women more privacy, made waiting lines work better—for instance, separate lines for men and women, since otherwise men just

push ahead of women—ensured a constant supply of water for washing, and made better provision for children. Many of them had separate children's toilets at the front so youngsters did not have to line up and wait.

Community management of the toilet blocks ensured that they could be maintained through user charges, with costs being much lower than conventional "public toilets." Typically, families paid a standard monthly fee. Caretakers and cleaners were identified from the local community. At first, local governments ignored or discouraged these efforts. Then the municipal commissioner in Pune, a city with over 2 million people, and eventually other city authorities recognized the poor quality of public toilets and the inadequate numbers built and supported this alliance in building community toilets. More than 500 toilet blocks have been built to date—mostly in Pune and Mumbai, but increasingly in other urban centers as well.

SOURCE: See endnote 23.

---

lowest-income urban populations.[23]

But it is difficult to make communal facilities work well, and especially difficult to make them safe and convenient for children and women to use at all times and to keep the toilets clean and well maintained. The costs of a regular water supply to such facilities and a high standard of maintenance have to be met—but this might imply higher charges than users can afford. Having enough toilets available to avoid waits at peak times increases costs—but lines discourage many people (especially children) from using the facilities. The difficulties of good communal provision are generally fewer if a toilet is shared among relatively few households who know each other—for instance, a toilet shared by those living around a yard.

Public or communal facilities are worth

considering where it is not possible to provide water connections and good sanitation to each household—but this decision has to made in consultation with those who cannot afford household provision. This will generally require an incremental approach, with good communal provision as a first step and with support for household provision thereafter. The possibilities of good household provision are often greatly increased if groups of households work together to help plan, manage, and finance the installation.

Increasing Private-sector Participation

In the 1990s, many international agencies—led by the World Bank—vigorously promoted private-sector participation in water and sanitation programs. This was in part a pragmatic response to the failure of public utilities to undertake the desired improvements during the 1970s and 1980s and in part a reflection of the broader decline of support for government planning in favor of private enterprises and markets. Public utilities were widely seen as inefficient, overstaffed, manipulated by politicians for short-term political ends, unresponsive to consumer demands, and—particularly in low-income settings—inclined to provide subsidized services to the urban middle class and leave the urban and rural poor unserved.[24]

Small water and sanitation enterprises in low-income urban settlements have been neglected in the debates on private-sector participation.

Proponents of increasing private-sector participation argued that by transferring utility management to private operators, under competitive conditions, it should be possible to overcome these problems and to increase investment in expanding water and sanitation provision. It was hoped that private concessions, such as those created in Buenos Aires and Jakarta, would point the way to a new form of public-private partnership, although a number of other contractual forms, including the management contracts that are now more popular, were also explored.[25]

Increased private-sector participation—or privatization, as it is more often termed by its critics—proved to be very contentious. This is perhaps not surprising. The fear of private companies gaining a monopoly over urban water systems and putting private profit ahead of public interest stretches back centuries. The controversy over contempo-

rary concessions has been heightened by the fact that consortia led by a handful of large foreign water companies initially won most of the major contracts. The notion that development assistance was being used to promote privatization also caused controversy, even in donor countries. With contemporary concessions, the ownership of assets has been kept in public hands and contracts have been signed that, at least in principle, prevent the companies from raising water prices unilaterally to secure excess profits. This has not been enough to allay the fears of critics, however.[26]

It is difficult to assess the impact of these developments on people without adequate access to water and sanitation, or even to determine whether the net effects have generally been positive or negative. The findings of the largest statistical reviews have been ambiguous. But when measured against the very optimistic claims that private-sector participation would expand provision, improve efficiency and accountability, and reduce corruption and political interference, the results must be viewed as disappointing.[27]

The early contracts did not produce the desired results, at least for the urban poor. The Buenos Aires contract was initially taken as a model of a well-designed concession, but it did not provide the basis for extending provision to the many unserved parts of the city where residents did not have secure tenure and it imposed connection costs far above what other unserved residents could afford. By the time such problems were being addressed, Argentina's economic crisis undermined the very basis for the concession. In Jakarta, the two concessions also ran into problems: soon after the contracts were signed the country was in economic crisis, President Suharto fell from power, and the close ties between the local partners and Suharto family business interests turned from being a necessity into a liability. These concessions were renegotiated, but many of the obstacles to extending provision to Jakarta's low-income residents still persist. Indeed, for a great many countries privatization has been more of a diversion than a solution to urban water and sanitation problems.[28]

Generally, governments that have problems managing public utilities also have problems with privately operated utilities. Indeed, many of the obstacles to public provision in the poorest areas remain when providers become private: uncertain or illegal land tenure, difficult terrain, complex and cramped plot layouts, large distances from existing water mains and trunk sewers or drains, illegal connections, difficulties collecting bills, corrupt practices. Moreover, as critics have pointed out, the private water companies have not invested nearly as much in the water sector as envisaged in the early 1990s.[29]

Understandably, private operators were most interested in securing concessions in very large cities that had an appreciable middle class. There was little interest in the smaller cities and towns where most of the ill-served and unserved actually live. Most of those without adequate provision today are not about to be connected to a functional water network and are nowhere near any sewerage network. At least in the near future, it matters little to most of them whether the large urban utilities are operated privately or publicly, as they will not be served in either case. Even in middle-income countries, many cities do not so much have water networks as they have a multiplicity of different systems, some more industrial and others more artisanal, some under more corporate control and others under more community control. In many low-income areas, the private sector is important, but primarily through the goods and services sold by small water and sanitation enterprises (including water vendors) rather than through utilities operated by multinational companies.[30]

In recent years, the international drive to privatize water and sanitation provision has stalled, and increasing private-sector participation is now rarely presented as the solution to water sector problems. In many countries the water sector is still undergoing internationally supported reforms, many of which involve shifts toward commercial principles and a regulatory environment that is more open to private-sector participation. Moreover, while the multinational water companies may be less active in pursuing concessions, other contracts and private enterprises are gaining in importance. In parts of Asia, for example, local companies are reportedly becoming stronger. Of 124 major water and sanitation contracts identified in a recent report as operational in Asia in 2004, for example, 42 had been awarded to national companies and a further 17 to companies run by Chinese entrepreneurs based in Malaysia or Singapore (with 84 of the 124 contracts being in China).[31]

Small water and sanitation enterprises have long had key roles in low-income urban settlements in Africa, Asia, and Latin America. They have been neglected in the debates on private-sector participation. Local governments have often suppressed them, despite the fact that the informal markets for small water and sanitation providers are closer to the free market ideal than the heavily regulated markets of water utilities are. There is growing recognition that utilities and governments need to acknowledge the strengths of these enterprises and work with them rather than against them. There are serious challenges to this, however. Many operate with technologies that do not meet official standards or sell at prices that exceed the controlled prices of the official utilities, making it difficult for governments to condone let alone support a role for them.[32]

Perhaps the mistake of the 1990s was to confuse good business principles on water and sanitation provision—keep down unit costs, be accountable to clients or potential clients, recover costs from users wherever possible—with the operations of large-scale private businesses and to assume that only large private enterprises could apply good business principles in the sector. In many ways, most of the local NGOs and community organizations mentioned earlier have done better at following these business principles. These are important, but there is no reason why they cannot be implemented by civil society organizations—and by government agencies.

Many case studies of cities with water shortages show that these situations are more often the result of poor management than of water scarcity.

As noted already, many public water and sanitation agencies provide good services and close to universal coverage, and they have done so by applying good business principles. There are also many others whose performance has improved greatly by doing the same. For instance, a government-owned water corporation in Uganda that serves Kampala and many other urban centers shifted from large losses to a surplus between 1998 and 2006; in these years, a series of initiatives tripled the number of households served, reduced unaccounted-for water from 51 to 29 percent, and lowered connection and reconnection fees. Customer relations were also much better. However, it is worth noting that although this can make good provision of water pay for itself, it does not provide the financial basis to address Kampala's huge backlog of people without adequate sanitation.[33]

Managing Water Resources Better

Improving provision of water and sanitation obviously depends on regular, sufficient supplies of fresh water, but it is important not to confuse this goal with the problem of water resource scarcity. Since the 1980s, there have been growing concerns about water scarcity and stress, driven primarily by land use changes and the large quantities of water used in food and biomass production. This growing water stress poses serious risks to existing aquatic and terrestrial ecosystems and to many agricultural systems. As of yet, however, there is no evidence that it is a significant factor in inadequate provision of water and any associated ill health and hardship.

Unfortunately, particularly in the more popular accounts of an impending water resource crisis, statistics on the number of people without improved water supplies and those suffering from water-related diseases have often been presented alongside estimates of the numbers living in water-scarce or water-stressed areas. The implicit suggestion is that water stress is why people fail to secure better water supplies and become ill. Statistically, however, there is no association between nations facing water stress and those with the most inadequate provision of water for rural and urban populations, even at similar income levels. Many large cities where provision of water and sanitation is quite inadequate have little or no overall shortage of freshwater resources, while cities facing severe water scarcity often still manage to achieve high levels of provision.[34]

Urban residents, particularly low-income groups, do sometimes face difficulties securing adequate water supplies as the result of water resource deficiencies, and these are not always evident in statistics on water provision. In Jakarta, for example, the poorest residents pay more than the wealthy for their water in areas where the groundwater is saline (and they are forced to buy high-priced vendor water) and less than the wealthy where the groundwater is suitable for drinking (after boiling). Neither the water vendors nor the unprotected wells they use can be considered improved water supplies, however. Indeed, in a somewhat perverse way the saline groundwater makes water coverage look better, since it motivates some households to connect to the piped water system. In Mexico City, on the other hand, water resource problems affect the piped water provision, especially in poorer, more peripheral parts of the city. And each dry season brings an increase in popular protests about water shortages and denunciations in the press. In this case, even the households facing supply breakdowns in the dry season would be considered as having improved supplies, again failing to detect the deficiencies resulting from water resource problems.[35]

Many case studies of cities with water shortages show that these situations are more often the result of poor management than of water scarcity—as in the case of Guadalajara, Mexico, and the drying up of Lake Chapala and in Beijing, where a great deal of water in and around the city is used inefficiently. Mismanagement can also lead to intermittent supplies from parts of the piped water system without any water resource scarcity whatsoever. In any case, the amount of additional water required to meet basic urban needs is small relative to the consumption for other purposes. The additional water required for adequate provision is, if well managed, unlikely make a significant difference to a country's overall water withdrawals. On the global scale, an additional 20 liters per day for

a billion people would still only amount to about 7 square kilometers a year, compared with aggregate water withdrawals on the order of 4,000 square kilometers.[36]

Thus, most solutions to water resource scarcity in cities lie in better local management. Often the cheapest way to increase available supplies of fresh water for the piped water system is to reduce leaks and introduce pricing structures that encourage larger users to withdraw less. As with the technologies for water supply and sanitation, there are numerous techniques through which water use can be cut or wastewater reused or recycled—or new sources tapped (for instance, through rainwater harvesting in homes, institutions, and public facilities). Where such measures are insufficient, it may still be possible to rehabilitate upstream infrastructure to achieve higher capacity utilization rates. In other cases, new sources of bulk water need to be identified. Addressing water supply issues in the piped water network thus requires a two-pronged approach: improving management and securing sources of bulk water for current and future needs.[37]

Overall, measures to improve water availability at the settlement scale are undoubtedly important in many urban centers, but they are rarely more than part of the solution to problems of inadequate provision for the worst-off residents. Indeed, it should be kept in mind that while water scarcity and infrastructure deficiencies are likely to hurt those with the fewest economic and political resources, measures to address scarcity and invest in new infrastructure will not necessarily help them. Only when these are combined with measures to improve provision in the more deprived areas and to secure more influence for such groups in the planning process will water resource management and infrastructure investments likely benefit those who need them the most.

Who Needs to Change the Way They Work?

The *Human Development Report 2006* characterized the way ahead for sanitation as more local development and more engagement with the unserved. This also summarizes the approach needed on water. As emphasized throughout this chapter, each urban center needs the best possible mix between good-quality convenient provision, what they can afford, and what can be managed locally.[38]

Local governments obviously have the central role in changing the approach—and in most examples of very good or much improved provision, the changes were underpinned by more competent and accountable local governments. But local governments are often weak and ineffective because higher levels of government want to keep them this way. And they are often unrepresentative of poorer groups' interests because of powerful local vested interests.

Success in water and sanitation rests not only in what local governments do, however, but also in what they encourage, support, and supervise. In many instances, small-scale water businesses (vendors and water kiosks), local NGOs, cooperatives, and community organizations already make important contributions to provision or to financing better provision. And there are many potential partnerships between local governments and these different service providers. Local government can do much to improve the provision of water and sanitation by other means—for instance, through measures to increase the supply and reduce the cost of land for new housing and through support for "slum" and "squatter" upgrading programs, as noted earlier. The many ways this can be done are often not recognized; for instance, extending roads and a good public transport system on the edge of a city's built-up core can reduce

land prices for housing (so that far more low-income households can afford to build their own homes) and cut unit costs for water and sanitation provision.

However creative and effective these community organizations, NGOs, or small water enterprises are, it falls to local governments to provide the framework in which these can work and, where needed, collaborate. Most local governments lack the capacity to manage or oversee a conventional model where formal agencies—whether public, private, or cooperative—provide good-quality water and sanitation to all buildings, with good management both "upstream" and "downstream." But most can provide a framework of support for a combination of public, small-scale private, NGO, community, and household provision.

It is also possible to combine conventional and unconventional models, with the conventional model used wherever possible (where users can afford to pay its full cost) and with alternative models meeting needs elsewhere. Within this, the conventional model can be expanded to cover growing proportions of the population. In addition, as shown by many examples in this chapter, household and community action can do much to improve provision within homes and neighborhoods. But they usually need the larger (external) systems from which to draw water and into which to dispose of liquid and solid wastes.

Promoting good hygiene is another important part of the solution—and the benefits of good hygiene practices are particularly large where sanitation provision is the worst. This falls within a more fundamental need for good relations between local governments, official service providers, and the unserved or ill served, because their cooperation and support is essential for improvements.[39]

The polarized debate about "public" versus "private" providers can obscure the fact it is the quality and efficiency of providers and their accountability to current and potential clients that is the key issue. Since most official water and sanitation providers in urban areas are still in the public sector, reform of these agencies is often the most effective local approach—including ensuring less political interference in setting prices, a more responsible attitude toward bill paying by public-sector water users, and a more business-like approach, such as increasing coverage through reducing costs and ensuring that costs can be fully recovered through user fees or secured from other reliable sources.[40]

In most nations, more effective and pro-poor local governments depend on support from higher levels of government. In many Latin American nations, improvements in water and sanitation provision have certainly been boosted by decentralization and by stronger local democracies, although perhaps less than might have been anticipated because many governments followed economic policies that reduced government roles and investments through much of the 1990s.[41]

Every nation also needs innovations of the kind shown by the Orangi Pilot Project in Pakistan and by the water engineers in Brazil who developed condominial systems for water and sanitation—or by federations of slum, shack, and pavement dwellers that have demonstrated far cheaper and more effective ways to develop new housing with good provision of water and sanitation. It is not so much what these groups did but the extent to which their innovations were rooted in local contexts and had as a priority reaching low-income groups. This allowed them to drive debates and discussions within their own districts, cities, and nations about different approaches—and provided innovative, working examples that politicians, civil servants, professionals, and community organizations

could visit. Innovations only work and spread if they are appropriate to local contexts.[42]

Enormously expanding and improving provision in urban areas in ways that reach low-income groups depends not so much on scaling up existing efforts as on supporting a vast multiplication of local initiatives. There are tens of thousands of urban centers where provision of water and sanitation is extremely inadequate; each needs locally driven processes to address the inadequacies.

It is the quality and efficiency of providers and their accountability to current and potential clients that is the key issue.

But supporting this approach is problematic for most international agencies. First, it means supporting work in urban areas; many of these agencies have long avoided working in cities or have restricted funding to urban projects in the belief that these areas attract an excessive share of development funds. They fail to see the inadequacies in provision of water and sanitation—and the scale and depth of poverty—in urban areas. And they fail to see how fast urban poverty is growing in most nations.[43]

Even if this changes, however, perhaps as great a problem is knowing how to make the large sums that international agencies can provide through governments support the diverse local processes needed to improve and extend provision. Official development assistance agencies are ill equipped and poorly structured to support a multiplicity of local processes implemented by local governments, NGOs, or community-based organizations, many of which require modest levels of funding and a strong engagement with the unserved. It is perhaps not surprising that most of the innovations in provision of water and sanitation in urban areas that reached low-income groups were not funded by these official agencies.[44]

If national governments provide the appropriate framework for local action, development assistance can support this. But it is the lack of national government support for local action that is the problem—especially in nations where provision of water and sanitation is the worst. Meanwhile, most international agencies are providing less support for water and sanitation initiatives as they shift funding to budgetary support (which is less staff-intensive than projects) or funding channeled through other agencies. And for the agencies that do still provide substantial funding for water and sanitation, it is often for high-cost infrastructure that delivers little to poorer groups.[45]

Low-income groups are the ones who have to drag the equivalent of six heavy suitcases of water back to their shelter every day.

As a recent United Nations report notes, it is time for external agencies to support local capacities to develop locally appropriate solutions, not impose their often inappropriate and costly solutions and often inappropriate conditions. Improving provision in urban areas does need substantial funding, especially for the big "trunk infrastructure" for water, sanitation, and drainage into which community or neighborhood improvements are integrated and for the larger water management upstream and waste management downstream. There is not much point in extending piped supplies to unserved communities if bulk water supplies are insufficient to cope with the new customers. But there is also not much point in providing funding for bulk urban infrastructure if city and local governments have no intention of improving and extending provision to most low-income groups.[46]

The bottom line in this discussion of water and sanitation challenges around the world is that low-income groups most need the world's help in their daily struggle for access to adequate supplies and facilities. They are the ones who have to drag the equivalent of six heavy suitcases of water back to their shelter every day. And who have to relieve themselves in the streets outside the shacks they call home.

The strain this brings to families is captured in the words of Chhaya Waghmare from Pune, India: "There are 280 families in our settlement.... Every day we get water brought to us in tankers. The delivery timings are not regular. We start queuing for water in the morning by putting our water containers in a line. If we have to go out we can leave the house only after we have filled the water. I have to go to work. My children are very young and cannot fill the water. So my sister stays at home and waits for the tanker. In order to be home when the tanker comes, she has stopped going to school."[47]

Shalini Sadashiv Mohite of Mumbai describes what it was like before the community organization of which she is a member built their own public toilet: "There was no toilet in this whole area. Men and women from the settlement squat along the road. Women do not go after six in the morning. They wait for the cover of darkness. We even eat less so that we do not need to relieve ourselves during the daytime."[48]

In the end, for most of those lacking adequate water and sanitation the issue is as much about better relationships with those in power and utility managers as it is about technology or pricing or infrastructure investments. Most examples of better provision in this chapter were underpinned by governments' recognition of the legitimacy of the needs of unserved groups—even groups who lived in illegal settlements. Many situations were also helped by urban poor organizations and local NGOs demonstrating to governments the possibilities of much improved provision, if they worked together. Yet most international funding agencies have yet to understand this—or even if they do, they have failed to restructure their funding to support it.

Collapsing Infrastructure

With more than 10 million residents, Lagos is literally bursting at the seams. This former capital of the Federal Republic of Nigeria and West Africa's leading seaport and commercial-industrial center received substantial federal government funding in road and bridge construction during the oil boom era of the 1970s. The city has consequently expanded phenomenally—right up to the margins of Ota, some 40 kilometers to the north, in Ogun State. A massive influx of people has placed unbearable pressure on transportation, housing, and water and electricity supplies in Lagos, creating a widening gap between demand and supply.[1]

While concerted efforts by the federal and state governments had ameliorated the situation up to the late 1970s, three developments aggravated the crisis. First, hostile, mainly military, federal governments since 1979 frustrated ambitious state government plans for infrastructure development, especially for an urban rail service. Second, the massive devaluation of the national currency and a steep decline in economic fortunes under the World Bank's Structural Adjustment Program, coupled with official corruption, denied the city much-needed resources for infrastructure development. Third, the transfer of the federal capital to Abuja in 1991 practically sealed the fate of Lagos. As architect Rem Koolhaas remarked, "Lagos was left to its own devices, then abandoned."[2]

Following the return to civil rule in 1998, all tiers of government can provide water, electricity, and roads. But the record of government service in these sectors is dismal. One new development has been the involvement of the private sector in service provision. The Lagos state government has initiated public-private partnerships in water, roads, and electricity supply. It has hired private-sector technocrats to run its water corporation and has started an independent power production scheme. Most recently, the federal government has proposed an ambitious scheme for developing the Lagos megacity.[3]

Water supply illustrates the challenges and possibilities of fresh initiatives in tackling infrastructure collapse in Lagos. The three major waterworks at Iju, Adiyan, and Isashi have a combined installed capacity of 119 million gallons a day, although their output is just 69 million gallons a day. Twelve mini and eight micro waterworks have a combined capacity of about 3 million gallons a day. While the major waterworks rely on surface water from the Ogun and Omo Rivers, the others depend on groundwater supplies. Yet large shortfalls have increasingly been experienced since British rule ended in 1960, due to population growth, dilapidated infrastructure (leaking pipes), illegal connections, poor maintenance, and inadequate access to the limited supplies. Official supplies meet barely half the demand, and coverage throughout the city is uneven.[4]

Private operators take up the deficit in official supplies. Water is brought in to high-income households in tankers, while hawkers sell to poorer people in high-density neighborhoods in four-gallon tins. The wealthy invest in private boreholes. Wells, which have been sunk across the city since the nineteenth century, are more widespread. But the quality of water from them is generally poor, given the low-lying terrain, the pollution of groundwater sources, and the shallowness of the wells. Since the mid-1990s, however, treated water sold in nylon sachets, known as "pure water," has emerged as a popular source of supply. "Pure water" is portable and satisfies the requirements of individual consumers as well as large religious, political,

Ganiyu Ajibola Aliyu/UNEP/Peter Arnold, Inc.

Lagos: rubbish pile in the foreground, commuters beyond

heightened concerns about the future of the megacity, the only one of its size without a functional rail mass transit system. Hostile intergovernmental relations have hampered a coordinated approach to the resolution of the infrastructure crises in Lagos. Agents of the federal and state governments have clashed over the control of roads in the city, and the federal government has withheld funds allocated to local governments in the state, in defiance of the Supreme Court. The national government has often frustrated initiatives by the state, such as independent electric power projects.[7]

and social gatherings. Unfortunately, the indiscriminate disposal of the nonbiodegradable sachets has compounded the problem of environmental sanitation.[5]

Meanwhile, the Lagos state government has attempted to privatize water supply, backed by a law passed in December 2004. This was in the face of a debt of $117 million incurred by the Lagos State Water Corporation (LWSC) by the end of 2004. Various stakeholders have opposed this scheme of "private-sector participation," which retains government ownership but hires private operators to manage the distribution of potable water. For the last two years the LSWC has been run by Olumuyiwa Coker, a chief executive recruited from the private sector, who admits that LSWC covers only 55 percent of the state.[6]

The inability of the state government to cope with collapsing infrastructure has

Relations improved, however, in 2006 following the appointment of a new Minister of Works. In July 2006, the federal government announced its intervention through a Lagos Mega City Region Development Authority, with a primary focus on the city's housing crisis. It also endorsed a 26-billion-naira ($200-million) International Development Association interest-free loan for the upgrade of infrastructure in the megacity. The renewed optimism this brought to the city is tempered by the realistic need to wait and see if all stakeholders meet their obligations.[8]

—Ayodeji Olukoju
University of Lagos, Nigeria

CHAPTER 3

Farming the Cities

Brian Halweil and Danielle Nierenberg

On the surface, Accra in Ghana, Beijing in China, and Vancouver in Canada seem to have little in common. They range in population from roughly 2 million in the metropolitan region of Vancouver to more than 14.5 million in Beijing. The per capita incomes are vastly different: about $700 a year in most of Ghana, about $2,200 in Beijing, and more than $32,000 in Vancouver. But take a closer look, digging a little deeper into the backyard and rooftop gardens, and you'll realize that these city folk share a preoccupation that has thrived since the first cities—raising food.[1]

Accra has a population of 6 million, including a steady supply of migrants from rural areas and immigrants who seek work in its factories. Because food is expensive, people farm anywhere they can: in backyard plots, in empty lots, along roadsides, and in abandoned dumps. These farmers grow a variety of crops for home use and sale, including exotic varieties like green peppers, spring onions, and cauliflower, as well as more traditional crops like okra, hot peppers, and leafy greens such as alefi and suwule.[2]

There are more than 1,000 such farmers in Accra. Their plots vary from just one tenth or one twentieth of a hectare (10 meters by 10 meters) to 20 hectares in the city outskirts. Among the biggest challenges they face is keeping their crops irrigated, since clean, affordable sources of water are not easy to find. Backyard farmers often use greywater— the waste water from bathrooms and kitchens. While sewage water can be a health hazard, farmers in Accra—and in cities all over the world—are finding that human waste can be a valuable fertilizer.[3]

In Beijing, city planners in the 1990s decided that urban agriculture was an important way to meet the city's food needs, preserve green spaces, and conserve the region's water and land resources more efficiently. They began offering courses and assistance for aspiring farmers, they surveyed existing land use to better understand the extent of urban farming, and they tried to incorporate urban farming into long-term city planning decisions.[4]

Today, urban and peri-urban agriculture

(farming in, around, and near cities) in Beijing not only provides residents with safer, healthier food, it also keeps farmers in business. Between 1995 and 2003, the income for farmers living just outside of Beijing doubled. The city includes tens of thousands of household farms and more than 1,900 agrotourism gardens for Beijing residents craving some rural experience. Although the share of the city's population involved in farming is currently very small—just about 1 percent—the municipal government plans to cultivate gardens on 3 million square meters of roof space over the next 10 years.[5]

Vancouver is known for being a popular destination for tourists. But what most visitors do not realize is that the city is a leader in encouraging its inhabitants to grow and buy fruits, vegetables, and other items produced in the city. According to a recent survey, an impressive 44 percent of Vancouverites grow vegetables, fruit, berries, nuts, or herbs in their yards, on their balconies, or in one of the 17 community gardens located on city property. Vancouver's mild temperatures and ice-free winters make it the ideal city to grow food nearly year-round. There, farming the city is part of a much larger movement that includes restaurants buying from local farms, buying clubs in which neighbors subscribe to weekly deliveries of produce, and the heavily attended Feast of Fields harvest festival twice a year on a farm outside the city that exposes city folk to rural life.[6]

A Rich History of Urban Farming

Growing food and raising fish and livestock in Accra, Beijing, and Vancouver—indeed, in cities all over the world—is nothing new. In some ways, these three cities are responding to the same challenges that urban gardeners have faced for millennia. The hanging gardens

in Babylon, for instance, were an example of urban agriculture, while residents of the first cities of ancient Iran, Syria, and Iraq produced vegetables in home gardens. This is partly because cities have traditionally sprung up on the best farmland—the same flat land that is good for farming is also easiest for constructing office buildings, condominiums, and factories—and partly because the masses of urban dwellers create a perfect market for fresh fruits and vegetables.

"In ancient times, the cost of transport was much greater," explains Jac Smit, head of the Urban Agriculture Network, "so the impetus for growing food in cities was greater." Of course, urban farmers continued to refine their craft. Centuries after the Incan residents of Machu Picchu raised food in small, intensive plots irrigated with the city's wastewater, enterprising Parisians developed bio-intensive production with steam-heated greenhouses and glass cloches that cover individual heads of lettuce; they sold their produce as far away as London. In China's cities, farmers developed complex cropping patterns and trellises that made use of every available square meter.[7]

But like the story of all local farming, a range of forces in the modern era—the Industrial Revolution, the evolution of the megacity, the invention of refrigeration—helped to render urban farming obsolete. In particular, when cities first combined industrial and organic wastes in one sewage stream at the end of the nineteenth century, they made wastewater too toxic for irrigation. And in many cities, urban agriculture became not only harder to practice but illegal as well, thanks to overzealous city officials and public health practitioners who wanted to eliminate urban livestock production.

Then during the 1970s, something changed. People working for the United Nations, the Peace Corps, and other development groups noticed the spontaneous appearance of home gardens and small retail farms in major cities throughout Asia, Latin America, and Africa. Rapid urbanization, inefficient and expensive transportation, and a greater demand for food made raising produce and livestock in cities possible and necessary. In other words, the same needs that had given rise to urban farming in ancient times had reappeared. And although cities in industrial countries might be able to compensate for traffic congestion and lack of local food with superior transportation and packaging, those in developing countries could not. Urban farming was posed to take off again.[8]

In fact, farming is ubiquitous in cities today. The U.N. Development Programme estimates that 800 million people are involved in urban farming worldwide, with the majority in Asian cities. Of these, 200 million produce food primarily for the market, but the great majority raise food for their own families. In a survey conducted for the United Nations, cities worldwide already produce about one third of the food consumed by their residents on average, a percentage that will likely grow in coming decades, given that the need for urban agriculture could be greater now than ever before.[9]

According to the U.N. Food and Agriculture Organization (FAO), the number of hungry people living in cities is growing. While malnutrition in rural areas is still a bigger problem in terms of actual numbers of people—of the 852 million people worldwide who are undernourished, 75 percent live in rural areas—urban residents, particularly children, also suffer from food shortages as well as micronutrient deficiencies. Urban agriculture can be one of the most important factors in improving childhood nutrition, by increasing both access to food and nutritional quality. Recent studies in the Philippines and elsewhere confirm this linkage between better childhood nutrition and the production of food in urban areas.[10]

Fortunately, urban politicians, businesses, and planners are beginning to regard urban agriculture as a tool to help cities cope with a range of ecological, social, and nutritional challenges—from sprawl to malnutrition to swelling landfills and the threat of attacks on the food chain. In this context, taking advantage of land in and around cities is essential and obvious. Unlike parks or other green space, which are generally financed by taxpayers, urban farming can be a functioning business that pays for itself. And for cities that use nearby farmland to filter wastewater, recycle garbage, and cool down the concrete jungle, farming is rapidly becoming something they can't do without.[11]

Replenishing Food Deserts

Local food takes on a very different meaning on a planet where half the people live in cities. As a greater share of the world's population resides farther from where food is grown, produce has to be moved across countries and sometimes around the world.

In 2001, FAO officials were concerned about the capacity of large cities in Asia, Latin America, and Africa to feed themselves. They found that by 2010 many of these cities will require massive increases in the number of truckloads of food coming into the area each year—increases that would overwhelm the capacity of these cities to distribute food. Bangkok will need 104,000 additional 10-ton truckloads each year, Jakarta will need 205,000, Karachi 217,000, Beijing nearly 303,000, and Shanghai just under 360,000. And while cities may never be able to meet all their food needs from local farmland, the tremendous infrastructure, energy, and cost required to shuttle food into densely populated areas argues for urban centers to secure as much of their food as possible from farmland within their borders or nearby.[12]

For the inhabitants, cities bring certain gastronomic advantages. A diversity of people and businesses means access to a wide range of cuisines compared with more-traditional fare in the countryside. Cosmopolitan commerce means that specialized stores and international supermarkets stock a variety of ingredients. At the same time, a more harried urban lifestyle often means that city folk have less time to cook or prepare meals from raw ingredients and that they opt for the convenience of processed, prepared, or even fast food. (Consumers in urban areas pay up to 30 percent more for their food than people in rural areas do, partly because they grow fewer of their own ingredients and partly because the food travels farther.)[13]

But the change in habits raises all sorts of nutritional and logistical concerns. Foods that are more processed require more refrigeration, clean water for preparation, and sophisticated transport lines. They also mean more sugar and fat in the diet, which combined with more sedentary urban lifestyles encourages diabetes and obesity. A study of 133 developing countries found that migration to the city—without any changes in income—can more than double per capita intake of sweeteners, simply because they are available cheaply. Traditional staples—whole grains, potatoes and other root crops, and some vegetables—on the other hand, are often more expensive in urban areas. For example, surveys show that recent migrants to Hanoi, Viet Nam, eat less rice, corn, vegetables, and beans than they used to and more meat, fish, eggs, milk, soft drinks, and canned and processed food. Home-prepared meals are gradually replaced by restaurant fare and street food.[14]

So this is the essence of the urban food quandary: People living in cities demand more food and a greater range of foods than their rural counterparts, but they live farther from the centers of food production. In response, people often start farming in the city simply because they cannot find an affordable and reliable source of the foods they crave from their rural roots or because they might not have the cash to buy food at all. As opposed to in the countryside, in cities a lack of money translates more directly into lack of food.[15]

In other words, growing food is not a hobby for most people, it is a necessity. Studies from several African cities have shown that families engaged in urban agriculture eat better, as measured by caloric and protein intake or children's growth rates. In terms of providing an essential source of food and income, urban and peri-urban agriculture is probably most important in sub-Saharan Africa. In the cities and towns in East Africa, a third of urban dwellers are engaged in agriculture. In West Africa, the number of households involved in urban agriculture varies from more than 50 percent in Dakar, Senegal, to roughly 14 percent in Accra, Ghana. In Dar es Salaam, Tanzania, 60 percent of the milk sold is produced right in the city.[16]

In densely populated Bangkok, home to roughly 10 million people, rising demand for aquaculture products such as morning glory, water mimosa, and freshwater fish is met primarily by an industry of peri-urban water farmers. Nearly one third of the nation's intensive urban aquaculture production comes from around Bangkok, and it generates about $75 million each year. Catfish farms in the northern part of the city produce more than 70 percent of the country's total output of this fish. And about 40 kilometers west of Bangkok, there are vast farms growing the aquatic plant morning glory (a staple of the Thai diet), while 20 kilometers south of the city tilapia and carp thrive in large ponds.[17]

Farms in the city can often supply markets on a more regular basis than distant rural farms can, particularly when refrigeration is scarce or during a rainy season when roads are bad. And local food production might be the best option for feeding urbanites neglected by the long-distance food chain. In both the industrial and the developing world, poorer urban households typically spend a greater share of their income on food than wealthier urbanites do. In some cases, poor urbanites spend 60–80 percent of their income on food, making them especially vulnerable to price changes.[18]

Nutritionists and sociologists have argued that many poor inner-city areas in industrial countries have become "food deserts" in recent decades. Supermarkets have left the inner cities to milk the more lucrative suburban markets, after pushing many of the independent mom-and-pop grocers out of business. Entire city neighborhoods have been left with only fast-food restaurants and convenience stores. This provides a good opportunity for farmers' markets and community-supported agriculture subscriptions, in addition to food co-ops and other locally owned stores. In the Anacostia region of Washington, DC, which has not had a supermarket for years, a new farmers' market is the first good source of fresh food for local residents.[19]

This reliance on city farmers is not always planned. Cuba depends heavily on urban farming—an estimated 90 percent of the produce eaten in Havana is grown in and around the city. The shift, however, was not entirely voluntary. In the early 1990s, the U.S. embargo and then the collapse of the Soviet Union left Cuba without agrochemicals, farm machinery, food imports, or petroleum, hobbling its capacity to produce food and ship it to cities. Confronted with massive food shortages, government officials set up a loose network of local extension offices that helped Cubans obtain vacant land, seeds, water, and gardening assistance.[20]

Cuba's main motivation was preventing a shortfall of food, but its support for city farming has also been a wise investment in jobs and crisis prevention. Urban farming in Cuba has created 160,000 jobs, including farmworkers, masons, vendors, herb dryers, and compost makers. Egidio Paez of the Cuban Association of Agricultural and Forestry Technicians notes that "the growth and spread of cities invariably creates many empty spaces…which often become trash-dumps that are sources of mosquitoes, rats and other disease vectors." Transforming these unhealthy spaces into farms and garden spaces creates jobs. Cuba's urban farmers raise food organically—without pesticides or chemical fertilizers—eliminating the health and environmental problems that result from agrochemicals.[21]

These welcome benefits of urban farming are needed in many countries. In Yaoundé, Cameroon, more than 70 percent of urban farmers do not have other occupations, a figure that rises to 85 percent in Abidjan, Côte d'Ivoire. In peri-urban Hanoi, agriculture still generates more than half of the incomes in some sections. In Kumasi, Ghana, the annual incomes of some urban farmers were estimated at $400–800, which is two to three times what they could make in rural farming.[22]

In many cases, farming is particularly suited to city folk without jobs or some outlet for developing skills for the working world. In Boston, Massachusetts, the Food Project trains inner-city youth how to do many of the jobs associated with a commercial catering business; the youngsters work on a farm, harvest the food, prepare it, and serve it at events. In Cairo, Egypt, teenage girls who are not allowed out of the house according to religious customs have found a calling—and generated their own income—by tending rooftop vegetable gardens. They use wastewater from their apartment buildings and have developed networks of friends and families for marketing the produce.[23]

Beyond providing jobs and good nutrition, urban farming can have a whole range of other health benefits. Research has connected gardening to reducing risks of obesity, heart disease, diabetes, and occupational injuries. For urban folks especially, working with plants and being in the outdoors can both prevent illness and help with healing. Some health professionals use plants and gardening materials to help patients cope with mental illness and improve their social skills, self-esteem, and use of leisure time. Horticulture therapy is relaxing and reduces stress, fear and anger, blood pressure, and muscle tension; it can also lessen patients' dependence on medications.[24]

Wayne Roberts of the Toronto Food Policy Council sees urban agriculture as the "new frontier in public health," benefiting health twice: first, by supplying urbanites with more foods and, second, by affording them the exercise involved in raising food. Roberts notes that obesity is epidemic in most wealthy nations and increasingly in Third World cities. Having food produced locally can radically change people's attitudes toward the produce. "Instead of pop and candy vending machines plastering the cityscape, people see fresh fruits and vegetables," notes Roberts.[25]

Because gardens can provide a social beacon in urban areas, their potential to educate extends beyond the basics of planting. In Lilongwe, Malawi, the Peace Corps has been using urban gardens as a way to raise medicinal plants and educate people about AIDS. According to Anne Bellows, a research associate at the Food Policy Institute in New Jersey, beyond physical health "urban gardens bring people together in public space, resulting in community growth, education, healthy lifestyles, and fun."[26]

Roberts also envisions a third way in which urban agriculture can benefit public health—by improving the social determinants of health, including the beauty and safety of neighborhoods and the strength of community ties and social interactions. Studies show that people at farmers' markets have as many as 10 times more conversations, greetings, and other social interactions than people in supermarkets. City planners are learning that farmers' markets can be used to bring people together in a central location, becoming a forum for politicians, activists, and other community leaders to raise awareness about local issues.[27]

A survey of community gardens in New York found that having a garden improved residents' attitudes toward their neighborhood, reduced littering, improved the maintenance of neighboring properties, and increased neighborhood pride. They also found that the presence of gardens was four times more likely to spur other community efforts in low-income neighborhoods than in high-income ones, due to a greater number of pressing community issues and a lack of meeting places. Add to this the other well-documented effects of community gardens—including greater consumption of fresh vegetables, reduced grocery costs, and the various psychological and health benefits associated with exercise in a natural setting—and it becomes clear that urban farming does a lot more than just replenish food supplies.[28]

Healing the Concrete Jungle

As cities strive to be more self-sufficient in food, several obstacles stand in their way. On the most practical level, tall buildings often obscure sunlight (although rooftop gardens provide one solution), and urban soils may be contaminated with the residue of past industries (although pesticide-laden rural soils are often not much cleaner). Raising livestock or fish close to dense human dwellings and the strain urban farming can place on an already tight city water supply present unique health and environmental challenges for cities. Handled appropriately, however, urban farming can actually serve to diffuse the potential public health issues and might even improve water quality.[29]

As fresh water becomes a more and more precious commodity in cities, using every drop more than once becomes important. Although urban farmers use rainwater and water from nearby rivers and streams to grow crops, many also use a source of water that is widely available in all cities—human waste. The International Water Management Institute (IWMI) estimates that wastewater is used on more than 50 percent of the urban vegetable supply in several Asian and African cities. (See Box 3–1.) In addition to contaminated water, urban farmers and consumers also have to worry about other sources of pollution in their food, including heavy metals and other toxins that can contaminate the soil.[30]

Despite these problems, urban farming can bring a bit of the country into the concrete jungle, creating benefits that reach well beyond a city farmer who makes more money or a city resident who has a more stable food supply. (See Table 3–1.) Urban farms bring some needed diversity to the urban landscape. They provide ground to help catch and filter rainwater; land for composting and reusing of organic wastes; city trees to create shade, reduce heat, and cut down on greenhouse gases; and even buffer zones for flood- or earthquake-prone areas. Shady, flower-filled plots might provide a respite for the weary urban soul, but these spaces can also heal the toxic city environment.[31]

In the broader sense, urban farming can also be an extremely efficient use of natural resources. Intensive production of vegeta-

---

Box 3–1. Urban Agriculture and Wastewater Use

It is hard to believe, but much of the food grown in cities in the developing world is irrigated with polluted water. Why? The reason is simple—wastewater from sewage systems and even raw urine and feces are low-cost, nutrient-rich sources of irrigation for the poor in urban areas. Worldwide, 3.5–4.5 million hectares of land are irrigated with wastewater. But wastewater contains a whole range of pathogens that can survive for weeks after being applied to fields, posing a public health threat.

Most wastewater irrigation is done informally. City authorities know it is taking place, but they lack the money or infrastructure to offer an alternative. In Ghana, there are few data on the extent of informal irrigation in the country, but in Kumasi (with a population of 1 million) at least 12,700 farmers irrigate more than 11,900 hectares in the dry season—more than twice the area with formal irrigation in the entire country.

In Accra, Ghana, 200,000 people a day eat salad from irrigated urban agriculture. While this production contributes to a diversified diet, it also gives a sense of the number of people potentially at risk from polluted water.

Governments and nongovernmental organizations, such as the International Water Management Institute, are working to educate urban dwellers about the risks and benefits associated with wastewater irrigation. Without it, however, millions of people would go hungry. Guidelines for using wastewater need to be flexible, reducing risks to public health and not punishing urban farmers for irrigating their crops.

—Pay Drechsel,
International Water Management Institute

SOURCE: See endnote 30.

Table 3–1. Multiple Uses and Benefits of Urban Agriculture

Use	City	Benefit
Sewage treatment and aquaculture	Beung Cheung Ek Lake, Cambodia	Thousands of families living around this sewage-contaminated lake cultivate water spinach—a local staple that thrives in nutrient-rich waters. For thousands of years Asians have been using aquaculture ponds enriched with human wastes to grow plants, rear fish, control floodwaters, and remove local pollutants.
Crisis prevention and food security	Cities in Cuba; Freetown, Sierra Leone	In response to the U.S. embargo, Cuban officials designed a network of urban gardens. In 1999 urban farmers produced on average 215 grams of fruits and vegetables per day per Cuban—in some cities harvests exceed the 300 grams per day target set by health ministers. A similar system exists in Freetown, where war forced residents, refugees, and schoolchildren to rely on urban agriculture.
Bioremediation and phytoremediation	New Orleans, United States	Hurricanes Katrina and Rita unleashed dangerous levels of DDT, arsenic, lead, and other soil toxins. But citywide plantings of sun flowers, wild mustard, oyster mushrooms, and compost are helping sequester and break down these toxins.
Creating equity and controlling crime	Los Angeles, United States; St. Petersburg, Russia	Teens in Los Angeles grow produce to sell at farmers' markets. St. Petersburg prisons use rooftop gardens to create income, pride, and a valuable sense of community.
Erosion and landslide prevention	San Salvador, El Salvador	One of the few remaining forested areas around the rapidly growing city is a 120-hectare parcel called El Espino. Known as the "lungs" of the city, it provides fresh air and groundwater replenishment for the city's water supply. Managed by a cooperative of coffee growers who tend their bushes in the forest's understory, El Espino has more than 50 species of trees and shrubs, which shelter 70 species of birds, including some not found elsewhere. In recent years, much of El Espino has been developed, however, and during Tropical Storm Stan in 2005, massive flooding caused huge landslides in areas that had withstood previous storms.

SOURCE: See endnote 31.

bles in cities can use less than a fifth as much irrigation water and one sixth as much land as mechanized rural cultivation. In Freiberg, Germany, officials subsidize farmers on the steep hillsides surrounding the city in order to reduce the risk of erosion. Coffee farms in the hills around San Salvador, El Salvador, serve the same purpose, and a proposed tax on the city water (which depends on the presence of forest-laden coffee farms in the watershed) would help keep these farmers in business.[32]

In the East Kolkata wetlands of India, farmers are helping protect the environment as well as earning a living. The wetlands cover 12,500 hectares and include 254 wild and farmed fisheries, space for agricultural production, and residences. And they happen to be on the coastal edge of one of India's most densely populated cities,

Kolkata (formerly Calcutta), where the wetlands are the primary means of absorbing wastewater from the city's sewers. In a unique system of recycling, the fish and vegetable farms extract nutrients from the city's wastewater; fish ponds covering about 4,000 hectares encourage a range of physical, biological, and chemical processes that help improve the quality of water before it empties into the Indian Ocean.[33]

Popularly known as the kidney of the city, the Kolkata wetlands produce roughly 18,000 tons of fish each year for sale and support around 60,000 residents through fishing, fish farming, fish processing, and related activities. Ironically, while such benefits have been an inspiration to coastal cities around the world, this has not kept speculators from increasing pressure on the government to develop these areas for residential and industrial purposes.[34]

Urban farmers are also adept at turning what some consider problems into solutions. "Despite the health and environmental risks posed by wastewater, its use in urban and peri-urban agriculture is a reality," says Gayathri Devi of IWMI. "If the city were to impose policies that restrict urban agriculture, they would not only be largely ineffective, but they would also likely cause significant socioeconomic problems for farmers and their families." For instance, Hyderabad, India's sixth largest city, boasts an emerging Internet and biotechnology hub and is widely recognized as the meeting place of northern and southern India. But 300,000 farmers and their families working 15,000 hectares within the city continue to rely for food and income on a decidedly ancient irrigation system: water from the nearby Musi River that is little more than untreated waste for much of the year.[35]

Technologically savvy cities can see farms in and around the city as allies in keeping water coming into the city clean and ensuring that it is not too dirty when it flows out. In Lima, Peru, treated wastewater is used to produce tilapia, and studies show that fish cultured in this way are acceptable to consumers and economically viable. The construction costs for this lagoon-based wastewater treatment facility were charged to the municipality; the local farmers who irrigated their crops with treated wastewater were happy to pay the land and operation costs, which were half of what some of them paid for groundwater.[36]

All the attention that urban farming is receiving can make people more aware of local pollution. In Hanoi, Viet Nam, concerns about how industrial runoff, garbage, and poorly maintained canals were affecting the safety and flavor of fish from surrounding fish farms prompted municipal authorities to retain large wetlands and lakes within the city boundaries for aesthetic and flood control reasons. According to a recent report, fish farming "is encouraged by the authorities as they believe the residents of Hanoi will equate food production with good environmental health, thus providing reassurance to consumers."[37]

For cities confronted with growing waste disposal problems—which includes virtually all cities—the strongest environmental argument for local farming is the opportunity to reuse urban organic waste that would otherwise end up in distant, swollen landfills. (See Box 3–2.)[38]

People have kept livestock in cities for centuries to help deal with urban waste as well as provide income and food. Farm animals recycle household refuse, agricultural waste, lawn cuttings, and other organic matter very efficiently, and the manure they produce can improve urban soils. Despite the common assumption that all of the world's pigs, chickens, cows, and other livestock are raised in idyllic country settings, more and more of the world's meat and animal products are pro-

---

Box 3–2. Mining Organic Waste

Efforts to turn urban organic waste into compost have generally been small, limited to the efforts of a few farmers collecting food scraps from hotels or vegetable markets or enterprising individuals who have begun to "mine" city landfills for organic matter. It is estimated that in Kano, Nigeria, 25 percent of the fertilizer needs of nearby farmers are met with municipal wastes. Among the barriers to greater recycling of city wastes are a lack of people interested in collecting it, high transportation costs, and the fact that most urban waste systems mix organic (food scraps, leaves, grass clips, newsprint) and nonorganic (plastic, metals, glass, hazardous chemicals) waste, which complicates the removal of the organic component. In most countries, the cost of dumping waste into landfills is so low that there is little incentive to look for alternatives.

In settings where the organic waste is easily separated and where farm and garden plots are located nearby, the transformation of urban wastes into fertilizer can be a lucrative business, particularly for poor urbanites. Eduardo Spiaggi of the University of Rosario in Argentina is training residents of the city's *villas*

miserias in composting techniques, since many of them already make a living from waste collection, classification, and recycling, although they often discard the organic part. In combination with training in small-scale gardening techniques, the participants—65–70 percent of whom are women—report more food for household consumption and some income from selling surplus food and compost.

For restaurants, hotels, supermarkets, and other businesses that generate large amounts of food waste, converting this "garbage" into compost can keep down disposal costs and even generate income. Projects from around the world have demonstrated the feasibility of collecting waste from an array of settings—supermarkets, restaurants, schools, hospitals—for composting on farms and spreading as fertilizer. In California, the Vons Companies Inc. and Ralph's Grocery Company supermarket chains, with more than 585 stores between them, have been able to reduce their waste stream by 85 percent and turned their scraps into profitable, branded products sold back to their customers.

SOURCE: See endnote 38.

---

duced in or near urban areas.

Consider this: People in developing countries now consume half of the world's meat, thanks to rising incomes and exploding urbanization. And people in cities in these countries are not just consuming more animal products, they are also becoming centers of production. In Bamako, Mali, for instance, 20,000 households keep livestock in the city. In Harare, Zimbabwe, more than one third of households keep chickens, ducks, pigeons, rabbits, and turkeys. In Dar es Salaam, Tanzania, 74 percent of people keep livestock, while in Dhaka, Bangladesh, the figure is 80 percent. Even in industrial countries people can be

found raising bees, worms, chickens, and other animals. (See Box 3–3.)[39]

But there can be too much of a good thing. Thanks to unregulated zoning and subsidies that encourage large-scale livestock production, massive chicken and pig operations are moving closer and closer to major urban areas, including in China, Bangladesh, India, and many African countries. This, says Michael Greger, a veterinarian with the Humane Society of the United States, is "bringing together the worst of both worlds—the congested inner cities of the developing world combined with the congested environment on industrial farms."[40]

Box 3–3. Bees and Worms: A City's Smallest Livestock

Until mid-July 2005, dozens of people in Vancouver, Canada, were engaged in a sweet but illegal activity—raising bees. But thanks to a dedicated group of apiarists, the city's health council changed the laws, allowing beekeepers to manage their hives legally. In London, beekeeping is also now a legal endeavor, with at least 5,000 registered beekeepers keeping beehives in their backyards and on rooftops. Beekeepers in New York City, however, have not been as lucky. They are restricted by a law that prohibits raising "wild animals," including honeybees. That hasn't stopped beekeepers there from producing some of the best-tasting honey in New York State.

Consumers may be reluctant to try honey produced in polluted cities, but because many cities have a huge variety of parks, private gardens, and even outdoor flower stands, the honey produced in urban areas is just as good—or better—than that produced in the rural countryside. In addition to producing honey, bees help keep urban gardens pollinated and biologically diverse.

And while urban beekeepers carry out their work primarily on rooftops, a growing number of cityfolk are keeping their livestock under kitchen sinks, in backyard bins, and even huge municipal waste dumps. Vermiculture—recycling organic waste with worms—can be an environmentally friendly alternative to more conventional waste disposal. For households, worm composting bins take up very little space and work quickly. Worms typically eat their weight in food daily—a kilogram of worms can eat a kilogram of food waste every day.

Vermiculture can also be done commercially. While organic waste—everything from carrots to bread to yogurt—can take years to decompose in regular landfills, worms can compost up to 90 percent of waste in little more than a couple of months. Although some communities may have a hard time adjusting to separating their organic and inorganic waste, many have been able to do it successfully. The Canyon Conversions Company near San Diego in California (a city of about 150,000 people) processes some 400 tons of municipal yard waste per year with 200 tons of worms.

And in Rosario, Argentina's third largest city, residents of the poor Empalme Graneros neighborhood are using worm compost made from discarded fruit and vegetable trimmings to nurture plots of vegetables, while selling worms to local fishers. This income stream is no small benefit in the city with Argentina's highest unemployment rate. The residents sort trash and separate out plastics, cardboard, metals, and glass for resale. By recycling organic waste into compost, the project reduced the quantity of dumped organic waste that posed a health threat.

SOURCE: See endnote 39.

While there needs to be a place for raising animals in cities, industrial farming within cities is an inhumane and ecologically disruptive way of producing meat. A 2005 report by the World Bank echoed this, noting that the "extraordinary proximate concentration of people and livestock poses probably one of the most serious environmental and public health challenges for the coming decades." Many experts are worried about the spread of diseases, such as avian flu, from animals to humans, and city officials are grappling with how to dispose of mountains of manure.[41]

In many parts of the world, including along China's eastern coast, in Thailand around Bangkok, and in Brazil near São Paulo, there is an "excessive concentration" of factory farms as well as animal manure. In fact, some provinces along China's eastern seaboard, near consumers and port facilities, have more than 500 livestock per square kilometer, which is five times as many animals as the surrounding land can handle.[42]

Even raising smaller herds of free-range livestock in cities can sometimes present waste management problems. In the city of Kisumu, Kenya, many residents rely on livestock for income and food. But there is little land available to absorb manure. According to a recent study, three fourths of the dung produced in Kisumu is not used as fertilizer for growing crops, nor is it a source of fuel for cooking and heating. And because there is no regular waste removal system in the city, the manure piles up and up, contaminating soil and water.[43]

But now some people living in Kisumu are using animal waste both as a source of fuel and as a money-making opportunity. With investment from Lagrotech Consultants, a private company, and from a development agency, Kisumu residents are turning dung into a safe, efficient source of fuel. The dung briquettes—made by mixing water, charcoal dust, straw, and other ingredients with animal manure—produce very little smoke (a health hazard of other fuels, particularly for women) and save residents from having to buy expensive commercial fuel. Livestock owners are also generating additional income by selling their excess briquettes.[44]

One way to prevent the problems that plague industrial livestock production is to discourage factory farming in or near cities. A recent FAO report suggests a combination of zoning and land use regulations, along with taxes, incentives, and infrastructure development that can encourage producers to raise animals closer to croplands, where manure can be used as fertilizer and where there is less risk of disease. According to FAO, figuring out where the best places are to produce livestock can help control land and livestock nutrient imbalances—in other words, raising livestock in areas that have enough land to handle waste. Thailand, for example, puts high taxes on large-scale poultry production within 100 kilometers of Bangkok, while giving farmers outside that zone tax-free status. Thanks to this, the concentration of poultry farms right outside of Bangkok has dropped significantly over the last decade.[45]

Planning Garden Cities

In the 1880s, Ebenezer Howard felt that the modern city was consuming itself and everything around it. Howard envisioned a different type of city, a "garden city," with parks and green spaces and suggested population and livestock carrying capacities. The city would include gardens to raise some of its own food, but it would also make room for deliveries of food from the nearby countryside. Howard realized that "people streaming into the city" not only represented a threat to the urban areas, it also could bleed the rural areas to death. He was not proposing a blending of the two into a homogenous suburb, but instead a symbiosis. "Town and country must be married," wrote Howard in *Garden Cities of To-Morrow*, "and out of this joyous union will spring a new hope, a new life, a new civilization." [46]

The remnants of Howard's garden city can be seen in greenbelt cities constructed during the Depression in the United States, in the postwar new towns of Great Britain, and in the parks that ring Portland, Oregon, today. Nonetheless, with few exceptions the marriage between town and country has not always been "joyous." The unchecked growth of modern cities, aided by freeways and mass transit that stretch ever farther from their core, remains one of the primary threats to the farmland that feeds them. As Howard suspected, the basic design of the modern city seemed to be inherently threatening to farming nearby. Rather than incorporating permanent farmland into urban design, planners pave it over, even as the growing urban population puts increasing demands on the remaining land. In the United States, 79 percent of fruit, 69 percent of vegetables, and 52 percent of dairy products are raised in metropolitan counties or fast-growing neighboring counties in the path of sprawl, which threatens to eliminate this form of urban agriculture.[47]

In Rosario, Argentina, seven farmers' markets and more than 800 community gardens sprouted up throughout the city.

Despite all that farming can do for the city landscape and the urban soul, politicians, businesses, and planners continue to regard food as a rural issue that does not demand the same attention as housing, crime, or transportation. This stubborn mindset partly explains the "piecemeal approach" to planning for city food systems, according to a study from the Department of Geography and Urban Planning at Wayne State University in Michigan. Urban planners around the world viewed gardens and farmland within city limits as an anachronism, not to be found in a "modern city." In many cities, farming has been outlawed. Policymakers would be wise to realize the nutritional, social, ecological, and economic benefits of reversing this mindset and putting programs in place to encourage cities to feed themselves.[48]

Planners interested in making room for farming in cities must look beyond farmers' markets and community gardens to much deeper issues of a city's design. An extensive light rail system that reduces the need for highways, or a municipal composting site that generates high-quality fertilizer, or city schools that serve local produce for lunch all represent important determinants of just how much a city can support the surrounding country.

Whether it is the English commonlands during the Middle Ages or the conservation reserves of today, keeping the countryside intact seems an essential ingredient in keeping urban life from destroying itself. Farmland is also less of a drain on public coffers than suburbs: research in the United States has shown that municipalities often spend several times more on public services for every dollar that new housing generates in tax revenue than they spend on services for every dollar generated by farms and open land. Advocates of farmland preservation point out that there is no shortage of creative policies at the disposal of interested municipalities. The scarce commodity has been the political will to confront powerful building and transportation lobbies.[49]

The location and design of food markets is vitally important for urban farming. In the absence of government leadership, the placement of food retailing outlets in cities is often haphazard and inefficient, and it ultimately ends up wasting food and driving up prices for poor consumers. For example, Edward Seidler of FAO's Marketing Group notes that of the five wholesale markets in Hanoi, a city of 5 million, only one was planned. The others all developed spontaneously and now find themselves deep in the inner city, where storage and waste disposal facilities are insufficient, food damage and losses are high, food quality is reduced, and traffic jams and parking are constant challenges for both buyers and sellers—all resulting in higher consumer prices. As many Third World cities begin to erect housing developments and transportation infrastructure to accommodate their rapidly growing populations, local officials who do not incorporate food shops and markets into their plans will force masses of residents to pay extra and travel long distances to buy food.[50]

Seidler suggests that city authorities consider establishing local retail markets that cater to low-income consumers, while simultaneously providing outlets for farmers, especially those who grow vegetables on the edge of cities. "In Dar es Salaam and Mbabane and Manzini in Swaziland, the local councils have established small retail markets to serve local clientele living in the suburbs," Seidler notes. "In Barbados and in many Caribbean countries, local councils have established small retailing facilities around local bus stops to provide services to hawkers who formerly sold their produce, exposed to the elements, on pavements blocking pedestrian traffic." [51]

In Rosario, Argentina, where farming in the city was initially a response to the nation's financial crisis, officials are trying to establish it as an integral part of urban life. They created the Programa de Agricultura Urbana (PAU), a cooperative venture that unites urban farmers, municipal officials, agricultural experts, and representatives of nongovernmental organizations. The PAU helped urban farmers secure and protect agricultural spaces, take advantage of value-added agricultural products, and establish new markets and market systems. Soon, seven farmers' markets and more than 800 community gardens—supporting some 10,000 farmers and their families—had sprouted up throughout the city. The cooperative also involved residents of Molino Blanco, a low-income housing project, in the design and construction of a large garden park that includes walking paths, soccer fields, and large designated areas where people can raise food. [52]

"Urban farmers tell me that they are not only pleased to have the opportunity to generate income and feed their families," said Raul Terille with the Centro de Estudios de Producciones Agroecologicas in Rosario and a member of the PAU. "But also, after years of feeling marginalized, they are making a genuine contribution to their city and are finally being recognized for it." [53]

From Cienfuegos in Cuba to Piura in Peru and Dar es Salaam in Tanzania, city officials are also taking inventories of available vacant land in the city through on-the-ground surveys by farmers and through geographic information systems and are analyzing the land's suitability for agricultural use. In a few cases, officials then demarcate certain areas to be permanently used for farming, which gives farmers the incentive to make long-term investments in the land. [54]

In the erosion-prone city of Villa María del Triunfo in Peru, where 83 percent of all urban farmers are women for whom this is the sole source of income, city officials surveyed the 70-square-kilometer urban landscape to determine what share was suitable for farming. By 2004 the municipality had established a dedicated urban agriculture office in its economic development branch; had earmarked money to subsidize seed, fertilizer, and other inputs for city farms; and encouraged an increase in local processing and marketing. The program helped create 399 family and community plots on formerly vacant land, examined sources of irrigation water, developed a municipal consolidation plan for urban agriculture through 2010, and formed a multistakeholder advisory group that supports ongoing implementation of the plan. [55]

Mapping can be used not just to find suitable farmland but also to track food availability, as the city of Philadelphia in the United States did several years ago. It found that a lack of healthy food options and high rates of diseases like cancers, diabetes, high blood pressure, and heart disease coincided in the low-income areas of the city. The Food Trust's Supermarket Campaign leveraged this information to create the Pennsylvania Fresh Food Financing Initiative, an $80-million public-private partnership that works to increase the number of grocery stores in underserved communities. Penn State University researchers are using National Institute of Health money to study the effects of this initiative on fruit and vegetable consumption patterns and on health.[56]

If the goal is to be more self-sufficient when it comes to food, city officials need to think creatively. Perhaps the most cutting-edge design innovation for bringing food back into cities is also the most sublime—rooftop gardens. At the midtown-Manhattan headquarters of Earth Pledge, an environmental organization hoping to lower New York City's temperature and reduce pollution, a green roof with an organic kitchen garden—filled with lettuces, tomatoes, eggplants, peppers, cucumbers, assorted herbs, and even sweet potatoes—thrives above the shade created by the skyline and is out of the way of ground-level pollution from cars.[57]

Rooftop gardens are springing up everywhere. City Hall in Chicago sports a green roof; in Tokyo, a new ordinance requires all new building plans with more than 1,000 square meters of floor space to cover 20 percent of their roofs with vegetation as a way to reduce energy costs and urban temperatures. (See also Chapter 5.) In Mexico, the Institute for Simplified Hydroponics has developed low-cost roof garden technologies that will help many more landless peasants in the world's expanding cities feed themselves and earn a living from urban farming. And in Morocco, students and community groups have built garden beds from old tires filled with compost and vermiculite on rooftops and achieved yields dramatically greater than conventional gardening. They collected and recycled water that drained through the bottom of the beds, reducing water use by 90 percent over standard gardening techniques—a critical factor for countries susceptible to drought.[58]

In some cases, urban food policy councils have been formed to help guide government decisions on food. These informal coalitions of local politicians, hunger activists, environmentalists, sustainable agriculture advocates, and community development groups allow food policy decisions to reflect a broad range of interests and tap possible synergies. For instance, hunger activists, senior citizens, and farmers might join to lobby for farmers' market coupons for the poor and elderly, so that hungry citizens could buy healthy food and farmers would have new customers.[59]

The Hartford Food System (HFS), for example, works to give people in Connecticut better access to nutritious and affordable food. The group has helped establish farmers' markets, distributes coupons to low-income households for use at these markets, created a grocery delivery service for homebound elderly people, and launched the Connecticut Food Policy Council—a body that helps guide Connecticut's decisions about food. HFS tracks prices at supermarkets and operates a 400-member community-supported agriculture program that distributes 40 percent of its produce to low-income people. It also educates the public about farmland preservation and lobbies for policies that preserve farmland.[60]

These local councils might have another policymaking advantage. "Only an entity on the ground that knows the community and knows the nuances of the local food system knows how to make the system work for local folks," says Mark Winne of HFS. Policies designed in the rarefied air of bureaucracies may not be relevant or effective for specific cities or communities. HFS interviewed hundreds of low-income Hartford residents to determine the main causes of hunger in the city. After finding a strong correlation between frequent bouts of hunger and poor access to transportation options, the group worked with city officials to modify bus lines so that routes connected low-income communities with supermarkets. HFS also helped to open several farmers' markets and a new supermarket in the same poorly served area.[61]

Without public participation, policies intended to support urban farming might actually harm it. In the 1980s, the government of Tanzania issued a policy encouraging people in cities to grow food. This policy built on years of farming and cattle raising in Dar es Salaam and other cities during colonial times. Much of that cow grazing, though, was in less populated parts of the cities owned by wealthy foreigners. The new policy meant animals were in the densest part of the cities, creating noise and manure problems.[62]

Bombarded with complaints, the city government in Dar es Salaam responded by being stricter about noise, dirt, and manure cleanup. Between 1985 and 2005, the number of animals kept in the city quadrupled, growing faster than the human population did, but the number of problems reported plummeted. The amount of land under cultivation has doubled, hundreds of jobs have been created, and the availability of locally grown food has increased dramatically. Women who keep cows or raise vegetables in their backyards report making two to three times as much per year as their husbands, inspiring people elsewhere in society. "Once national and municipal leaders understood the on-the-ground reality of urban agriculture, they were convinced of its economic value—especially for poor families and women," says George Matovu, Regional Director of the Municipal Development Partnership in Tanzania.[63]

One way to make it easier for cities to feed themselves is to slow the flow of people into them from the countryside. Policymakers in rural areas have to make living there a healthy, viable option for the world's poor, so that they are not forced to move to cities. In just the last 50 years, some 800 million people have moved from the countryside to urban areas in search of higher incomes and a better way of life.[64]

Investing more in rural agriculture can help ease this migration, according to a 2006 study by FAO. The report found that governments and policymakers are largely unaware that if "properly managed," agriculture can not only produce food but also have a positive impact on poverty alleviation, food security, crime control, and protection of the environment in both cities and the countryside. In particular, improving roads and rural infrastructure, access to credit, and social services in rural areas can help curb the rate of people leaving the country and ease pressure on urban centers. People leaving the country often gravitate to a nation's capital or a few large cities, a strain that can be lifted by medium-sized towns that embrace urban agriculture as farming and farming-related industries flourish.[65]

The most cutting-edge design innovation for bringing food back into cities is also the most sublime—rooftop gardens.

A low-cost option for growing food in cities might be even more important than ever before. The migrations that prompted Ebenezer Howard to demand a new pattern of city development are minuscule compared with the changes under way in the Third World today. "On the longer term, urban agriculture will be sustainable especially if its potential for multifunctional land use is recognized and fully developed," noted René van Veenhuizen, editor of *Cities Farming for the Future* by Resource Centers on Urban Agriculture and Food Security. "The sustainability of urban agriculture is strongly related to its contributions to the development of a sustainable city: an inclusive, food-secure, productive and environmentally healthy city."[66]

CITYSCAPE:
FREETOWN

Urban Farms After A War

Freetown, the capital of Sierra Leone, evolved as a farming settlement. From the mid-1700s to the early 1800s, apart from being one of Africa's best natural harbors, this area offered safe drinking water, fresh grapes, apples, lettuce, spinach greens, potatoes, goats, sheep, chickens, and ducks to European explorers and traders en route to India. West Africans recaptured on their way to slavery and those freed in Europe and North America often resettled in Freetown.[1]

Farming and trading emerged as the main socioeconomic activities under the British colonial administration from 1808 to 1961. After the Second World War, a freehold land tenure system emerged, which was distinct from the communal ownership found in the hinterlands. By the end of British rule, most agricultural lands had been sold for urban use: residences, industries, and stores. Freetown's urban center grew in built-up land area and population after independence. The Greater Freetown Area now covers about 8,100 hectares. The population nearly quadrupled between 1963 and 1985—from 127,917 to 469,776 people. Economic life and food security deteriorated rapidly. It took a war before agriculture began to resurge in the city.[2]

In March 1991, a retired military corporal launched the Revolutionary United Front (RUF) rebel war from the eastern part of the country. In 1992, the military defending the country against the RUF overthrew the ruling All People's Congress Party, forming the National Provisional Ruling Council (NPRC). In 1996, through a democratic process, the NPRC handed power over to the current ruling party, the Sierra Leone Peoples Party, which gradually ended the war with the help of local and international organizations.

The war destroyed lives and property.

More than 2 million people were displaced, and major economic activities, such as farming, mining, and forestry, were disrupted. People flooded into Freetown, increasing the demand for food.[3]

Urban agriculture returned to the city, as many public sector workers became unemployed. Some of their spouses entered the informal sector, cultivating leafy vegetables and marketing fruits and vegetables within and near the Freetown municipal boundary. Young displaced people and women returning to the city turned to farming locally and international trade in vegetables between Conakry in Guinea and Freetown. Those who could not afford to travel operated as go-betweens and served as local retailers between wholesale importers and consumers in Freetown. Other women joined the urban agriculture marketing chain by preparing fast food for the growing numbers of unemployed, single, separated, or divorced family members.

During the war, Njala University College (now Njala University) relocated to Freetown in early 1995, and researchers there started studying technical, health, and institutional issues among farmers in the city. For instance, farms in Freetown are rainfed from April to October and irrigated from November to March. During the irrigated months, farmers work with old, worn-out containers that leak. Those who get wet are prone to falling ill with arthritis and pneumonia. Of particular concern to health and energy experts are school-age children, who during the weekends often help their parents prepare seed beds, transport household refuse, water, and market the crops.[4]

Urban farming grew even more important during the nine months of the Armed Forces Revolutionary Council (AFRC) mili-

Thomas R. A. Winnebah

Women selling vegetables from Guinea in Freetown

tary junta in 1997–98. As a result of an international trade embargo against the AFRC regime and the rebel blockade of trade between Freetown and its hinterland, food shortages became acute. Farming within and near Freetown became a basic survival mechanism.[5]

After the war ended in late 2002, the first post-war census revealed that Freetown's population had increased to 772,873 and now contained 16 percent of Sierra Leone's population. To help feed the burgeoning city, since 2005 the Ministry of Agriculture and Food Security (MAFS) has promoted urban farming associations and training under the U.N. Food and Agriculture Organization Special Project for Food Security's Farmers Field School. For instance, drip irrigation technology was extended to some of these plots.[6]

In June 2006, the MAFS teamed with the International Network of Resource Centers for Urban Agriculture and Food Security to launch the Freetown Urban and Peri-Urban Agriculture Project (FUPAP). The project has trained people in multistakeholder processes for action planning and policymaking for urban agriculture. Stakeholders trained include representatives of the Freetown Municipality, the Ministry of Lands and Country Planning, the MAFS, Njala University, the National Commission for Environment and Forestry, the National Farmers Association of Farmers, the Decentralization Secretariat, and a variety of nongovernmental organizations.[7]

The growing population, particularly in the city's east ward, has increased the demand for housing. Dwelling units are being constructed within stream valleys where farming takes place. Farming is being pushed to peri-urban lands. With the re-establishment of the Freetown City Council, the newly constituted multistakeholder city team under FUPAP is working toward the full integration of agriculture into city development plans. This may help redress access to land as another constraint to farming.

Farming—the foundation of life in Freetown since pre-colonial times—was marginalized by nonfarm land uses during urbanization, but it resurged in importance during the recent war. The current attempts at placing agriculture at the core of urban planning suggest that farming will always exist in Freetown as long as there are mouths to feed.

—*Thomas R. A. Winnebah,*
Njala University, Sierra Leone
—*Olufunke Cofie, International Water*
Management Institute, Ghana

CHAPTER 4

Greening Urban Transportation

Peter Newman and Jeff Kenworthy

The meeting was over, a motion to remove the rail tracks was passed—and yet a rail revival seemed almost certain. Friends of the Railway (F.O.R.) in Perth, Australia, was trying to restore the Perth-Fremantle railway in 1982 after the government had closed it in 1978. Perth was a modern city and the car was king. The government wanted the railway reserve to become a freeway, and buses were to replace the trains. But since the rail's closure, buses had lost 30 percent of the riders who had used the train. Further, the first signs of a global oil crisis had hit Australia in 1979, and a transportation system entirely reliant on oil-fueled cars and buses did not seem sensible.[1]

After four years of intense lobbying, F.O.R. was gaining public support. The group had called a public meeting; the government's political party had responded by busing in 600 supporters two hours before the start of the event, preventing any one else from entering. After the government supporters passed their "unanimous" vote, the media were merciless in their coverage of a "bankrupt government."

Within a year the government was voted out of office and the train was restored. Now Perth has 180 kilometers of electric rail, with 72 stations covering every corridor of the city. The rail system is often the fastest way to get around the city, and patronage grew from 7 million passengers a year to 47 million in 2005, with sharp increases again since fuel prices skyrocketed.

Perhaps more important, land near stations has become the preferred site for living and working. As the city has focused new development around stations, these areas have become hives of activity. Perth has pioneered a travel demand management system called Travel Smart that has educated people about transportation options. As a result, more than

Peter Newman is Professor of City Policy and Director of the Institute for Sustainability and Technology Policy, and Jeff Kenworthy is Associate Professor in Sustainable Settlements at the Institute, at Murdoch University in Perth, Australia.

15 percent of car trips have changed into public transport or bicycle trips. Extensive cycleways now lace the metropolitan area, including a "veloway" for bicycles along rail lines. Funding for public transportation and bicycles now outstrips road funding and includes a hydrogen fuel cell bus trial.

Although still powerful, the car is not quite the king it was in Perth. Politicians there can now bathe in the glory of having had the foresight to prepare their city for an oil-constrained future. The F.O.R. campaign and hundreds of others in cities around the world demonstrate that modes other than the car can be facilitated by popular governments. For this to happen, however, civil society must develop a vision of what this will involve for each city. Turning that vision into reality will unavoidably involve politics.

"Greening" urban transportation, as Perth has tried to do, is about finding ways to make walking, bicycling, other nonmotorized modes like rickshaws, buses, light rail (modern trams), and heavy rail more competitive with the car as well as about reducing the need to travel in the first place through land use changes, for example, that put people's homes and jobs closer together. This chapter looks at the rationale for doing this, at the policies that can move cities in that direction, and at some cases where this is happening.

Transportation and Land Use in Cities

While each city has its own transportation story, the past 30 years have seen an explosion in the growth of cars in cities worldwide. In 1970 there were 200 million cars in the world, but by 2006 this had grown to more than 850 million—and the number is expected to double by 2030. Heavily marketed, highly successful in its political campaigns, and now the symbol of success for any

aspiring person from Boston to Belgrade to Beijing, the car seems unstoppable.[2]

Cities everywhere have been filling with cars—leading to congestion, road accidents, and bad air quality. Traditionally the response has been to build more road capacity, which displaces transit (public modes of transportation), bicycling, and walking and which causes a city to spread outwards along highways. Travel distances get longer and travel times favor cars. This sets up a spiral of more cars, more roads, and more sprawl that seems never-ending.

Cars are useful, and the desire to own one is so great that every city has to work out how to deal with this. Yet ownership of cars may not be the problem, as the extent to which cars dominate a city varies enormously. Cities are shaped by their economies, their culture, and their transportation priorities in a synergistic way.

This chapter uses data on urban areas from the Millennium Data Base prepared for the International Union of Public Transport and conducted by the Institute for Sustainability and Technology Policy at Murdoch University in Australia, the most comprehensive urban transportation and land use data set available. Data for 84 cities in 1995 (the latest year available) are used to establish a global perspective (available at www.sustainability.murdoch.edu.au), while 15 cities typical of the situation in industrial and developing countries are presented here to illustrate the main differences. The indicators discussed include transportation energy, transit use, population density, nonmotorized transport (NMT), relative speed of transit to traffic, and length of freeway.[3]

The variation in the amount of fuel used in private passenger cars in the 15 cities is shown in Figure 4–1. Not surprisingly, Americans lead the world in the use of cars and fuel, although there are significant differences between Atlanta at 2,962 liters of gasoline annually per person and New York at 1,237 liters. People in Australian, Canadian, and New Zealand cities are next, with 700–1,200 liters per person. European cities use around 450 liters per person, while people in East European cities use about 100–240 liters. The figures in wealthy Asian cities are also extremely low, averaging 275 liters per person. Cities in developing countries are mostly at the lower end of this array, ranging from around 70 to 300 liters per person. Ho Chi Minh City—with 27 liters per person—is hardly measurable on the same scale.[4]

Cities vary in their ability to afford cars and to provide roads for them, but there is surprisingly little correlation between car fuel use and city wealth. Cities like Tokyo and Hong Kong are very wealthy, for example, but their residents use 10 times and 25 times less gasoline than Atlanta. European cities in general are among the wealthiest in the world, but people there use six times less fuel than people in Atlanta. It seems that cities invest in either the ability to travel by car or the ability to travel by other modes such as transit.[5]

Figure 4–2 shows the proportion of motorized transportation on transit in the 15 typical cities. These show an even greater spread. U.S. cities have vanishingly small levels of transit, although 9 percent of New York City's motorized transport is on public transportation. Australian, Canadian, and New Zealand cities are just a little better, varying from 5 percent in Perth to 14 percent in Toronto. Most European cities are over 20 percent transit, although a few have less than 10 percent. East European cities, in contrast, are all around 50 percent transit. The wealthy Asian cities are very high in transit, with Hong Kong at the top with 73 percent. Figures for cities in developing countries are highly scattered, from Mumbai at 84 percent to Ho Chi Minh City at 8 percent and

Riyadh at 1 percent. Again, these patterns do not seem to follow per capita wealth levels. Some cities appear to invest in transit. Others do not.[6]

The third indictor of significance is non-motorized transport—mainly walking and biking, but also some rickshaw use. Figure 4–3 shows that the cities with high car use and low transit generally have low walking and biking: North American, Australian, and New Zealand cities average 8–16 percent (Atlanta is just 3 percent) whereas European, Latin American, and Asian cities are around 30 percent (although Shanghai registers 78 percent and Ho Chi Minh City 44 percent). The density of these cities helps explain the differences, as distances in compact cities are short enough for walking. The political priority assigned to facilitating NMT is also important, which explains Bangkok's relatively low figure. Copenhagen and Amsterdam have high bicycling rates—over 30 percent—due to the exceptional provision of cycling infrastructure. For Chinese cities, the bicycle remains the largest contributor to urban transport, but this is being challenged by rapid motorization. (See Box 4–1.)[7]

The fourth important indicator is density of population in people per hectare of developed urban land. (See Figure 4–4.) In general, higher-density cities have the most walking/biking and transit use, while low-density cities have the most car use. This is seen clearly in a larger sample of cities. (See Figure 4–5.) Although there are some exceptions—Curitiba and Krakow are a little lower in density than expected for their low car use, for instance, and Bangkok is higher density than expected for the city with the highest car use in Asia—the link between urban form and transport seems to be quite clear. Some analysts argue that transport patterns are mostly caused by other factors such as income and gasoline price, but Figure 4–6 shows the same

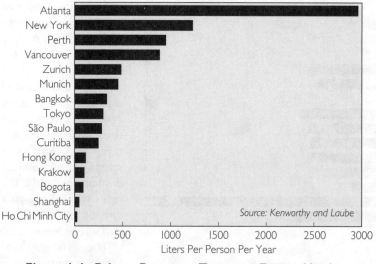

Figure 4–1. Private Passenger Transport Energy Use in 15 Cities, 1995

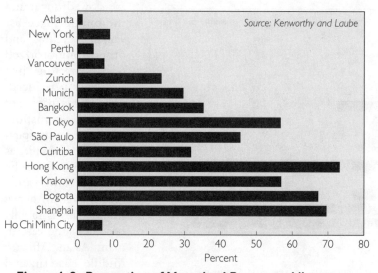

Figure 4–2. Proportion of Motorized Passenger-kilometers on Public Transport in 15 Cities, 1995

large variation in transportation energy with density across Sydney's suburbs. The density of activity is very high in the center (City of Sydney), where the fuel use is similar to Shanghai or Krakow, whereas usage in the inner

suburbs is like Munich and in the outer suburbs it is like Houston. This also shows that it is not income driving these patterns, because Sydney—like all Australian cities—declines uniformly in wealth from the center outward.[8]

From these transportation and land use data, it is clear that there is one set of problems associated with low-density cities—sprawl-based car dependence—and a different set associated with high-density cities—congestion-based car saturation.

Cities built around cars sometimes use automobiles 10 times as much as other cities, and they have land use patterns that suggest there is little alternative. Many of these urban areas have a pattern of low-density land use that is so car-dependent that it reinforces the downward spiral of greener modes of transportation. They have traffic problems not only on the freeways that lace them but also in their centers, which are often dominated by car parks and traffic conflicts. But they also have no other viable options for transportation. It appears that only cars can get

69

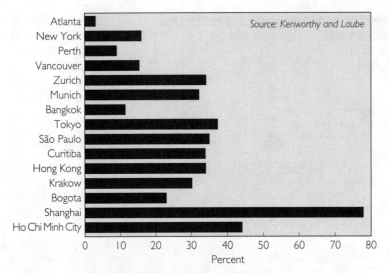

Figure 4–3. Proportion of Total Daily Trips by Nonmotorized Modes in 15 Cities, 1995

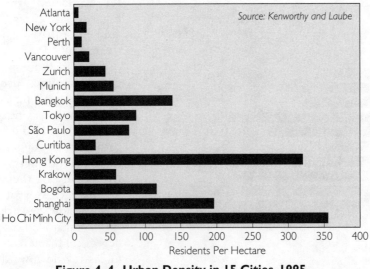

Figure 4–4. Urban Density in 15 Cities, 1995

use generally support more transit and walking/biking. But they also have less space for vehicles on their streets. Thus in many of these denser cities, cars, scooters, and three-wheel taxis are rapidly filling the streets—the traditional domain for public life in a city—leading to a decline in air quality and more noise and traffic congestion. Worse, as public space is converted to use by private vehicles, buses are slowed down and the public space for walking, cycling, and other nonmotorized transportation like rickshaws is reduced. The result is that use of greener transport modes begins to spiral down in dense cities as well. Cars become the quickest and safest means even in heavy traffic. This is rapidly becoming the case in many Asian, African, Middle Eastern, and Latin American cities.[10]

us to destinations in a reasonable time. This is mainly the case in U.S., Australian, New Zealand, and Canadian cities as well as increasingly on the fringes of European cities and in transitional cities that are allowed to sprawl.[9]

Cities that have retained denser land

Car-dependent and Car-saturated Cities

Cities are shaped by many historical and geographical features, but at any stage in a city's history the patterns of land use can be

---

Box 4–1. Is the Motorization of Chinese Cities a Threat to the World?

Some commentators fear that the world will not be able to cope with the growth in Chinese cities as people there are buying cars so rapidly. Yet the 200 million Chinese who have moved into cities over the last 10 years use around 50 liters of transport fuel per person—in total, the 200 million use less in a year than Atlanta's 4.1 million people use and one quarter as much as Sydney's 3.5 million people.

Of course, if the Chinese choose to build freeways and sprawl their cities like Atlanta, they will end up using cars like Atlanta does. But there is almost no possibility of that hap-

pening, despite the government's recent attempts to build freeways. Chinese cities are built of high-rise towers, so they house some 150–200 people per hectare. Space for highways becomes the limiting factor in high-rise cities, and the current building phase appears to be reaching that limit in cities like Shanghai. Atlanta, at 6 people per hectare, is the lowest-density city in the world. Freeways there are constantly being expanded to enable movement, and there is still a lot of space for such roads.

SOURCE: See endnote 7.

---

changed by altering its transportation priorities. Italian physicist Cesare Marchetti has argued that there is a universal travel time budget of around 1 hour on average per person per day. This "Marchetti constant" has been found to apply in every city in the Millennium Data Base's global sample as well as in data on U.K. cities for the last 600 years and defines the shape of cities:[11]

• "Walking cities" were and remain dense, mixed-use areas that are no more than 5 kilometers across. These were the major urban form for 8,000 years, but substantial parts of cities like Ho Chi Minh City, Mumbai, and Hong Kong, for example, retain the character of a walking city. Krakow is also mostly a walking city. In wealthy cities like New York, London, Vancouver, and Sydney, the central areas are predominantly walking cities in character.

• "Transit cities" from 1850–1950 were based on trams and trains, which meant they could spread out 20–30 kilometers, with dense centers and corridors following the rail lines and stations. Most European and wealthy Asian cities retain this form, as do the old inner cores in U.S., Australian,

and Canadian cities. Many developing cities in Asia, Africa, and Latin America have the dense corridor form of a transit city, but they do not always have the transit systems to support them, so they become car-saturated.

• "Automobile cities" from the 1950s onward could spread 50–80 kilometers in all directions and at low density. U.S., Canadian, Australian, and New Zealand cities that were developed in this way are now reaching the limits of the Marchetti constant of a half-hour car commute as they sprawl outwards and hence are redevloping.

In some cities, like growing megacities or rapidly sprawling ones, the travel time budget for an increasing proportion of people can be exceeded. Invariably people will adapt by moving closer to their work or finding a better transportation option. The search for better options can form the basis of social movements that seek to provide greener transportation.

There are many reasons to overcome car dependence and car saturation. (See Table 4–1.) Perhaps the best way to see it overall is the lack of resilience in a city when cars begin

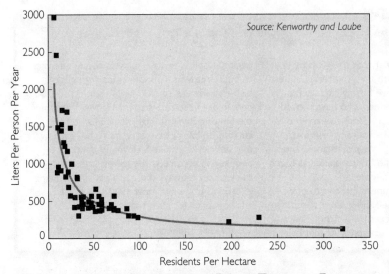

Figure 4–5. Urban Density versus Private Transport Energy Use in 58 Higher-Income Cities, 1995

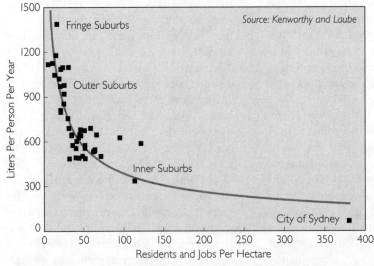
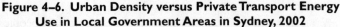

Figure 4–6. Urban Density versus Private Transport Energy Use in Local Government Areas in Sydney, 2002

stranglehold of the single "car-only" option for cities and providing instead a wide range of options. This can build resilience into the city, especially when it faces crises like climate change and the peaking of world oil production as well as the variations in economic and social functions that transport needs to address.[12]

Fuel is a major problem for car-dependent cities if predictions about global oil production peaking come true. Cities that are not prepared with greener transportation options may find the era of oil depletion more difficult than cities that have struggled to combat car dependence. However, fuel is only one of the issues listed in Table 4–1. Too many cars in a city will continue to be a problem no matter what fuel is used. But even fuel problems in vehicles are not easily dealt with by simply improving efficiency. Table 4–2 sets out the fuel efficiency performance of transport options in the global cities sample.[13]

Urban car travel is on average nearly twice as energy-consumptive as average urban bus

to dominate. A city needs many kinds of transportation and land use options, not just one type. This is what Eric Britton, an American transport planner living in Paris, calls the New Mobility Agenda—breaking the

Table 4–1. Problems in Cities Related to Cars

Environmental	Economic	Social
Photochemical smog	Costs from accidents and pollution	Oil vulnerability
Toxic air emissions	Congestion costs	Loss of street life and community
High greenhouse gas contributions	High infrastructure costs in new sprawling suburbs	Loss of public safety
Loss of forest and rural land	Loss of productive agricultural land	Access problems for those without cars and those with disabilities
Greater stormwater problems from extra hard surface	Loss of urban land to asphalt	Road rage
Traffic problems—noise, neighborhoods cut up by roads	Obesity and other health impacts	

travel, 3.7 times more than typical light rail or tram system travel, and 6.6 times more energy-intensive than average urban electric train travel. Light rail and tram systems typically involve a lot more stopping and starting than heavy rail, with stations much closer together. So although their average passenger load is similar to heavy rail, their energy efficiency is a little poorer. Diesel rail is only a little more fuel-efficient on average than an urban bus. Average occupancies on trains are roughly equal across types—generally more than twice that of buses and about 20 times higher than cars. Overall, these data re-

Table 4–2. Average Fuel Efficiency and Occupancy by Mode in 32 Cities, 1990

Mode	Average Fuel Efficiency	Measured Average Vehicle Occupancy
	(megajoules per passenger-kilometer)	(number of occupants)
Car	2.91	1.52
Bus	1.56	13.83
Heavy Rail (electric)	0.44	30.96
Heavy Rail (diesel)	1.44	27.97
Light Rail/Tram	0.79	29.73

Note: Rail mode occupancies are given on the basis of the average loading per wagon, not per train. The average occupancy of cars is a 24-hour figure.
SOURCE: See endnote 13.

emphasize the importance of developing a good backbone of electric rail in cities if energy conservation is to be enhanced. Cities without such systems are the ones with very high gasoline use.

Many commentators look at the theoretical potential in the various modes to see if actual passenger loads could be increased. Bus loadings vary enormously, although these are not easily improved if, for example, a bus route is through highly dispersed suburbs. The main target is the ridership level in cars, which could carry four people instead of the average 1.52. This average number is so universal, however—it ranges from a low in Geneva of 1.20 to the highest in Manila at 2.5—that it appears unrealistic to expect much higher occupancy. Car pooling has little potential in this area as it offers participants only a limited timetable. But public transport has a much higher range—from 1.8 in bus jitneys in Manila to 239.4 in Mumbai per carriage (often with many people on the roof). So there is a real potential for change in transit occupancy levels.[14]

The other way to improve vehicle fuel efficiency is through improving the vehicle technology itself. Since the first oil crisis in the early 1970s, the

73

solution has been cast as one of creating more technologically efficient cars and trucks. This is still seen by many commentators as the most important response. Energy entrepreneur Jeremy Leggett of Solarcentury, for instance, suggests that U.S. dependence on Middle Eastern oil would be reduced to zero if fuel efficiency were improved in the U.S. vehicle fleet by just 2.7 miles per gallon. This assumes that people do not drive more if they need less fuel per kilometer, yet the past few decades have seen people driving more and more in the United States.[15]

Vehicle engines have become more efficient in recent decades, but this has been offset in the United States and Australia by an increasing proportion of heavier sports utility vehicles in the overall fleet. Thus U.S. vehicle fuel efficiency improved from 1975 to 1987 but has dropped from 26.2 miles per gallon then to 21.4 in 2006, and Australian fleet averages are now below their levels in the 1960s. In contrast, European and Chinese standards on vehicles are being increased, which is likely to have ripple effects throughout the vehicle industry. Huge technological advances are available in vehicle fuel efficiency; they just need to be mainstreamed.[16]

Air quality data from across the world's cities indicate continuing problems in this area as well. Vehicle and fuel technology improvements in most industrial-country cities have been able to hold overall ambient air pollution levels steady or reduce them despite increases in driving. But this is not the case in cities in most developing countries. Yet this is even more important in these high-density cities, where air pollution threatens the health of millions. In cities like Mexico City, Delhi, Mumbai, Kolkata, and Dhaka, most traffic consists of three-wheeler taxis with two-stroke engines (in Mexico City, for instance, 55 percent of traffic is mini-taxis). These are cheap to run but dirty, causing air pollution that is sometimes three to five times the level recommended by the World Health Organization.[17]

In Delhi, one study suggested that 10,000 people were dying every year from air pollution. Anil Agarwal and Sunita Narain from the Centre for Science and Environment there first raised this issue in the early 1980s; in 1985, environmental lawyer M. C. Mehta sued the government of India for inaction on the issue. In 1998, after much consideration, the Supreme Court mandated that all buses in Delhi use compressed natural gas (CNG) as fuel by 2001. By 2006, some 80,000 CNG vehicles could be found on Delhi's streets, including all public buses and mini-taxis. Air pollution dropped by 39 percent. Now other Indian cities have followed, and Beijing wants to do likewise.[18]

To deal with another major problem in car-saturated cities—congestion—municipal governments often turn to freeways. These are usually proposed to help free up congestion. Speeding up traffic will save time (at least for awhile) and is thought to save fuel and emissions because vehicles are involved in less of the stop-and-start driving that wastes fuel. Traffic planners use benefit-cost analyses based on these ideas to justify the large capital cost of freeways. But the data do not support these contentions.

Is congestion associated with higher fuel use in cities? No, cities with higher congestion have lower fuel use. (See Figure 4–7.) Those with the least congestion use the most fuel. Although individual vehicles in less-congested cities are moving more efficiently, they are being used much more and for longer distances; at the same time, people in these cities are not using more fuel-efficient modes of transportation as much.[19]

When road capacity increases, car use does too, to fill the space created. In a study of U.S.

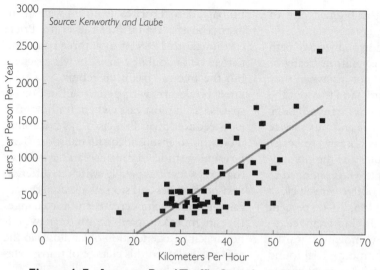

Figure 4–7. Average Road Traffic Speed versus Private Transport Energy Use in 58 Higher-income Cities, 1995

cities over the past 30 years, the Texas Transportation Institute found that there was no difference in the levels of congestion between cities that invested heavily in roads and those that did not. Thus it is possible that a policy of road-building designed to reduce traffic can actually have the opposite effect and increase it.[20]

The building of freeways varies considerably across the world's cities. Data on the amount of freeway per person in cities generally indicate that "if you build roads, cars will come." (See Table 4–3.) West European cities and wealthy Asian cities place very little emphasis on such roads. These cities are among the wealthiest in the world. Latin American and Chinese cities have 50 times less freeway per capita than U.S. cities do. When freeways are expressed in terms of meters per dollar of wealth in the city, then sadly the cities that are investing most heavily in freeways

relative to their wealth are those in Africa and the Middle East. This would indicate a muddled development assistance priority system.[21]

Limiting car use in order to reduce congestion is a more sustainable solution than building freeways. This has been done in Greater London. Mayor Ken Livingstone did what many people thought was impossible—he levied a congestion tax that proved to be popular. For many years transport economists have recommended that cities tax the use of cars to reduce congestion and to pay for motor vehicles' external costs, such as pollution and accidents. Although Singapore and Oslo had already set up these

Table 4–3. Freeways in 84 Cities, Summary by Country or Region, 1995

Cities within Region or Country	Freeways per Population	Freeways per City Wealth
	(meters per thousand population)	(meters per $1,000 gross regional product)
United States	156	4,970
Australia and New Zealand	129	6,520
Canada	122	5,850
Western Europe	82	2,560
Asia high-income	20	650
Eastern Europe	31	5,260
Middle East	53	11,880
Latin America	3	620
Africa	18	6,410
Asia low-income	15	3,990
China	3	1,170

SOURCE: See endnote 21.

systems, London was the biggest city to attempt it.

Officials ringed London's inner city with sensors that let people pay automatically or that fined those who did not pay when they crossed into the central city. And they put the money raised back into better transit. Traffic was reduced by 15 percent and bus service improved dramatically, as buses were better able to meet their schedules and the city supplied more of them. People who continued to drive appreciated the 60,000 fewer vehicles a day entering central London; 50–60 percent of those who stopped driving changed to transit. Stockholm is now moving toward a congestion tax after a six-month trial found a reduction in traffic of 25 percent in the morning rush and 40 percent in the evening. About half the commuters moved to transit, with a 4.5-percent increase in its use.[22]

Some cities faced with congestion have decided not to build freeways— and have turned out to be global leaders in green transportation.

Some other cities faced with congestion have decided not to build freeways—and have turned out to be global leaders in green transportation. Copenhagen, Zurich, Portland in Oregon, and Vancouver and Toronto in Canada all faced considerable controversy when freeways were initially proposed. The city governments decided instead to provide greener options—light rail lines, cycleways, traffic calming, and associated urban villages. All these cities had citizen groups that pushed visions for a less car-oriented city and a political process that allowed and encouraged their inputs.[23]

Today many other cities are trying to demolish freeways that have blighted central urban areas. San Francisco removed the Embarcadero Freeway from its waterfront district in the 1990s after the Loma Prieta earthquake in 1989. It took three ballot initiatives before public consensus was reached, but the freeway has been rebuilt as a tree-lined boulevard with pedestrian and cycle spaces. As in most cases where traffic capacity is reduced, San Francisco has not found it difficult to ensure adequate transport. Traffic calming studies for many years have shown that cities adapt as people switch to different transport modes and stop taking unnecessary car trips and as cities create land use changes that are more compatible with the new traffic capacity. Regeneration of land uses in the area has followed this change of transportation philosophy.[24]

Seoul, South Korea, has removed from its center a large freeway that had been built over a major river. The freeway had become controversial because of its impacts on the built environment as well as the river. After a mayoral contest in which the vision for a different kind of city was tested politically, the new mayor began a five-year program that saw the freeway dismantled, the start of a river rehabilitation process, the restoration of a historical bridge over the river, the restoration and rehabilitation of the river foreshores as a public park, restoration of adjacent buildings, and extension of the underground rail system and a bus rapid transit (BRT) system to help replace the traffic. The project has been very symbolic for Seoul, as the river was a spiritual source of life for the city. Now other Asian cities, especially in Japan, are planning to do similar projects, and Aarhus in Denmark has removed a major arterial road covering a small river and created a beautiful foreshore regeneration.[25]

What these projects have shown is that we should, as David Burwell from Project for Public Spaces, says, "think of transportation as public space." Freeways, from this perspective, become very unfriendly solutions as they are not good public spaces. But boulevards with space for cars, cyclists, pedestrians, and a busway or light-rail system—all packaged in good design and with associated land uses that create attractions for everyone—these are the gathering spaces that make green cities good cities. The Aarhus River now attracts people and investment to the area that was once blighted by the busy road. The Demos Institute, a U.K. public policy research group, has shown how public transport helps create good public spaces that in turn help define a city.[26]

Traffic engineers are increasingly aware of this new paradigm for transportation planning. Andy Wiley-Schwartz of the U.S. public policy group Project for Public Spaces says that "road engineers are realizing that they are in the community development business and not just in the facilities development business." He calls this the "slow road" movement. A similar movement in Denmark spearheaded by the urban designer Jan Gehl has stressed the importance of making all public spaces, especially roads, into people-oriented spaces where the priority is for pedestrians and cyclists.[27]

Both car-dependent and car-saturated cities thus need a combination of transportation and land use options that are more favorable for greener modes—that is, they must save time compared with a car. First, public transportation needs to be faster than traffic down each major corridor. Cities where transit is relatively fast are the ones with a reasonable level of support for it (see the data on the 84 cities on the Web site). The reason is simple: public transport in these cities saves time.[28]

Second, more people need to live and work where they have greener transportation options. Densities in central cities, regional centers, and neighborhood centers need to increase so that whatever mode people choose they can travel less. Transit needs densities over 35 people and jobs per hectare of urban land (though preferably 50 people), cycling also begins to be effective over 35 people per hectare, while walking requires densities over 100 people and jobs per hectare.[29]

And third, transit service levels and connectivity need to allow time savings to occur. Transit must be available at 15-minute intervals or less and be provided at night and on weekends. A city with inadequate public transportation can be restructured into a series of transit cities where local bus services link to a faster service down corridors. These local bus services can then go across the corridors by linking into stations in a coordinated way from many directions. Vancouver and Sydney have restructured in this way, and Denver has a rail project planned that will transform that city's transit system.[30]

Rebuilding Cities with Transit

To reverse sprawl-based car dependence, it is almost impossible to redevelop a whole city so that transit is faster than the traffic. But it is possible to make transit faster than traffic down all main corridors. The European and Asian cities with the highest ratio of transit to traffic speeds have achieved this with rail. Rail systems are faster in all 84 cities in the Millennium Data Base by 10–20 kilometers per hour (kph) over bus systems, which rarely average over 20–25 kph. Busways can be quicker than traffic in car-saturated cities, but in lower-density car-dependent cities it is important to use the extra speed of rail to establish an advantage over cars in traffic. This is one of the key reasons railways are being built in more than 100 U.S. cities.[31]

Rail, together with proper zoning, can induce density around stations. The Metrorail system in Washington, DC, built in 1976, has grown to 168 kilometers of track with 86 stations and has become a key factor in shaping housing and employment patterns. This process—called transit-oriented development, or TOD—is becoming a guiding philosophy for planners, politicians, and developers as it not only reduces car use, it also saves money on infrastructure and helps create community and business centers.[32]

Centers or TODs need to be planned along every public transport system so that transit cities are built as an antidote to automobile cities. In these centers—especially major regional centers—walking and cycling need to be given priority so that short journeys can be done quicker. The walking city is just as functional in today's economy as in any other period of history. A combination of transit cities and walking cities can be built within an auto city to make it more resilient. Most Australian and Canadian cities are being planned and redeveloped with this concept in mind, as are Portland (Oregon), Chicago, and Denver in the United States.[33]

Less than 1 percent of the trips in Denver are by transit, yet the city is rebuilding itself around six new rail lines. Inspired by neighboring Boulder, with its transit and bicycle-friendly center, Denver groups like the Transit Alliance began a campaign to transform the transit and land use system in the mid-1980s. A partnership was forged between political leaders, business, nongovernmental organizations (NGOs), and neighborhood groups. In 2004, voters approved a proposition to raise $4.7 billion to provide 192 kilometers of new rail track with 70 new stations, 30 kilometers of bus rapid transit, and a plan to focus development around the transit system.[34]

Greener transportation has begun to transform Vancouver, Canada. The city's population, like that in many North American downtown areas, began declining in the 1970s and 1980s. In response, the City Council created quality urban spaces, good cycling and walking facilities, reliable electric rail (the Sky Train) and electric trolley buses, and high-density residential buildings with at least 15 percent public and cooperative housing. The population has grown by 135,000 in the last 20 years, and transportation patterns have changed. Car trips in the city declined by 31,000 vehicles per day between 1991 and 1994 (35 to 31 percent of all trips), while cycling and walking increased by 107,000 trips daily, rising from 15 to 22 percent of trips. By 2006, walking and cycling had gone up to 30 percent of trips.[35]

Families are moving into the city, and schools, child care centers, and community centers are now crowded, while the number of cars owned in the city has dropped to less than it was five years earlier—probably a world first, especially in a city undergoing an economic boom. The city requires each development to provide public spaces and social facilities equal to 5 percent of the cost of the project. These funds improve the city's walkability as well as local community facilities. Vancouver also has redeveloped many areas around the Sky Train stations.[36]

Paris, like many European cities, has a strong transit system and a walkable central area, but over recent decades it has given more and more space to the car. Now, in a bid to reclaim its public spaces, it is implementing a series of policies to reduce the number of cars in the city. The new measures include 320 kilometers of dedicated bike lanes, conversion of one-way express routes for motor vehicles into two-way cycle lanes with additional street trees, and the removal of 55,000 on-street parking spaces a year. The city also plans a new light rail system linking a dozen subway and express train lines, providing cross-city linkages, and 40 kilometers of dedicated busways that enable buses to travel at twice their normal speed, with bus stops that provide real-time information. Mayor Bertrand Delanoë says 80 percent of Parisians support these innovations.[37]

Bus rapid transit is filling a niche between rail and conventional buses. (See Box 4–2.) Its main features are dedicated lanes, pre-paid fares, level boarding, frequent service, large capacity, signal priority, and intelligent control systems. As BRT can fit onto existing roads, it is cheaper than rail. Ottawa, Curitiba, and Bogotá were the first cities to demonstrate BRT on a large scale.[38]

The strength of BRT is that it can be like a rail system in bypassing road traffic. In many cities in developing countries, thousands of minibuses have crammed the streets. With its extra speed, some BRT systems can carry 20,000 people an hour. Buses in traffic can rarely carry 8,000 passengers per hour. BRT offers a greener option that is faster than traffic or minibuses. Buses have emissions and noise problems, so they are less able to attract dense development around their main stops, though this can be overcome with emissions regulations (especially by favoring CNG) and by noise insulation in buildings.[39]

Paris, Los Angeles, Pittsburgh, Miami, Boston, Brisbane, Mexico City, Jakarta, Beijing, Kunming, and Chengdu are all now developing BRT systems. Describing the BRT trend in Chinese cities, the International Energy Agency notes: "If Chinese cities continue the momentum they have gained in the past few years, transport will serve city development, the strangulation by smaller vehicles seen elsewhere will be avoided and Chinese cities will move a large step towards sustainability."[40]

Transformation of many car-saturated cities in the developing world can occur through a well-placed BRT or rail system, as so much activity is already built into each corridor. The question of whether to use BRT or rail depends on the number of people to be transported and the space available in the central area. In Mumbai, where 5 million people a day arrive by train, the space is so confined that only rail could work. In high-density cities, transit-oriented development nodes are already in place, they simply need the transit. Bangkok has begun to provide a rail rapid transit system through and above its congested streets that has an average traffic speed of 14 kph and a transit (bus) speed of 9 kph.[41]

All these hesitant moves toward greener urban transportation received a boost from higher oil prices during 2006. According to the American Public Transportation Association (APTA), use of public transport in the United States in the first quarter of 2006 was more than 4 percent above a year earlier. Ridership on light rail was up 11.2 percent, and buses carried 4.5 percent more passengers. Bus use has seen some astonishing growth, especially in smaller cities, notes APTA president William Millar. Tulsa, Oklahoma, has seen bus travel jump 28 percent in one year, for example. Many systems are running their buses on natural gas or cooking oil, to save money and draw in green riders. But it is light rail that is booming across the United States. Phoenix, Charlotte in North Carolina, and Oceanside in California are building light-rail lines from scratch. Denver, Dallas, St. Louis, and many others are racing to extend existing systems, sometimes along old railway tracks.[42]

Facilitating Walking and Biking

Nonmotorized transport—bikes, rickshaws, walking—needs to be given as much priority as transit if it is to be facilitated. The large numbers of people who walk or bike or who would do so if safe routes existed provides plenty of justification for cities to set aside whole streets and parts of streets.

Box 4–2. Bus Rapid Transit: The Unfolding Story

Since 1974, when the bus rapid transit system in Curitiba, Brazil, opened, engineers have known it was possible to provide the speed, capacity, and comfort of subways at a fraction of the cost. Curitiba was able to move roughly 12,000 passengers per direction at peak hour (pphpd) at average speeds greater than 20 kph. Although most metrorail or subway systems have capacities above 20,000 pphpd and speeds around 30 kph, Curitiba's system was closer to those levels than any bus transit had achieved.

Curitiba built beautiful tube-shaped stations with platforms at the same level as the bus floor. By paying before entering the tube, passengers could quickly enter all four bus doors. With a dedicated right of way and signals that gave priority to the bus corridors, buses avoided traffic. Buses were switched from direct routes to trunk and feeder services, which cut the number of them on the roads. Businesses enjoyed the decreased traffic, reduced air pollution, and high-speed passenger access.

The share of trips in Curitiba taken by public transport increased with the BRT and remained above 70 percent for over two decades, countering the iron rule that transit use declines with economic development. The system's low cost made it possible to expand bus service to keep pace with metropolitan growth. When BRT expansion stopped in the early 1990s, the share of trips by public transit began to decline. Today it is about 54 percent—albeit still high for a city with motor vehicle ownership of around 400 cars per 1,000 people.

For more than two decades, BRT failed to thrive outside Curitiba. Brazilian cities such as São Paulo, Belo Horizonte, and Porto Alegre built bus lanes superficially resembling Curitiba's but without the key elements: prepaid platform-level boarding stations, restructured bus routes, and bus priority through the city center. When experts considered why no city could replicate Curitiba's success, they noted that its Mayor, Jaime Lerner, had been appointed during a dictatorship and had military backing

to force private bus companies to reform. The municipal transit agency collected fares and paid bus companies by the kilometer. Bus operators in other Latin American cities blocked such changes.

In 1998, Quito built the first real Curitiba-style BRT, routed boldly through the city's historical core on narrow streets. Its speed and capacity are slightly lower than Curitiba's, as it goes through such a dense city center. In 2006, the Quito BRT lane was briefly reopened to mixed traffic, however, significantly compromising the system and illustrating one of the risks of BRT.

Bogotá's TransMilenio BRT system was built by Brazilian and Colombian engineers who had analyzed the Curitiba and São Paulo systems. The main bottleneck in Curitiba is the bus stop. During rush hour, buses back up waiting to discharge passengers. TransMilenio's innovation was a passing lane and multiple bays at each stop. Up to three buses, not just one, can allow passengers to board and alight at once. The passing lane also allowed express bus services. TransMilenio achieved an operating capacity of 35,000 pphpd and speeds around 28 kph. With overcrowding, TransMilenio moves 53,000 passengers per direction per hour, comparable to all but the highest-capacity metros. With redesigned streets, bicycle lanes, sidewalks, and public spaces, traffic accidents on the corridor dropped, bicycle and pedestrian trips increased, air pollution decreased, and the quality of life improved.

The rapid dissemination of BRT since the late 1990s owes much to Bogotá's charismatic Mayor at the time, Enrique Peñalosa. Also, development agencies and NGOs began to recognize motorization-related problems, from air pollution to climate change. Many private bus companies that had opposed BRT became threatened by growing private motor vehicle use and paratransit minibus services; when they visited profitable bus operators in Bogotá and Curitiba, they became BRT supporters.

Jakarta opened the first Curitiba-style BRT

in Asia in January 2004. Three corridors of TransJakarta are completed, and three more are under design. While capacity (5,000 pphpd) and speed (18 kph) are not as high as in Curitiba, correcting design flaws such as having only a single door on each bus and at each station could more than double capacity.

In December 2004 Beijing followed suit, with support from the Energy Foundation. Seoul, South Korea, also built an extensive bus priority system, though it does not have all the BRT features. Mexico City followed in 2005 with Metrobus, with early World Bank involvement and support from EMBARQ at the World Resources Institute. New systems have opened in Hangzhou, Guayaquil, and a half-dozen other cities. Today, new BRT systems are under development in dozens of cities in both developing and industrial countries. The systems in Brisbane, Ottawa, and Rouen in France are among the best examples in industrial countries.

Yet obstacles to the spread of BRT remain. There is a risk that the misapplication of the lessons of Bogotá's TransMilenio will lead to suboptimal, largely failed systems. Even U.S. cities have branded many marginal improvements in bus services as BRT, though these systems lack most of the features that made Curitiba and Bogotá such a success. And Metro and light rail interests have organized against BRT in cities such as Hyderabad. Much of this competition is healthy, but for many large Asian megacities, integrated metro and BRT systems could offer a much denser, faster network for less than a metro-only system.

—*Walter Hook*
Institute for Transportation
and Development Policy

SOURCE: See endnote 38.

Cities need to create a network of "green streets" where motor vehicles are banned. This is happening in Denmark and the Netherlands, where cities have bicycle infrastructure such as cycleways. Whole streets are set aside for bicycles, and the number of people using bikes is increasing. In Copenhagen, for instance, 36 percent of residents biked to work in 2003 (27 percent drove, 33 percent used transit, and 5 percent walked). In Chinese cities, which have a long history of cycling, many local mayors began to remove bicycle and pedestrian rights in favor of cars. Now that the national government has outlawed anti-bike measures, the green streets of Chinese cities are to be restored. Yet the tension between car drivers and cyclists is likely to remain, as it does in all cities.[43]

Providing green streets was also a goal of Indonesia's Kampung Improvement Program, a slum-upgrading initiative. In several Indonesian cities, such as Surubaya, narrow streets within the kampungs were kept closed to cars, and traditional social life was maintained. By banning motorized vehicles, the government helped protect the revitalizing kampungs from gentrification. The only modes of transportation used within the kampungs are foot, bicycle, and becak, the traditional Indonesian trishaw.[44]

Every city desiring to have greener transport will need to favor its public transport and NMT system over road building. This is fundamental to greening urban transportation. It all comes down to priorities in the planning of a city. (See Box 4–3.) Two former governors from different parties in the United States, Democrat Parris Glendening and Republican Christine Todd Whitman, agree: "If you design communities for automobiles you get more automobiles. If you design them for people you get walkable, livable communities."[45]

Box 4–3. São Paulo Bicycle Refuge

Cycling is not an easy option in traffic-choked São Paulo, a sprawling metropolitan region of 19 million. Motorized vehicles are responsible for 90 percent of São Paulo's smog, which often hangs over the city in a dark veil, contributing to severe respiratory problems. Many of the wealthiest residents forgo the streets; the city's huge helicopter fleet is second only to New York City's.

But a one-of-a-kind bike parking lot in the São Paulo suburb of Mauá is prompting city officials to rethink the role of cycling. Cyclists wanting to ride to the Mauá rail station had nowhere to park until 2001, when station manager Adilson Alcantâra created ASCOBIKE (Association of Bicycle Riders). For a $5 monthly fee, members park their bikes and receive regular maintenance. The initial 700 spaces filled quickly, 1,800 members have joined, and the association is planning to expand.

Laura Ceneviva of the Environmental Department of the City of São Paulo and the head of the city's bicycle working group said "users pay a low fee for a good service, and jobs are created as well. There is no reason why we could not reproduce this successful and efficient service throughout São Paulo."

—Jonas Hagen
Sustainable Urban Transport Consultant

SOURCE: See endnote 45.

The Economics of Urban Transportation

Many cities built around nineteenth-century rail in Europe and parts of the United States, Australia, and Latin America remain in their transit city form. As transit systems need to keep pace with urban growth, new rolling stock and lines into car-dependent suburbs are required to keep ahead of the avalanche in car use. To fund public transportation, cities have had to find innovative solutions, such as the congestion tax in London or the use of land sales around stations for Copenhagen's new light rail.

Car-dependent cities without adequate transit need to build new systems because traffic growth and urban sprawl are accelerating car dependence. In recent decades this has meant new train systems. As indicated earlier, rail can offer a faster option than cars (often with average speeds over 50 kph) and can facilitate the development of walkable centers. But transportation funding sources have tended to favor buses, as these fit into the main priority of building roads.[46]

Political intervention has led to the revival of rail when funding has been opened to local choice. For example, the U.S. federal transportation funding process was freeway-oriented from its inception in 1956 until the 1990s. Then the Surface Transportation Policy Project (a coalition of over 100 NGOs) drafted the Intermodal Surface Transportation Efficiency Act of 1991, which requires that local choices be made by cities. Funding for greener transportation has grown, and the act has been reauthorized twice with large congressional support.[47]

Car-saturated cities can model their transit systems on Hong Kong and Tokyo, where transit is funded almost entirely from land redevelopment. In poorer cities, the use of development funds for mass transit is clearly justified because transit helps transform the economy through better planning and efficiency in the city.[48]

Transit-based cities spend around 5–8 percent of the GDP of the city's region on transportation, but in heavily car-based cities the figure is more like 12–15 percent (and it reaches 18 percent in Brisbane). Why? It appears to be due to the costs of car travel and the sheer space required for cars. Car travel is estimated to cost more than transit (about 85¢ per passenger-kilometer versus 50–60¢), even though motorists routinely think it is cheaper—no doubt because they fail to take into account such costs as depreciation and insurance. Data from U.S. cities show that transportation costs between the 1960s and 2005 have risen from 10 to 19 percent of household expenditures (before the recent rise in gasoline prices); cities with the highest level of car dependence have the highest percentages and those with better transit, the lowest.[49]

The biggest economic impact of cars on cities is the sheer space they take for roads and parking. Freeway traffic carries 2,500 people per hour, a bus lane carries 5,000–8,000, a light rail or BRT can carry 10,000–20,000, and a heavy rail system carries 50,000 people per hour—20 times as many as a freeway. It is no wonder that freeways fill so quickly. Likewise, most car-dependent cities require five to eight parking spaces for every car. All this space costs money and is simply unproductive land.[50]

Two calculations for Sydney illustrate the uneconomic, space-hungry nature of car dependence. If downtown Sydney closed its rail system, the central area would be required to build another 65 lanes of freeway and 1,042 floors of multistory car park. In reality, business would just scatter instead, as it does in most heavily car-dominated city centers. This is also a big economic issue in today's global economy, as the new interactive kind of jobs that are becoming available seem to favor city centers where a large concentration of people can meet. Another relevant calculation: if the next million additional people in Sydney were located in transit-oriented development so that each household had one less car, the city would save around

$18 billion in capital opportunity costs due to space saving and $3–4 billion in annual driving costs. And this does not include the reduced external costs due to pollution reductions or the health savings due to less obesity and depression associated with excessive car use. Greener transportation is a healthier economic proposition.[51]

The biggest economic impact of cars on cities is the sheer space they take for roads and parking.

The economics of greening transportation have been assessed by the Center for Transit-Oriented Development in Oakland, California. Based on a detailed survey across several states, the staff calculated that people in 14.6 million households wanted to live within half a mile of a TOD. This is more than twice the number who live there now. The market for TOD is based on the fact that those living in such centers now (who were found to be smaller households, although the same age and the same income level on average as those not in a TOD) save some 20 percent of their household income by not having to own so many cars. People in TODs owned 0.9 cars per household compared with 1.6 cars for other people. This freed up on average $4,000–5,000 per year. In Australia, a similar calculation showed this would save some $750,000 over a lifetime.[52]

The economic benefits of greening urban transportation are even beginning to be seen by some parts of the normally car-oriented conservative side of politics. According to the Washington lobby group Free Congress Foundation: "Conservatives tend to assume that transit does not serve any important conservative goals. But it does. One of the most important conservative goals is economic growth. In city after city, new rail transit lines have brought higher property values, more customers for local businesses and new development."[53]

Needed: Political Leadership and a Vision of Greener Options

Cities are organisms that work as a whole regional system and as a series of local parts. They need viable governance systems at both the regional and the local level to create green transportation options for more resilient and sustainable cities.

Visionary master plans and regional governance structures enabled governments to build the urban freeways of the auto city. Today, cities need new visionary plans to generate the political momentum and find the funds for greening transportation, as well as the governance structures that can carry out the plans. Regional governance structures for transportation exist in most cities in Canada and Australia and, to a lesser extent, the United States. Bogotá and Curitiba relied on regional governance systems, city money, and World Bank support to build their transformative transit systems. Mumbai and Kolkata also have a regional governance system that manages their massive transit systems. Megacities without such a system will find regional transit difficult.[54]

Cities need new visionary plans to generate the political momentum and find the funds for greening transportation.

But regional transit systems cannot work unless local systems feed into them. Local governments, residents, and businesses need to come up with their own green transport plans, since each part of a city has different economic functions. The cities that have done best at building regional transit systems—Zurich, Munich, Hong Kong, Singapore, and Tokyo—also have active local transport planning processes. Zurich has each canton choosing the timetable they require for their transit.[55]

Surveys constantly show that people want to see greener transportation options given higher priority in their cities. In Perth, people were asked if they saw a need for more transit, walking, and biking over cars, and 78 percent agreed they did. Then they were asked if they would transfer road funding to pay for these greener modes, and 87 percent agreed with that idea. In Porto Alegre, Brazil, a people's budget approach asked citizens to assign priorities for expenditures; the vast majority of neighborhoods put greener modes of transportation above the need for more roads. In Milwaukee, Wisconsin, a survey showed that bus and rail projects were favored by 70–85 percent of those surveyed, whereas "more highway capacity" came in last, with 59 percent support. A gasoline tax was the preferred way to pay for such improvements. And in Oregon, the Transportation Priorities Project showed similar sentiments about transit over freeway options.[56]

In the past, transportation priorities have generally been set by engineers, not the public. But in the United States, at least, voters are sending the clear message that they want better options and are willing to pay for them. Between 2000 and 2005, voters in 33 states approved 70 percent of transport ballot measures, generating more than $70 billion in investments—much of it for public transportation.[57]

Even in Atlanta—the city with the fewest people per hectare and the most car use per person—there are signs of hope. A $2.8-billion Beltline loop has been proposed that will link unused railroad rights-of-way with the city's existing transit lines and will create walkable communities joined together by green trails and public transport. The project has lifted the sights of the city, bringing together a coalition of interests. "It's very important that cities and communities go for a big vision," says Peter Calthorpe, an urban planner based in San Francisco, "Cities need these bold moves and elements to make them exciting places to live. It's exactly the kind of thing that will differentiate a city from the suburbs. Suburbs are the sum of a lot of little ideas." [58]

Determining the views of the public on how they want their city to function in the future can involve deliberative democracy techniques, such as town meetings as occurred in New York after the September 11th terrorist attack. In Perth, 1,100 people were brought together to plan the city; the result was a strong endorsement for greener transport in a greener city. In the United Kingdom, a foresight technique has been used to look at how cities can be envisioned. The greening of urban transport options proved to be the one the community desired the most. [59]

As described in this chapter, reaching that goal requires:
- a transit system that is faster than traffic down all major corridors;
- viable centers along the corridors that are dense enough to service a good transit system;
- walkable areas and cycling facilities that can mean easy access by nonmotorized means, especially in these centers;
- services and connectivity that can usually guarantee access without time wasting;
- the phasing out of freeways and phasing in of congestion taxes that are directed back into the funding of transit and walk/cycle facilities as well as traffic-calmed boulevards; and
- continual improvement of vehicle engines and fuels to ensure that emissions, noise, and fuel consumption are all reduced.

Cities need visions about how they can be transformed from car dependence and car saturation to greener modes. And they need political leaders who can overcome the various barriers that prevent these visions from coming true.

CITYSCAPE: LOS ANGELES

End of Sprawl

Los Angeles is known around the world as the mother of sprawl, thanks to countless aerial photographs portraying vast landscapes of monotonous suburban houses. It may come as a surprise to learn that Los Angeles will soon be known instead as the birthplace of the post-suburban city.

Los Angeles is a city, a county, and a region—an immense mosaic of continuous development in Southern California. The city's 1,300 square kilometers are inhabited by 3.8 million people (according to the 2005 U.S. Census estimate), translating to a density of 30 persons per hectare or 2,930 persons per square kilometer. Compared with other global cities, Los Angeles is unquestionably "suburban" if the term means low density. Los Angeles has half the density of London, one quarter as much as São Paolo, and one tenth as much as Hong Kong or Mumbai.[1]

Seemingly unbounded expansion has been part of the city's geographic pattern for over 200 years. The first ring of single-family, detached dwellings was followed by an explosion of suburbs following World War II, when immense residential developments like Westchester, Lakewood, much of Orange County, and the San Fernando Valley were built up. Like their contemporaneous if more famous cousin Levittown, on Long Island, New York, these subdivisions came to define the image of suburban sprawl.

From the mid-1940s to the mid-1960s, tens of thousands of home sales were financed by the federal government through various means such as mortgage interest deductions, Federal Housing Administration loans, and infrastructure subsidy. Federal housing policy had several political uses: it was an employment program for the vast construction industry, it provided housing to the middle class, and it served conservative national interests (based on the belief that no homeowner, with so much work to do, would have enough spare time to be a revolutionary). The housing policies were also segregationist, since nearly all early suburbs were restricted to whites. Zoning regulations for housing, first tested in Los Angeles, instituted further restrictions against mixed-use or multifamily housing.[2]

Los Angeles was the engine behind an emerging, nationally organized real estate lobby at the beginning of the twentieth century—an industry dominated by residential builders who monopolized the ever-widening periphery. In an effective marriage between the government and housing construction interests, 58 percent of Californians became homeowners by 1960 and the figure never climbed higher.[3]

Los Angeles continued its unabated growth as long as geography, policy, economics, and the environment were amenable. It is only in the new millennium that definitive research corroborated Angelenos' impression that "sprawl has hit the wall." There is no more land for easy suburban growth, the commute times exceed viability, and water—long a determining resource in Southern California—is in short supply. Yet Los Angeles expects to add 6 million new inhabitants—the population of two Chicagos—by 2020. The mother of sprawl must now transform herself into the mother of invention, and there are signs that she is doing just that.[4]

The city has slowed its outward growth and has begun to fill interior gaps in the city fabric. In 2000, the Los Angeles urban area was more densely populated than San Francisco, New York, or Washington, DC. Unlike the dramatic landscape transformation wrought by postwar suburbia, the next metropolitan era is creeping into the city more

© Bettmann/CORBIS

New homes in Lakewood, Los Angeles County, in 1950

stealthily, as existing neighborhoods fill in and grow denser. Residents notice changes that may not seem linked: traffic increases, housing prices skyrocket, parking is more difficult, children need to be bused to outlying schools, new lofts are built downtown, communities grow more ethnically diverse, neighborhood organizations become more protectionist, older houses are torn down and replaced by multifamily housing where permitted, and older housing stock is upgraded.[5]

Some of this infill is improving the quality of life in Los Angeles. Yet in certain areas the results are a looming disaster. For example, the core of urban poverty in the region is a vast, 272-square-kilometer zone that cuts across municipal boundaries and holds nearly 40 percent of the county's poorest households. These neighborhoods are growing more overcrowded with shadow housing, as new generations of immigrants double and triple up in existing apartments or turn garages into semi-habitable quarters. Inadequate infrastructure and services exacerbate problems of poverty and inequity.[6]

Still, within that urban core of poverty as well as in the inner-ring suburbs and some of the early postwar suburbs, there is a flowering of infill projects that are adding to the housing stock and to neighborhood quality. Formerly industrial sections of downtown are witnessing reinvention as young urban dwellers move into the new market-rate housing. Streets that had been abandoned are now filled late at night with people walking dogs and dining out. Creative nonprofit housing corporations are demonstrating ways to build affordable, multifamily housing on scattered sites throughout the region. Mixed-use developments are supporting local retail and services along with housing. Smaller public schools are being established in older neighborhoods so that children can be educated locally.

Los Angeles has a more than adequate supply of sites for future infill, with residents of every income and race willing to live at higher densities, provided they get the housing and services they need. Local policymakers are struggling to create stronger public guidance, so that the next Los Angeles will be characterized by greater community equality. The lessons this city of nearly 4 million holds for metropolitan futures may stem the tide of exporting the tired U.S. suburban model to the rest of the world and may simultaneously cultivate a new, more compact Los Angeles.[7]

—*Dana Cuff*
University of California, Los Angeles

CITYSCAPE:
MELBOURNE

Reducing a City's Carbon Emissions

Melbourne's Queen Victoria Market, a nineteenth-century shopping complex, has a modern facade: 1,300 solar panels—the largest single urban grid-connected solar installation in Australia. Regularly classified as one of the world's most livable cities, Melbourne has 60,000 residents and a 660,000 daytime population, is the heart of a metropolis of 3.6 million, and has emerged as a national environmental leader.[1]

The Melbourne City Council is promoting partnerships between business and government that encourage business growth while promoting environmental quality. The city intends to build new export-oriented industries and jobs, with a target of a 60-percent growth in environmental management capacity. The council has adopted important environment strategies on carbon emissions, water consumption, and waste management.[2]

The ambitious Zero Net Emissions by 2020 effort, on reducing carbon emissions, is backed by comprehensive policies and programs. As the first Australian city to achieve all five carbon-reducing milestones of the Cities for Climate Protection Campaign organized by the ICLEI–Local Governments for Sustainability, Melbourne has turned the serious threat of climate change into a triple-bottom-line opportunity by combining market mechanisms and regulation. While some observers claim that fighting climate change requires costly action, Melbourne is showing that businesses can reduce operating costs and improve their competitiveness with energy-efficient design.[3]

The City Council, which has already cut its own carbon dioxide emissions by 26 percent, has decided to lift its 2010 target from an initial goal of a 30-percent cut to 50 percent. Its new office complex, Council House 2, is the first in Australia to achieve the maximum Green Star rating of six. Lord Mayor John So wants the building to set a benchmark for sustainable design: "We hope that CH2 will change the way that buildings are designed and constructed in Melbourne, Australia and round the world."[4]

Solar-powered louvers on the building's facade track the sun, and automatic "night purge" windows allow fresh air to cool the building after dark. Wind turbines, solar panels, and a gas-fired cogeneration plant provide power. A water mining facility mines water from a neighboring sewer, treats it to Class A standard, and uses it to flush toilets and run the cooling towers. The new building uses 87 percent less energy than the old one and 72 percent less water, while providing occupants with 100 percent fresh air.[5]

These types of innovations are being adopted citywide, as the new Melbourne Planning Scheme requires that all new office buildings improve energy efficiency, reduce emissions, use passive solar design, use solar energy or heat pump technology, collect and reuse rain water, recycle wastewater, encourage waste recycling, and have no impact on the solar collecting of adjoining buildings.

For existing buildings, owners are encouraged to audit and reduce their own energy and water use with the GreenSaver program, which subsidizes participants' energy and water audits, as well as products such as low-flow showerheads, efficient light bulbs, and draft sealers. A quarterly Melbourne Forum holds commercial green building discussions with the real estate, property development, and architecture sectors to drive the adoption of sustainability principles. A municipal Savings in the City program helps hotels reduce greenhouse gas emissions and the impacts of water use and waste. City-run pilot projects are targeting high-rise apartment buildings.[6]

David Hannah/City of Melbourne

The Council House 2 building

Other Council initiatives include purchasing green power for street lighting and Council buildings and forming neighboring councils into a green power bulk buying group. The city is promoting a voluntary carbon market to allow businesses flexibility in emissions management and is investigating joining the Chicago Carbon Exchange.

The city is also promoting carbon sequestration through urban tree planting and a pilot investment in rural conservation plantings to offset emissions from the Town Hall. Melbourne's car fleet has signed up with Greenfleet, which plants 17 trees per car. When Mayor So attended the 2005 World Environment Day in San Francisco, emissions from his travel were offset by tree planting.[7]

The city has also created a $5-million Sustainable Melbourne Fund that has been working with Investa Property Group, Australia's largest listed owner of commercial property, to finance a Greenhouse Guarantee program that delivers energy savings to commercial tenants. The fund also invests in water-saving infrastructure. Businesses, universities, and other organizations can undertake a water audit to determine possible savings. The fund will finance the purchase and installation of water-saving technology. The recipient then has a lower water bill but pays the fund the difference between the old and new bills until the investment plus interest is repaid.[8]

The 2006 Melbourne Commonwealth Games set new benchmarks in reducing the environmental impacts of a world-class event. The state and the city cooperated to plant a million trees to offset emissions generated in transporting athletes to Melbourne, provided free public transport during the event, and housed athletes in "green" buildings.[9]

Melbourne's leadership role has most recently been demonstrated by its membership in the Large Cities Climate Leadership Group that is supported by the Clinton Foundation's Climate Initiative. Mayor So linked the grand plan with the practical "agreement to create a buying club for member cities to get lower prices for sustainable products."[10]

Melbourne's commitment to combating climate change has the support of Labor, conservative, independent, and Green councilors. The two major Melbourne-based banks are active members of the U.N. Environment Programme's Finance Initiative, and the National Australia Bank highlighted its award-winning green headquarters in its *2005 Corporate Social Responsibility Report.* The State Government and the private-sector Property Council have also been closely involved in these efforts.[11]

Elected councilors, senior staff, and community and business supporters have combined to set achievable and ambitious targets and then raise and expand them. The goal is "an Environmentally Responsible City which seeks to actively increase natural assets through the decisions it takes, the development it chooses to pursue and the benefits and impacts these have on the natural world."[12]

—*The Honorable Tom Roper*
Former Minister, Victoria Government, Australia

CHAPTER 5

Energizing Cities

Janet L. Sawin and Kristen Hughes

At night, Earth's cities are visible from space as stars or chains of light in a sea of blackness. Many appear to be vibrant and to exude energy. People are drawn to the "lights on Broadway," for example. But a closer examination reveals complex webs of streets, enormous buildings, vehicles, and burgeoning populations—all of which need energy to build, use, and sustain. To meet these needs, cities draw energy from the world around them, providing local benefits—but with health, security, and environmental consequences for all.

The portrait of urban energy use today contains real differences in consumption and resulting ecological footprints among the world's cities—differences that reflect the vast financial wealth separating the world's most industrialized, rich urban areas from the poorer cities just now experiencing rapid economic growth. Indeed, millions of people who live in or around the world's poorest cities do not have access to modern energy services.

Even as industrializing and poor nations seek to expand their economies to levels nearer those of rich nations, Earth's atmosphere and ecosystems are demonstrating real limits to our ever-increasing consumption of resources. Partly in recognition of this, hundreds of cities around the world are working to reduce their ecological footprints.[1]

The re-visioning of urban life reflects a critical moment in the history of cities, as many unfavorable factors have converged to make present trends insupportable. Over the past 150 years, cities have become increasingly reliant on dirty and distant energy sources, leaving them vulnerable to supply disruptions and destroying community-based notions of environmental protection. In the next few decades, the vast majority of expanded energy supply will be to meet the needs—direct and indirect—of cities. Increasingly, cities will need to play a more active role in planning and

Kristen Hughes is a research associate and doctoral candidate at the Center for Energy and Environmental Policy, University of Delaware.

building their own energy futures.

Cities in the future will bear the brunt of many challenges related to today's unsustainable energy systems—from air and water pollution to climate change—particularly as growing populations put increasing pressure on resources. As cities continue to expand, the enormity of their contributions to major social and environmental problems is only expected to rise. Yet this same enormity of scale offers cities the potential to make beneficial changes with significant local and global impacts. As this chapter describes, cities hold the key to mitigating problems through urban planning, building design, and choice of end-use products and energy resources and technologies. Around the world, numerous cities are already improving their efficiency and producing more of their energy locally and sustainably, and many of these efforts can be replicated elsewhere to reduce environmental impacts and improve the quality of life for urban and rural dwellers alike.

Urban Energy Needs and Constraints

It took millennia to make the transition from human muscle power to draft animals and then to primitive machines that tapped renewable energy flows from wind and water. In contrast, the Industrial Revolution came along in the mere blink of an eye. In the span of a few generations, cities were transformed from dense areas of narrow streets with small, low dwellings to skyscrapers and sprawling suburbs. Over time, urbanites traded horses for streetcars and, eventually, private vehicles. Rarely very large before the advent of steam engines, urban populations began to soar as opportunity drew waves of immigrants to cities and as cleaner streets reduced death rates. Energy use surged as well, and the

advent of the fossil fuel age—which provided power for elevators, electric lights, and motor vehicles—enabled cities to become what they are today.[2]

Direct energy consumption per person in industrial-country cities is often lower than in rural areas due to the greater density of living and commuting spaces. Urban residents in Japan, for example, use less energy per capita than rural residents do. In older cities, designed before the widespread use of private cars, energy use per person is lower than in sprawling modern cities. The dense environment of Manhattan more than compensates for its massive, often old and inefficient buildings, making New York City one of the most resource- and energy-efficient places in the United States.[3]

In developing countries, where many rural people lack access to modern energy services, the reverse is often true. The one third of India's population who live in cities consumes 87 percent of the nation's electricity. And in China, urban residents typically use 40 percent more commercial energy than their rural counterparts. (People in rural China actually use more total primary energy, mainly in the form of biomass, but most of this is lost during inefficient combustion.)[4]

Cities require energy to build infrastructure, to light, heat and cool buildings, to cook, to manufacture goods, and to transport people. The infrastructure itself, including streets, buildings, bridges, and other urban features, represents large quantities of embodied energy—the energy invested in these structures during their lifetimes from the cradle of raw materials, to city block, to eventual grave. (See Box 5–1.) Urban residents also consume large amounts of energy indirectly in the food and other goods they import.[5]

Most if not all of the energy used directly in cities must be imported as well, raising a host of significant costs and challenges. Pipes

Box 5–1. Reducing Construction's Environmental Impact

The construction industry accounts for more than one third of global carbon dioxide (CO_2) emissions and produces nearly 40 percent of all human-generated waste. That waste traps enormous amounts of embodied energy. For example, concrete (composed of sand, aggregate such as crushed stone, and a cement binder) contains about 817,600 Btus per ton. A ton of steel contains 30 million Btus—about half the annual energy consumption of a typical home in San Francisco.

While steel contains a higher amount of embodied energy per ton, concrete accounts for the largest portion of construction waste. Although cement represents only 12 percent of the average concrete mixture, it accounts for 92 percent of concrete's embodied energy. In 1997, manufacturing 1.5 billion tons of cement worldwide emitted more CO_2 than Japan did that year. The industry's impacts could be reduced significantly with more use of fly ash—a fine waste powder produced during coal combustion that is toxic if inhaled. Increasing the generally accepted rate of 15 percent fly ash content in cement to a feasible 65 percent could avoid emissions of the equivalent of Germany's entire annual contribution to climate change. In mid-2006, researchers dis-

covered a way to produce lighter, stronger bricks and build aggregate entirely from fly ash. Completely replacing cement with fly ash may also be possible. Increased use of fly ash can reduce the need to manufacture cement and crushed stone, while also sequestering a toxic substance.

Using local and traditional materials for construction, such as stone, wood, clay, and plant materials, can reduce costs, provide local jobs, and improve occupants' comfort and health while minimizing the embodied energy of construction materials. Energy inputs for production are often lower, and transportation needs—which account for 12 percent of concrete's embodied energy—are reduced substantially.

Some people are taking the next step by opting to reuse construction "waste." The demolition of Boston's Central Artery, a stretch of freeway replaced during the infamous Big Dig project, generated 20,000 tons of waste concrete and over 38,000 tons of waste steel, most of which went to landfills. However, Big Dig engineer Paul Pedini had a different vision for the old highway: he took some concrete slabs and steel beams and built himself a house.

—Stephanie Kung

SOURCE: See endnote 5.

carrying gas, for example, pose serious safety and environmental threats in urban areas, where leaks or explosions can cause injuries and deaths. Electricity generally comes from large, central power plants via transmission and distribution (T & D) systems that are often inefficient and unreliable. The centralized grid permits blackouts to cascade throughout entire regions. The August 2003 blackout in the northeastern United States and Canada, for example, which was caused initially by a fallen tree, affected 50 million people and cost the region $4.5–10 billion. A month later, another tree hit a high-volt-

age transmission line in Italy, leaving 57 million people in the dark.[6]

Transmission bottlenecks are particularly pronounced in large metropolitan areas, which require vast amounts of power to traverse great distances through a limited number of lines. The share of electricity lost along the way ranges from 4–7 percent in industrial countries to more than 50 percent in parts of the developing world, where much of the loss is due to people tapping lines illegally. In parts of New Delhi, electric cables are caught in a tangle of hooks and wires as slum dwellers, small factories, Hindu temples, and even wealthy businessmen siphon off 36 percent of the city's power.[7]

One of the greatest challenges of the current system is getting energy services to all urban residents. Nearly one fifth of the estimated 1.6 billion people worldwide who lack access to electricity and other modern energy services live in the world's cities. Because access is defined as areas with grid extensions, the actual number truly without access could be higher. About one third of Africans live in urban areas, and at least one quarter of city dwellers on the continent do not have access to electricity.[8]

Too many of the world's people thus must struggle daily to afford or find energy resources—most often wood, charcoal, dung, or other biomass. Indoor air pollution caused by burning these inferior fuels results in millions of deaths annually. Heavy reliance on biomass has also increased the destruction of forests around cities, exacerbating local air pollution and soil erosion. In India, Sri Lanka, and Thailand, wood harvesting by the urban poor has produced a halo of deforestation around cities, towns, and roads. And a radius of some 400 kilometers has been cleared around Khartoum in Sudan.[9]

For those with access to modern energy services, the predominant fuel used for non-transport energy is coal, which accounted for nearly one fourth of total global energy use in 2003; the International Energy Agency (IEA) projects that coal consumption will continue to rise significantly through at least 2030. Energy from coal and other conventional sources comes with high costs, including soil and water pollution resulting from resource extraction and use, air pollution from burning, and associated health problems. In China alone, coal use causes the death of 100 miners weekly on average, significant urban air pollution, and acid rain damage to more than a third of the country.[10]

Furthermore, heavy reliance on fossil fuels, particularly in cities, is the primary driver of global climate change. Cities now house just shy of half the world's population, but they are responsible for the vast majority of greenhouse gas emissions from human activities.[11]

Reducing Demand without Dimming the Lights

Much of the energy that people pipe, wire, and truck into cities is used by and in buildings—constructing and operating them as well as making their occupants comfortable. Globally, buildings account for more than 40 percent of total energy use. When the energy required for materials, transportation, and construction is included, buildings devour more than half the energy used in the United States each year.[12]

As cities become more populated, more and more of the world's buildings are found in urban areas. In 2005, Shanghai constructed more building space than exists in all the office buildings of New York City. Every month, China adds urban infrastructure equal to that found in Houston, Texas, simply to keep up with the masses of people migrating from rural areas to cities.[13]

The advent of cheap and readily available energy let the modern building work in spite of nature rather than with it. Yet around the world there is a small but rapidly growing movement to make buildings "green"—lowering their energy needs, for example, through efficiency improvements, embodied energy reductions, and the use of on-site energy resources. Green buildings incorporate designs and technologies often considered new and innovative; in reality, many of these ideas have been around for centuries. Today, architects, planners, and others are rediscovering traditional ways to light, heat, and cool indoor spaces and adapting them for modern uses.

Lighting accounts for nearly 20 percent of total electricity consumption worldwide. Much of this occurs when the sun is shining. Thus energy use could be reduced dramatically with simple design techniques such as natural daylighting, mirrors and reflective paints, and light shelves—horizontal fins at windows that act as shading devices, reduce glare, and allow daylight to penetrate deep into buildings. Technology has improved to the point where glass transmits light while reflecting unwanted heat. These techniques and materials not only offset some of the lighting load, they also lower the significant heat gains associated with lighting, reducing air conditioning needs.[14]

The Accord 21 Building, opened in 2000 in China, uses 70 percent less energy than standard buildings.

Once design changes have lowered the need for artificial light, energy demand can be reduced further with modern technologies like motion sensors—which turn lights, appliances, or machinery off when they are not needed—and energy-efficient bulbs and lamps. Conventional incandescent bulbs convert about 10 percent of energy to light and the remainder to heat. In contrast, compact fluorescent bulbs and light-emitting diodes (LEDs) use far less energy to produce a comparable amount of light while producing a fraction of the heat. These alternatives cost more upfront, but they save energy and money over their lifetimes.[15]

Heating water and space also requires significant amounts of energy. Better insulation, proper building orientation, and the use of solar heating and other techniques can dramatically lower energy demand and associated costs, as can reducing the scale of buildings.

In developing countries, one of the most cost-effective ways to increase thermal comfort for the urban poor is to install ceilings beneath their roofs in order to reduce heat loss; energy savings from such programs in South Africa have exceeded 50 percent.[16]

"Waste" heat that is vented in conventional large-scale power plants or in small systems like microturbines or fuel cells can be captured for heating, cooling, or additional power generation. Such combined heat and power systems improve overall efficiency levels dramatically. The Verdesian, a new building in New York City's Battery Park, captures heat from a natural gas microturbine to produce hot water, increasing overall energy efficiency to 80 percent or higher, compared with the 25–35 percent efficiency of a typical fossil fuel power plant.[17]

The means used to distribute such heat can also improve efficiency. Radiant floor heating, for example, is generally more energy-efficient than conventional alternatives, and today's systems can operate with fossil or renewable fuels. Rediscovered early in the twentieth century and now commonplace in much of Europe and the United States, radiant heating was devised by the Romans, who placed terra cotta pipes beneath stone floors to heat villas with flue gases from wood fires. A more modern technique is used in the Hewlett Foundation building in Menlo Park, California, where air is circulated through a raised floor, heating and cooling workers rather than the space above them and allowing individuals to control temperatures.[18]

During hot months, space cooling is becoming increasingly important to keep cities running. The concrete and asphalt jungles that replace natural life absorb heat and raise urban temperatures further, creating what is known as the "heat island effect." In China's major cities, air conditioning accounts for 40 percent of the public's summer energy demand and is the primary cause of power shortages that began in 2003. And in Tokyo, a modeling study found that waste heat emissions from air conditioning are responsible for 1 degree Celsius of warming during the summer, exacerbating the heat island effect. A similar study of Houston, Texas, found that total waste heat emissions were responsible for warming of up to a half-degree Celsius in daytime and 2.5 degrees at night.[19]

For at least 2,000 years, people in the Mediterranean region have passively cooled buildings with a variety of techniques. These include cross ventilation over the surface of a pool of water, open buildings, tree shading, careful placement and sizing of windows, and the use of massive, thick walls and floors as insulation from summer heat. Some of these techniques are being revitalized today, along with options not available two millennia ago.[20]

Natural ventilation—the use of outdoor air to cool buildings—reduces the need for air conditioning in some climates. Studies show that effective night ventilation, adapted to local conditions, could reduce the cooling load in office buildings by 55 percent or more. And the U.S. Environmental Protection Agency (EPA) estimates that careful placement of trees can reduce the energy required for cooling by 7–40 percent, depending on the extent of tree canopy.[21]

Another way to reduce energy demand for cooling is to top buildings with reflective surfaces—such as white paint or metal shingles that act as radiant barriers. An EPA-funded study that considered both cooling benefits and heating penalties of such "cool roofs" found significant net savings in energy use for 11 major U.S. cities.[22]

Green roofs and walls reduce heat gain in summer and they also insulate buildings from cold in winter. Temperatures on conventional roofs can be 50 degrees Celsius (90 degrees Fahrenheit) higher than the ambient temperature; atop a "green roof," the temperature on a hot day can actually be below ambient. A study of an eight-story residential building in Madrid, Spain, found that adding a green roof cut annual energy use by 1 percent, while reducing the peak cooling load on upper floors by 25 percent. With enough "cool" or green rooftops throughout a city, substantial reductions in the urban heat island effect are possible, with the added benefit of less urban smog. Green roofs also filter and retain storm water, reducing urban runoff problems, and they create habitat for birds and recreational space for people.[23]

Each of these features alone provides significant savings. The integration of intelligent design with several efficiency measures can reduce energy use to half or less that in a comparable conventional building. Some experts believe savings of up to 80 percent are possible. As peak loads for lighting, heating, and cooling decline, the required size of boilers, fans, and other machinery does also, providing greater savings in energy and construction costs. The Accord 21 Building, opened in 2000, was the first internationally certified green project in China. It uses 70 percent less energy than standard buildings, causing astonished inspectors to return repeatedly to check that energy meters are functioning properly.[24]

There are good economic reasons for constructing more-efficient buildings: they generally have healthier and more-comfortable occupants, higher worker productivity, reduced tenant turnover, and better performing students in schools. The Internationale Nederlanden Bank headquarters in Amsterdam uses about 10 percent of the energy of its predecessor and reduces worker absenteeism by 15 percent, for a total savings of $3.4 million annually. In the United States, the average premium for a "green" building is 2–5 percent, but studies find that the associated financial benefits over 20 years are more than 10 times the initial investment. And the costs of green buildings are falling with design and construction experience.[25]

Although the marginal cost of improving efficiency is lowest when buildings are constructed, retrofits can be highly cost-effective as well. Simple strategies like daylighting, efficient lighting, and glazing can pay for themselves in as little as one year. More than 300 retrofit projects—from insulation to water system improvements—undertaken in China in recent years had an average payback period of 1.3 years.[26]

Such advances can also provide important benefits for the world's poor. In industrial nations, maximizing efficiency through design and cost-effective end-use technologies can ensure that poor residents are not forced from their homes by rising energy costs. In the developing world, efficiency advances can bring dramatic quality-of-life improvements by making energy services more affordable to the poor. LEDs, for example, provide an estimated 200 times more useful light than kerosene lamps. At $55 each, solar-powered lamps with LEDs could brighten the nights of the poor. In Tembisa, a shantytown of Johannesburg, South Africa, a survey found that almost 10,000 households spend more than $60 each for candles and paraffin every year; with access to microcredit (see Chapter 8), such families could afford cleaner, better lighting freely powered by the sun.[27]

In ancient Greece, many cities were planned in grids so that every home had access to the sun for warmth and light in winter; the ancient Romans went so far as to pass "sun-right laws," forbidding builders from blocking access to the winter sun. Green roofs date back thousands of years, the most famous being the Hanging Gardens of Babylon, constructed around 500 BC. The lessons of these ancient practices, combined with state-of-the-art technologies and materials, provide today's cities with powerful tools to achieve dramatic efficiency improvements.[28]

Powering Cities Locally

When Thomas Edison installed his first electric systems in the late nineteenth century, he envisioned an industry with dozens of companies generating power close to the point of use. Such a system would be particularly suited to densely populated urban areas. Initially, the industry evolved along these lines, with many companies producing power on site and capturing the waste heat. But by the mid-1930s most industrial countries had established monopoly industries, driven greatly by the economic benefits of ever-larger generating stations matched with transmission and distribution systems. It was not until the 1980s that efficiency limits were met—which, combined with a variety of economic and environmental challenges, led many experts to realize that bigger is not always better when it comes to energy production.[29]

Small-scale, locally installed power equipment, also called distributed generation (DG), could enable cities to meet much of their own energy needs once again. Today, DG remains more expensive per unit of energy output than conventional, centralized generation, but costs continue to fall and associated benefits are significant. Distributed generation reduces the need for expensive transmission and distribution infrastructure while lowering grid losses. By bypassing the T&D system, DG also improves reliability and reduces vulnerability to accident or sabotage. Because they are modular and can be installed rapidly, distributed small-scale generators can expand to keep pace with demand as a city grows, deferring or preventing the need for new central power plants. This is particularly important in developing countries, where migration is rapidly raising urban numbers as well as energy demand. And distributed systems provide local control and ownership of energy resources, encouraging community-level economic development. (See Chapter 8.)

Most DG today comes from inefficient diesel generators or natural gas turbines. But several new options are emerging, with technological progress on a variety of fronts. For example, advanced technologies such as high-performance microturbines and fuel cells promise reliable, efficient alternatives. Fuel cells require minimal maintenance and can be sited in crowded urban centers because they are clean, quiet, and highly flexible. Several fuel cell technologies are under development, with many already producing power for modern office buildings and hotels; advanced fuel cells could soon generate enough energy to supply a large proportion of the electricity and heat needed to power a city and warm its buildings.[30]

Today fuel cells or advanced microturbines must rely primarily on natural gas that has to be piped into cities. But alternatives already exist: methane from a local landfill will soon drive a fuel cell in the city of Vaasa, Finland, supplying heat and power for 50 homes. Eventually, fuel cells can use hydrogen produced from a variety of renewable sources.[31]

Far beyond feeding turbines and fuel cells, renewable resources can provide energy for cooking, lighting, heating, cooling, and even transportation in the world's cities and beyond. Renewables already meet the energy needs of millions of people around the globe, and renewable energy markets are experiencing exponential growth. Wind and solar power are the fastest-growing electricity sources, and biofuels are the world's fastest-growing fuels; all are experiencing double-digit annual growth rates.[32]

Green roofs date back thousands of years, the most famous being the Hanging Gardens of Babylon, constructed around 500 BC.

Wherever the sun shines, buildings—whether shacks or skyscrapers—can become mini-power or heating stations. Solar photovoltaics (PVs) generate electricity directly from sunlight, often at precisely the time when power demand is greatest and electricity is most costly. PV technology has advanced to the point where it can literally be integrated into structures—in roofing tiles and shingles, outer walls, and glass windows—generating not only electricity but also shade and insulation. When used for building facades, PVs can be cheaper than granite or marble. Building-integrated PV (BIPV) is now widely used in Europe and is spreading to other regions as well. The IEA estimates that BIPVs could meet nearly one fifth of annual electricity demand in Finland, more than 40 percent in Australia, and about half of the total in the United States.[33]

Solar thermal systems, which use the sun's warmth to heat water and space, adorn rooftops from Freiburg in Germany to Jerusalem in Israel and can pay for themselves in just a few years through fuel savings. Shanghai and other Chinese cities are becoming hotbeds for solar energy, driven by the need to reduce coal and oil consumption. China now leads the world in the manufacture and use of solar thermal systems. Solar power and heating offer enormous potential in other developing-country cities as well, where they could provide electricity, heat, and hot water for families and communities in informal settlements that currently have no access to the electric grid or other modern energy services—and for far less than it would cost to extend the grid.[34]

A new district with 1,000 dwellings in Malmö, Sweden, meets 100 percent of its electricity needs with solar and wind power.

Cities can also tap the insulating properties of the ground beneath them. Heat pumps use the near-constant temperatures of Earth or groundwater as a heat source in winter and a heat sink in summer to heat and cool water and space. The U.S. military replaced individual space heating, cooling, and water heating systems with ground-source (also called geothermal) heat pumps in more than 4,000 housing units in Fort Polk, Louisiana, eliminating nearly one third of the community's electricity use and 100 percent of the natural gas previously required for heating and cooling. In the world's largest residential application of this technology to date, the Beijing Linked Hybrid Project will use heat pumps to heat and cool almost 140,000 square meters (1.5 million square feet) of new apartments.[35]

There is evidence that high-temperature geothermal water was used to heat buildings in ancient Pompeii. Today, such sources are tapped for district heating systems in cities in France, Iceland, the United States, Turkey, and elsewhere. Paris has the largest such system in the European Union.[36]

Although cities have little land available for energy crops, they have an enormous potential resource for biomass energy: urban waste. New York City, for example, produces 12,000 tons of garbage per day. The waste must be shipped as far away as Ohio,

and disposal costs the city more than $1 billion annually. In industrial- and developing-country cities alike, per person generation of municipal waste is increasing with population and lifestyle changes. Due primarily to a lack of resources and disposal sites, as much as 90 percent of the waste in some developing-country cities is not collected; instead, it is burned or left to rot in the streets, creating heavy smoke and fumes, water pollution, and disease.[37]

But one person's trash is another's black gold, and urban waste can be used to produce everything from cooking fuel for individual households to grid-based electricity for office buildings and homes or biofuels for modern vehicles. Where waste does make it to landfill sites, methane can be extracted to generate electricity, reducing release into the atmosphere of a greenhouse gas (GHG) that is 21 times more potent than carbon dioxide. Landfill gas produces electricity in many U.S. cities, in São Paulo in Brazil, and in Riga in Latvia, and it meets nearly two thirds of power demand for lighting in Monterrey, Mexico.[38]

Waste can also be treated in anaerobic digesters, which break down almost any organic material—from paper and yard waste to garbage and municipal sewage—into compostable solids, liquid fertilizer, and a gaseous fuel that can be carried or piped to stoves, heaters, electric turbines, and any device fueled by natural gas. Most poor people in the developing world spend at least 20 percent of their monthly incomes on fuel for cooking. But low-cost, household-sized digesters fed with feedstock readily available in urban areas can displace dung or firewood, reducing pressure on local forests while providing families with a smoke-free and healthier environment. And a Tanzanian study found that biogas could save five hours of household labor daily, giving women and children more time for productive activities.[39]

On a larger scale, many industrial-country cities—including Frankfurt, Vienna, and Zurich—are converting waste to gas for energy. In early 2006, San Francisco launched a pilot project to produce power from dog waste after finding that it accounted for nearly 4 percent of the residential garbage collected. Oslo, Norway, has perhaps the largest system in the world that uses raw sewage to produce space and water heating. Heat is drawn from the sewer and transferred to a network of water pipes that feed thousands of radiators and faucets throughout the city. And the Swedish coastal city of Helsingborg runs its buses on biogas made from local organic wastes. New technologies can convert even inorganic materials—from hospital and industrial wastes to car tires—into electricity and transport fuels.[40]

Although the potential is limited in urban areas, even wind and water can provide some cities with much needed energy. Wind energy, in particular, faces visual and resource siting constraints, but these challenges have not always discouraged its use. Tokyo has installed 2.5 megawatts of wind turbines along its waterfront, and in May 2005 an electricians' union installed the first commercial wind turbine in Boston, which will provide electricity for its regional training center. Cities along coastlines or large water bodies can tap local resources from new directions, helping to alleviate transmission constraints. The Middelgrunden Windfarm off the coast of Copenhagen meets 4 percent of the city's electricity needs and is the world's largest cooperatively owned wind power project.[41]

Both New York and San Francisco have proposed projects to use marine energy for power. And some cities are literally tapping local water sources for cooling. Paris pumps water from the Seine River to run air-conditioning systems, and Toronto uses the deep, frigid waters of Lake Ontario for district cooling. Toronto's system has enough capacity to cool 3.2 million square meters of office space, or the equivalent of 100 office towers.[42]

Although few cities will meet all their energy needs with distributed renewable resources in the foreseeable future, some urban areas are already doing so. A new district with 1,000 dwellings in Malmö, Sweden, meets 100 percent of its electricity needs with solar and wind power, gets its heat from sea and rock strata and from the sun, and fuels its vehicles with biogas from local refuse and sewage. The planned Chinese eco-city on Dongtan Island will tap similar resources for an expected population of 500,000 by 2040.[43]

Energy efficiency improvements in building design, proper orientation and materials, and more-efficient end-use technologies facilitate the use of renewable energy for two reasons. First, because the scale becomes more manageable, renewables can meet a city's energy needs more easily; second, as a city reduces its demand for energy, it is in a better position to bear the higher costs per unit of output that come with many renewable technologies today.[44]

While renewable energy technologies are capital-intensive, they have low to zero fuel costs, reducing exposure to fluctuations in fossil fuel prices. They have far lower impacts on air, soil, and water and, as a result, on human health than conventional fuels and technologies. And they can provide a reliable and secure supply of power. An analysis of the 2003 blackout in the U.S. Northeast found that a few hundred megawatts of PV generation strategically placed in and around the major cities involved would have reduced the risk of the power outages dramatically.[45]

Renewables also provide local control over energy supply and generate valuable tax revenue and local jobs—one of the most pressing concerns of city mayors, according to a 1997 U.N. Development Programme survey. Approximately 170,000 new jobs in Germany are attributed to the renewable energy industry. About 250,000 Chinese are employed in the solar heating industry, and the biogas industry has created more than 200,000 jobs in India. Further, renewables can provide energy services where many conventional technologies do not or cannot go—into the homes and communities of the very poorest people.[46]

Pioneering Cities

While cities face formidable challenges in reforming energy generation and use, many are taking bold steps in this direction—ranging from daily municipal operations to special events and gatherings. (See Box 5–2.) Their actions demonstrate at practical levels which policies have proved most effective in a variety of conditions of economic wealth, natural resource endowment, and cultural and political heritage. They also indicate the vital role that cities can play in reducing greenhouse gas emissions and averting climate change.[47]

In Barcelona, Spain, after the Green Party won in city council elections it introduced strong policies to support renewable energy and reduce reliance on nuclear power. The primary focus has been on developing the city's solar energy potential—which is 10 times as large as its total energy demand. From 1995 to 1999, demonstration projects and stakeholder consultations took place to develop policy and a realistic timeline for industry compliance.[48]

In 2000, the Barcelona city council mandated that solar water heating provide 60 percent of hot water in new and substantially refurbished buildings. Less than four years

---

Box 5–2. "Greening" Special Events

Some 9,000 international gatherings take place around the world every year, giving cities a prime opportunity to address climate change in a very public way. For example, the Olympic Village constructed for the 2000 Games in Sydney, Australia, represented the biggest solar-powered residential development in the world at the time.

As part of Beijing's successful bid for the 2008 Olympics, city leaders are working to improve local air quality. With assistance from the U.S. Department of Energy, the city is trying to reduce coal consumption and to increase the use of solar energy for both electricity and pool heating.

In Germany, a series of Green Goal targets for the 2006 World Cup Games included a 20-percent decrease in stadium energy use and energy generation from renewable sources. These efforts reflect municipal desires to attract prestigious and lucrative special events while avoiding strains on local infrastructure and resources as well as the global commons.

SOURCE: See endnote 47.

---

after enactment of the Solar Ordinance, installed solar capacity in Barcelona had grown nearly twelvefold; by April 2004, the city's solar water heating systems saved the equivalent of almost 16 megawatt-hours of energy a year, reducing CO_2 emissions by 2.8 tons annually. The city has since extended the requirements to even more buildings. By early 2006, more than 70 Spanish cities and municipalities had adopted solar water heating ordinances; following their lead, the national government has enacted a similar policy.[49]

In other cities where governments encourage increased local reliance on green power, one popular mechanism is quota systems, which require that a growing amount of municipal or community energy be obtained from renewable resources, with market forces competing to identify the most economical projects. Often referred to as renewable portfolio standards, these policies can apply to public or private energy utilities. The publicly owned Sacramento Municipal Utility District in California—building on its long-running commitment to green energy—aims to derive 23 percent of its electricity supply from renewable resources by 2011. And to encourage local PV installations by residential, commercial, and industrial customers, the utility offers incentive payments for every watt installed.[50]

Cities served by privately owned utilities or other actors over which the municipality has little control must often follow other strategies. In 1995 and 1999, Chicago sweltered under serious heat waves that brought rolling blackouts and hundreds of local deaths. Following a $100-million settlement with the private utility ComEd due to the outages, the city chose to apply the funds it received toward greater sustainability in local energy use in order to reduce the likelihood and impact of future blackouts. In 2001, Chicago negotiated a new power purchase agreement with ComEd, requiring the utility to provide 20 percent of the city government's electricity from renewable sources by 2006 (although that was later changed to 2010).[51]

Through these and other initiatives, Chicago has started a campaign to become "the most environmentally friendly city in America." As of 2004, new or substantially refurbished public buildings must meet Leadership in Energy and Environmental Design (LEED) certification as defined by the U.S. Green Building Council. Retrofits of municipal buildings totaling 1.4 million square

meters (15 million square feet) could save the city $6 million in energy costs annually.[52]

Chicago's vision for change is not only bearing economic fruit, it is also altering the very texture of the urban environment. Green roofs have sprouted to life atop City Hall and on more than 232,000 square meters (2.5 million square feet) of residential and commercial structures. Some 250,000 trees planted over the last decade offer shade and beauty to local neighborhoods. In effect, a city long known for its industrial heritage is preparing to seize the next wave of global economic opportunity—one linked explicitly to "green" and "clean" development.[53]

Less than four years after enactment of the Solar Ordinance, installed solar capacity in Barcelona had grown nearly twelvefold.

Another option for cities with private utilities is evident in the growing movement for governments to help a collection of communities meet their energy needs. In the United States, for example, cities and towns in California, Massachusetts, New Jersey, Ohio, and Rhode Island are now authorized to do this for local government, area homes, and businesses, thanks to recent regulatory changes. In turn, localities may shop among a range of energy options. This community aggregation may allow cities to set more-stringent rules for energy efficiency and renewables than federal or state standards as a condition of utility contracts.[54]

Beyond the issue of municipal control and local utility ownership, some cities seek clean local power as a way to keep pace with the demands of an industrializing society. Since 2000, Daegu in South Korea has pursued increasingly comprehensive urban planning that links renewable energy with local economic development. During the 1997–98 Asian economic crises, the devaluation of South Korea's currency contributed to a doubling of energy prices due to the nation's large reliance on imported energy. Against the backdrop of high population density and rapid urbanization, this focused attention on Daegu's need to alter its energy model.[55]

Daegu has established a goal of local renewables meeting 5 percent of its total energy demand by 2010, with long-term targets set through 2050. In addition, the Center for Solar City Daegu, a joint effort of the municipality and Kyungpook National University, is working to disseminate green technologies. These include PV and solar water heating installations at schools, on the university campus, and at sewage and water treatment facilities. To help homeowners install solar roof systems, the city and national government are funding up to 80 percent of installation costs. Strong citizen participation has been reinforced by municipal leadership in Daegu.[56]

The need to address environmental threats while widening social access to critical energy services are driving efforts in Mexico City—home to 20 million people in the metro area—where a cloud of haze relentlessly shrouds views of surrounding mountains. In 1998, the World Resources Institute named Mexico City "the most dangerous city in the world for children" because of its poor air quality, and the city remains among the world's most polluted urban areas.[57]

In 2002, officials finally addressed this situation when they enacted a range of policies that are now organized under Mexico City's Proaire initiative for climate protection. Energy efficiency improvements are being achieved through the installation of advanced light bulbs in 30,000 new residential units and 45,000 existing homes. Solar heating systems are due to be installed in some 50,000 residences. Financial supporters of Proaire include local electric and water utilities, the World Bank, corporate foundations, the Chicago Climate Exchange, and nonprofit organizations.[58]

Since 2003, Cape Town in South Africa has sought to advance energy efficiency and renewable energy as a way to bring basic electricity service to poor, underserved neighborhoods and to reduce the impact of a national power shortage that is expected to begin in 2007. The municipal government aims for 10 percent of its energy to come from renewables by 2020 and has begun energy audits and efficiency retrofits at public facilities. In the Kuyasa region of the city, a pilot project under the Clean Development Mechanism (CDM) of the Kyoto Protocol, which aims to reduce GHG emissions in developing countries, has insulated ceilings and provided residents with solar water heaters and compact fluorescent bulbs. The GHG reductions earned Kuyasa Gold Standard CDM recognition in 2005 for exceptional standards in sustainable design.[59]

Numerous other cities are adopting goals and programs that support sustainable energy systems. (See Table 5–1.) And many cities have united to form larger networks that can pursue green energy development for both climate protection and urban quality of life. In many ways their collaboration—as well as the actions of regional and state governments—reflects an effort to act in place of national governments and the international community, which to date have largely failed to resolve major problems associated with conventional energy use.[60]

Examples of these networks include the U.S. Mayors' Climate Protection Agreement, which encourages cities to lobby the federal

Table 5–1. Selected Municipal Energy Targets

City	Target
Beijing, China	Reduce energy intensity of the city's economic output by 32 percent between 2004 and 2010
Berlin, Germany	Reduce energy use in public buildings 30 percent by 2010; incorporate solar water heating into 75 percent of new buildings annually
Copenhagen, Denmark	Energy audits required for buildings exceeding 1,500 square meters; all new buildings must rely on district heating (electric heating banned)
Freiburg, Germany	10 percent of all public and private electricity must come from renewable sources by 2010
Leicester, United Kingdom	Reduce municipal building energy use 50 percent from 1990 level by 2025
Melbourne, Australia	Increase municipal use of renewable energy by 50 percent from 1996 levels and private use by 22 percent by 2010
Oxford, United Kingdom	10 percent of homes must use solar hot water or PV by 2010
Portland, Oregon, United States	100 percent green power for municipal government by 2010; all new city-owned construction to meet LEED Gold certification
Tokyo, Japan	Minimum 5-percent renewable energy use in large municipal facilities starting in 2004; renewables proposed to supply 20 percent of total energy by 2020

SOURCE: See endnote 60.

government for a national climate change policy, and the Cities for Climate Protection Campaign of ICLEI–Local Governments for Sustainability, which focuses on the design and use of climate-related policies among some 650 participating local governments. Through such partnerships, city officials are able to share best practices and encourage ongoing municipal leadership. And a few governments are now stepping forward to reinforce these efforts. For example, the Australian government has funded a national independent ICLEI office, which involves 216 councils representing 87 percent of Australia's population.[61]

The International Solar Cities Initiative, created to address climate change through effective actions in cities, has devised an explicit target to guide "pathfinder" cities toward major GHG emissions reductions. The target was established by estimating how much greenhouse gas each person on Earth can emit annually without overwhelming the ability of the atmosphere and biosphere to absorb it. The target for 2050 is about 3.3 tons CO_2-equivalent per person. This is about as much as the average person in China or Argentina emits today.[62]

Lighting the Way

Cities have great potential to influence change. This power comes not only from the more manageable scale of local population and energy use but also from their role as national and regional seats of political power. Cities also frequently represent centers of political and technological innovation, where constituents are closer to these seats of power and thus retain more influence over policymakers. And because powerful industries do not wield the same influence at the local level as at national or regional levels, cities can provide a more even playing field for all. Under such conditions, supporters of clean power and related alternatives may find it easier to introduce groundbreaking changes in cities.

Given that local renewable energy development can yield significant benefits, what is standing in the way of change? One major obstacle is the limited resources available to pursue local initiatives. As noted, there are numerous options for minimizing energy use and increasing reliance on clean power, but cities need financial, technical, and administrative support to pursue these strategies. Although this is more commonly a problem in the developing world, it is also a constraint among municipalities in industrial countries.

Many sustainability goals can be pursued through policies that do not increase taxpayers' costs.

Investment priorities deserve particular attention in the world's poorest urban areas. To help achieve more balanced, sustainable economic development that simultaneously meets people's needs, nongovernmental organizations (NGOs) and community groups can encourage governments to link clean energy access to poverty alleviation. Bilateral and multilateral program funding must also move more quickly from fossil fuels toward renewables. Initiatives under the CDM and related global programs could be used more frequently for energy projects that reduce GHG emissions.[63]

The second fundamental challenge is posed by national and international politics. For decades, conventional fuels and technologies have received the lion's share of global investment in energy infrastruc-

ture. In 2002, the World Council for Renewable Energy noted that the $300 billion of energy subsidies spent every year on nuclear power and fossil fuels is four times as much as has been spent promoting renewable energies in the last two decades. This trend is all too evident, for example, in the Bush administration's push for next-generation nuclear and "clean" coal technologies, in efforts to boost nuclear power in India and China, and in subsidies used by some developing countries to support fuels like kerosene and diesel, which make renewable energy less competitive. Countering these developments is going to require a political commitment to clear, mandatory targets for renewable energy use and for technology research and development.[64]

A third barrier is market pressures that ignore environmental and social costs and benefits in energy prices. As a result, development of green energy remains at a disadvantage beyond the most immediately profitable niches, such as wind generation as a hedge against volatile natural gas prices. This is particularly clear in areas where the electricity sector has been privatized over the last decade, where governments have often found it necessary to impose firm renewable energy goals for retail electric providers in order to ensure green power's continued advance. Such actions highlight national governments' crucial role in correcting for prices and market structures that fail to signal the true costs of conventional fuels.[65]

The effect of market pressure is also apparent in the priorities of most electric utilities, which focus on expanding supply rather than conservation to meet customers' needs. "Negawatts"—electricity that is never actually produced or sold—would be a viable energy service to consumers if more governments introduced regulations that encouraged utilities to pursue conservation.[66]

The issue of pricing and costs also plagues the building sector. Although developers in cities like Chicago now have trouble finding the requisite "anchor" tenants if a new building does not meet certain voluntary green standards, this is rarely the case in other municipalities. Energy costs often represent only a small share of overall business or household expenses, and cost savings from efficiency measures are not always reflected in conventional accounting. As a result, price signals fail to drive change.[67]

Another fundamental challenge involves altering the common skepticism that even a large number of small-scale, local renewable systems combined with conservation and efficiency will ever be able to produce enough energy to meet the demands of a large city. To some extent, such mindsets are starting to change, as evident in the growing movement toward more sustainable cities and in recent efforts by former President Bill Clinton to encourage climate protection in some of the world's largest urban centers.[68]

Yet a great deal remains to be done in cities. As one example, despite some policy efforts to encourage or require green construction, the typical new U.S. home still remains highly energy-inefficient, requiring 30–70 percent more energy than new "advanced" green homes. This gap points to the need for larger awareness of the long-term gains, both ecological and economic, that can be achieved through more ambitious mandates for sustainable practices. In effect, a paradigm shift is needed—one that embraces radical improvements in energy efficiency, with the remaining demand met primarily by renewable energy.[69]

Relevant actors and institutions—from all levels of government to the finance sector—must consider new ways of evaluating the life-cycle costs and benefits of renewable energy and of building design that considers local conditions and uses local knowledge. This will mean involving the authorities that have the most power to mandate new requirements and monitor enforcement. It can also ensure the institutional capacity—in the form of financing for "green" home improvements, for example—to assist people who participate in efficiency and renewable energy programs.[70]

Contrary to some people's perceptions, many sustainability goals can be pursued through policies that do not increase taxpayers' costs, as in Chicago, where green buildings receive expedited permitting. City planners can incorporate the "new urbanism"—which involves building for people rather than cars—and related planning approaches for mixed-use communities that combine residential and commercial space. This can minimize energy use and suburban sprawl while making city life more sustainable and enhancing the overall quality of life.[71]

In addition to education and public awareness campaigns, political pressure must be brought to bear against powerful forces that favor the status quo. Positive changes in the energy sector, particularly in the world's poorest urban areas, will require action from not only municipal authorities but also regional, provincial, and national governments as well as NGOs and aid and lending institutions. (See Table 5–2.)[72]

The challenge lies in moving beyond local voluntary partnerships toward strong intergovernmental and societal commitments for change. Wider civil society involvement will be critical and has already figured prominently in many recent movements for more-sustainable energy use in cities. Citizens' groups can do more by calling for national and international changes in investment priorities and can work with private financial institutions favoring clean energy as a profitable strategy for minimiz-

Table 5–2. Roadmaps for Powering Cities Locally

Obstacle	Strategic Response
Lack of control over energy sector	Municipal government can set targets for its own green energy use, procure goods and services made with local green power, aggregate customer demand, and form power purchase agreements with utilities.
	Municipal government can target energy efficiency and conservation in public and private buildings by requiring energy audits and mandating use of specific technologies and construction practices, through city planning and permitting.
	Citizens can form cooperatives for local energy development or purchase green power.
Lack of widespread access to energy service (particularly common in low-income cities)	Governments can support pricing reform and commit to replanting trees to ensure wider availability of fuelwood and other biomass resources.
	Legalized secondary power arrangements can give urban dwellers access to power sources "owned" by other individuals, thereby avoiding or reducing otherwise prohibitive upfront fees (the utility can set basic technical standards to enhance safety of energy delivery, while the de facto electricity distributor determines rates).
	Reduced lifeline electricity tariffs (available to low-income users for lower levels of use) can spread out upfront fees (such as grid connection charges) into future payments over time.
Lack of funds or expertise to identify and undertake projects	Local actors (public or private) can partner with energy service companies or, in low- to moderate-income cities, bundle projects to leverage microfinance or multi- or bilateral assistance for the lease or purchase of solar water heaters, PV systems, and safer and more efficient stoves and smoke hoods.
Lack of awareness or understanding of benefits of local green energy or how to use technologies	Municipal government can work with local trade organizations, private-sector champions, and citizens' groups on information campaigns, product labeling, professional training, and school curricula.
	NGOs and community groups can sponsor demonstration projects.
Lack of utility involvement or of regional, national, or international emphasis on renewable energy development, energy efficiency, conservation, and GHG reductions	Municipal or grassroots efforts can coordinate lobbying across locales for changes in political priorities (toward regional or national targets and commitments) to include mandates for both public and private utilities.
	States and cities can develop and implement their own policies and band together in multi-state or multi-city agreements to set "de facto" policy.

SOURCE: See endnote 72.

ing business risks from climate change.[73]

Today cities have an unprecedented opportunity to change the way they supply and use energy. New eco-cities such as Dongtan in China may show the way, even as existing cities turn to technologies rooted in the past—from adobe architecture to passive solar heating. When complemented by conservation, more-efficient technologies, and new decentralized, small-scale energy services, these efforts can help cities confidently navigate the forthcoming peak of cheap oil and natural gas production while reducing the impact of climate change. Energy transformation in cities can be the doorway to security and vitality in urban life.

Solar-Powered City

Buildings in Rizhao, a coastal city of nearly 3 million on the Shandong Peninsula in northern China, have a common yet unique appearance: most rooftops and walls are covered with small panels. They are solar heat collectors.[1]

In Rizhao City, which means City of Sunshine in Chinese, 99 percent of households in the central districts use solar water heaters, and most traffic signals, street lights, and park illumination are powered by photovoltaic solar cells. In the suburbs and villages, more than 30 percent of households use solar water heaters, and over 6,000 households have solar cooking facilities. More than 60,000 greenhouses are heated by solar panels, reducing overhead costs for farmers in nearby areas. In total, the city has over a half-million square meters of solar water heating panels, the equivalent of about 0.5 megawatts of electric water heaters.[2]

Kouguan Town Primary School is one of the satisfied users of solar energy in Rizhao. Since the wall-hanged solar heat collectors were installed in 1999, the school has been relying on solar energy for all classroom heating in winter and hot water supplies for the entire school all year. After more than a decade, the system is still functioning well.[3]

The fact that Rizhao is a small, ordinary Chinese city with per capita incomes even lower than in most other cities in the region makes the story even more remarkable. The achievement was the result of an unusual convergence of three key factors: a government policy that encourages solar energy use and financially supports research and development, local solar panel industries that seized the opportunity and improved their products, and the strong political will of the city's leadership to adopt it.

As is the case in industrial countries that promote solar power, the Shandong provincial government provided subsidies. Instead of funding the end users, however, the government funded the research and development activities of the solar water heater industry. Mayor Li Zhaoqian explained: "It is not realistic to subsidize end users as we don't have sufficient financial capacity." Instead, the provincial government invested in the industry to achieve technological breakthroughs, which increased efficiency and lowered the unit cost.[4]

The cost of a solar water heater was brought down to the same level as an electric one: about $190, which is about 4–5 percent of the annual income of an average household in town and about 8–10 percent of a rural household's income. Also, the panels could be simply attached to the exterior of a building. Using a solar water heater for 15 years costs about 15,000 yuan less than running a conventional electric heater, which equates to saving $120 per year.[5]

A combination of regulations and public education spurred the broad adoption of solar heaters. The city mandates all new buildings to incorporate solar panels, and it oversees the construction process to ensure proper installation. To raise awareness, the city held open seminars and ran public advertising on television. Government buildings and the homes of city leaders were the first to have the panels installed. Some government bodies and businesses provided free installation for employees, although the users pay for repairs and replacement. After 15 years of effort, it seems the merit of using a solar heater has become common sense in Rizhao, and "you don't need to persuade people anymore to make the choice," according to Wang Shuguang, a government official.[6]

Rizhao would not be the city it now is

Rizhao City Government

Rizhao solar-powered street light

without the clear vision and innovative thinking of its leaders. Although the program was started by his predecessor, Mayor Li Zhaoqian has a special interest in continuing it. Before becoming mayor, Dr. Li was vice president and professor at Shandong University of Technology and served as vice director general of the Economic and Trade Commission of Shandong Province, where he helped industries improve solar energy production technology and efficiency.[7]

Widespread use of solar energy reduced the use of coal and help improved the environmental quality of Rizhao, which has consistently been listed in the top 10 cities for air quality in China. In 2006, the State Environmental Protection Agency designated Rizhao as the Environmental Protection Model City.[8]

Rizhao's leaders believe that an enhanced environment will in turn help the city's social, economic, and cultural development in the long run, and they see solar energy as a starting point to trigger this positive cycle. Some recent statistics show Rizhao is on track. The city is attracting a rapidly increasing amount of foreign direct investment, and according to city officials, environment is one of the key factors bringing these investors to Rizhao.[9]

A good environment also brings more people to Rizhao. The travel industry in the city is booming. In the last two years, the number of visitors increased by 48 and 30 percent. Since 2002, the city has successfully hosted a series of domestic and international water sports events, including the International Sailing Federation's Grade W 470 World Sailing Championship.[10]

The favorable environmental profile of Rizhao is changing its cultural profile as well, by attracting high-profile universities and professors to the city. Peking University, the most prestigious one in China, is building a residential complex in Rizhao, for example. More than 300 professors have bought their second or retirement homes in the city, working and living in this new complex at least part of the year. Qufu Normal Unviersity and Shandong Institute of Athletics have also chosen Rizhao for new campuses. Again, one of the reasons they cited was the environmental quality in the city.[11]

—*Xuemei Bai*
Commonwealth Scientific and
Industrial Research Organization,
Australia

CITYSCAPE:
MALMÖ

Building a Green Future

As people drive across the Øresund Bridge from Copenhagen in Denmark to Malmö in Sweden, their eyes are drawn to the Turning Torso in Malmö's Western Harbour. This apartment tower's white marble-clad walls twist 90 degrees as they rise 54 stories above a city of just under 300,000 people. Architect Santiago Calatrava's design mimics the human body in motion. The building, completed in 2005, can be seen as symbolizing Malmö's efforts to move from its recent history of industrial pollution and unemployment to an ecologically, socially, and economically sustainable future.[1]

Malmö, Sweden's third largest city, is an international port that has survived many transitions. Founded in 1275 as a Danish city, although its earliest settlements date to 10,000 BC, Malmö became part of Sweden in the seventeenth century. Today, one quarter of Malmo's residents are foreign-born. For many of the last 150 years, the city was a prominent shipbuilding center. Kockum's shipyard, established in the mid-nineteenth century, employed more people than any of the city's other industries, which included leather, textiles, and food processing. After the shipyard closed in the mid-1980s during a recession, some 35,000 people left Malmö within a couple of years.[2]

In the 1990s, Malmö began to forge a new vision of itself as a "sustainable" city. Among the first steps: transforming the Western Harbour from abandoned industrial sites into a model of ecological design to host the 2001 European Housing Exhibition. The government invested in cleaning up contaminated land. The city took responsibility for public spaces and infrastructure. Sixteen development companies chosen to participate were in charge of everything inside their plot boundaries.[3]

Malmö partnered with E.ON Sweden, a subsidiary of the largest privately owned energy company in Europe, to get 100 percent of the area's energy from local renewable sources: wind, sun, water, and gas from garbage and sewage. The Swedish government granted roughly 250 million krona ($34 million) to offset costs for environmental investments.[4]

A 2-megawatt wind turbine, supplemented by photovoltaic sun shields on one building, provides virtually all electricity to homes in the Western Harbour and powers heat pumps that supply hot water and district heating. The pumps extract heat from seawater in the city's canal, from solar collectors installed on rooftops, and from an innovative aquifer storage system. All apartments in the Turning Torso have units that grind organic waste, which is collected to produce biogas for cooking and to fuel vehicles. Residents in the new district receive training in how to use the environmentally friendly technologies.[5]

The system that supplies the district with renewable energy is attached to the city grids for electricity and district heating. Malmö's 50-year-old district heating system covers about 90 percent of the residential area. The Western Harbour exports energy when production is high and demand is low and imports it when more energy is needed. Over the course of a year, the local renewable supply matches local demand.[6]

Malmö aims to decrease carbon dioxide emissions citywide on average by 25 percent in 2008–12 compared with 1990 levels, which will mean bringing them 10–15 percent below 1999 levels. Malmö has joined three other cities—Dublin in Ireland, Hilleröd in Denmark, and Tallinn in Estonia—in a three-year project for Sustainable Energy Communities and Urban Areas in Europe, or SECURE, to seek good energy

Marcus Österberg

The Turning Torso under construction

solutions and produce local energy action plans. Malmö's strategy involves extending its district heating system and increasing the use of natural gas for electricity. The city is also educating citizens about their role in climate change. Billboards around Malmö show techniques for lowering energy use.[7]

Malmö seeks to learn from setbacks in the Western Harbour redevelopment. The first homes that were built did not achieve the targets set for energy efficiency. Engineers and architects have studied these to avoid repeating the mistakes. Even though the plan was to offer various housing types—for the elderly, large families, people seeking either collective housing or less expensive half-completed flats, for example—in the first stage only student housing and fully equipped flats, on the upper end of housing prices, were built. Although this created a low-diversity housing area, currently viewed as an area for the rich, many people who live in other areas use the public spaces and services in the Western Harbour area. As there are now roughly 1,000 residents and the target is to one day house 10,000 people, there is still considerable opportunity to add greater diversity to future housing stock.[8]

The city's environmental initiatives have sparked change. Malmö University was founded in 1998 with a focus on the environment, conservation of natural resources, ethnicity, and gender. Since 1997, many information technology and telecommunications companies have opened offices. Malmö hosts a growing number of cultural activities, sports events, exhibitions, and shows. New shops, including famous retail chains, have set up business. Housing construction is steadily increasing, with 1,265 new dwellings built in 2005. Joblessness among adults is decreasing: from 16 percent in 1996 to 8.8 percent in 2005. Today, 39 percent of people in Malmö have a university or postgraduate education—double the figure in 1990.[9]

On the negative side, the city is grappling with a rise in crime and has yet to integrate its large immigrant population. Just 3 kilometers from the newly built Western Harbour and close to the city center, the Rosengård area is home to 21,000 immigrants. Some 84 percent of this area's inhabitants have both parents foreign-born, and 50 different languages are spoken. The area was built in the 1960s and 1970s as part of a massive social housing program. Unemployment is high, and the language barrier makes inhabitants even more isolated from the rest of Malmö.[10]

Residents have taken note of the city's rapid transformation. Jeanette Andersson, a young eco-toxicologist said: "Moving to Malmö, as a young student, offered me several advantages such as good housing and a cheap living. After just a few years the situation had changed and suddenly people started fighting about apartments. You could tell by just walking through the city that the population and atmosphere of Malmö was about to change."[11]

—Ivana Kildsgaard
IVL Swedish Environmental
Research Institute

CHAPTER 6

Reducing Natural Disaster Risk in Cities

Zoë Chafe

On a winter night about 2,000 years ago, residents of a majestic Greek city were no doubt shaken awake by the massive earthquake. In quick succession, soil below the buildings liquefied, the city dropped below sea level, and everything and everyone within it was subsumed by a tsunami. The year was 373 BC; classical Greek civilization was just reaching its peak, and this city—Helike, well known for its Temple of Poseidon, a tribute to the god of earthquakes and the sea—disappeared overnight. For 2,000 years, Helike's legacy would survive only through Plato's legend of Atlantis, which was inspired by Helike's demise.[1]

Within the past decade, while excavating despite another lethal earthquake in Greece, archeologists located ancient remains buried by the earthquake: rock walls, coins, and pottery. Imagine their surprise as they simultaneously uncovered a nearby town, this one 4,000 years old, that had met a fate similar to Helike's some 2,000 earlier and in almost exactly the same spot.[2]

The legacies of Helike and its historic neighbor remind us that natural hazards are—and always will be—a part of our lives. Yet our understanding of disaster risk and our actions to perpetuate or reduce it can change. By breaking apart the ingredients of disaster—natural hazards, vulnerability, risk, and risk management—the ways we can protect ourselves and our cities become more obvious.

Large natural disasters, such as the earthquake at Helike or the 2004 Indian Ocean tsunami, garner media attention, inspire action, and remain emblazoned in our memories. But the suffering caused by chronic small-scale urban disasters (such as local flooding, water contamination, and landslides) often escapes the limelight. While cities are increasingly home to both types of disaster, they can also be great places to tackle the underlying issues that leave people vulnerable to the risk of disaster.

What hazards create particular impacts on cities, and what can be done to prepare for the onslaught of disasters that appears to lie ahead? As human populations swell and our environment continues to change, these basic questions should guide the planning and development of cities in our increasingly urban world.

The Rising Toll of Disasters

Although natural disasters are often presented as rare and unexpected tragedies, the reality is that they now occur more frequently, affect more people, and cause higher economic damages. The number recorded each year fluctuates, and reporting gaps make data analysis difficult, but there has unquestionably been a general upward trend in the number of disasters recorded each year. In 2005, the Center for Research on the Epidemiology of Disasters (CRED) recorded 430 natural disasters, which killed 89,713 people and affected 162 million others worldwide. For comparison, CRED recorded an average of 173 natural disasters each year during the 1980s and an annual average of 236 during the 1990s. The definition of a natural disaster varies between reporting sources, but for CRED it means any incident that kills 10 or more people, affects 100 or more people, or necessitates a declaration of emergency or call for international assistance. Unfortunately, no reliable data source catalogues the global incidence of natural disasters solely in urban contexts.[3]

Spurred by population growth, rapid urbanization, environmental degradation, and climate change, the number of people affected by natural disasters has increased remarkably over the past 20 years. In the late 1980s, about 177 million people were affected by natural disasters each year—roughly equal, at the time, to the entire population of Indonesia or the world's 13 largest cities. (See Figure 6–1.) Since 2001, the annual average has risen to 270 million—an increase of more than 50 percent. Now the average number of people affected each year is on par with the

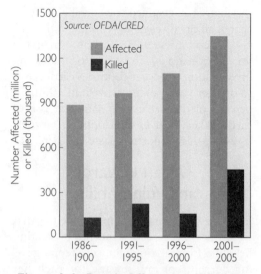

Figure 6–1. People Affected or Killed by Natural Disasters Worldwide, 1986–2005

current populations of the world's 18 largest cities. (CRED counts in the total number affected all those injured, made homeless, or requiring immediate assistance to meet basic survival needs during an emergency.)[4]

Natural disasters have a disproportionate impact on low-income countries. Over the past 25 years, a stunning 98 percent of the people injured or otherwise affected by natural disasters were living in the 112 countries classified as low income or low-middle income by the World Bank. These countries account for about 75 percent of the world's population, including 62 percent of the world's urban dwellers. They were also home to 90 percent of the people who lost their lives to natural disasters during the same time period. This means that less than 10 percent of the individuals who died from natural disasters lived in the 96 richest countries.[5]

Urbanization is proceeding at a quick pace in poorer countries. (See Chapter 1.) In 1980, just under 50 percent of all urban dwellers lived in the 112 poorest countries. By 2005, this figure had changed considerably: 62 percent of all city dwellers were found in the 112 poorest countries and 38 percent were in the 96 richer nations. The trend of increasing urbanization in poorer countries carries over to disaster vulnerability as well: in 1950, 50 percent of the urban population at risk of earthquakes lived in developing countries; by 2000, this proportion passed 85 percent.[6]

Urban Hazards and Vulnerability

Here are some images of modern cities that we do not like to see: a mother protecting her child from driving rain, sidestepping rising currents of raw sewage to reach her home; teenagers standing atop the skeleton of a felled house, surveying the heaps of brick and mangled metal that formed their neighborhood before the earthquake struck; a family watching helplessly as flames ravish crowded rows of makeshift shelters, including their home. But through a combination of poor urban governance, economic poverty, inadequate urban planning, and inappropriate building styles, these scenes of disaster are plausible endings to any day for many city dwellers.

Disasters are not simple chance occurrences, as often portrayed in the media. They are the product of an ever-changing relationship between natural events (hazards), social and physical conditions (vulnerabilities), and the risk management systems that exist—or all too often do not exist—to protect us. (See Boxes 6–1 and 6–2.) With few exceptions, people are not killed by high winds or seismic waves; rather, they are killed by the effects that these natural hazards have on their houses, their schools, their offices, and their surroundings. Response efforts can save lives and property but, conversely, a lack of prior planning and proper communication can constitute the final ingredient in a devastating disaster.[7]

Risk is created through a complex set of interactions between our built and natural environments. We often consider whether the food we eat is safe and nutritious, the water we drink clean, or the neighborhood we live in safe. But we seldom stop to consider the health of our immediate environment or

---

Box 6–1. Defining Disasters

Disaster: A rare or abnormal hazard that affects vulnerable communities or geographic areas. Causes substantial damage, disruption, and possible casualties. Leaves the affected communities unable to function normally and requiring outside assistance.

Natural hazard: A geophysical, atmospheric, or hydrological event that has a potential to cause harm or loss: death or injury, property damage, social and economic disruption, or environmental degradation.

Risk: The likelihood of a hazard occurring in a specific location and its probable consequences for people and property.

Disaster risk management: A range of activities to cope with risk. Includes mitigation, prevention, and preparedness (actions to minimize disaster risk, preferably taken before disaster occurs); relief (actions taken immediately after disaster strikes); rehabilitation (restoring normal activities within two years of disaster); and reconstruction (long-term work to restore infrastructure and services).

Vulnerability: The potential to suffer harm or loss. Increased susceptibility to the impacts of hazards from physical, social, economic, and environmental factors.

SOURCE: See endnote 7.

Examples of Natural Hazards:

Earthquake	Wave/surge
Flood	Landslide
Volcano	Wildfire
Windstorm	

Examples of Vulnerability Factors:
Poorly planned development
Deforestation and erosion
Precarious buildings and shelters
Lack of insurance or banking services
Prohibitively expensive health care
Inability to get information
No access to emergency services

Examples of Risk Management:
Write and enforce building codes
Improve access to sanitation
Offer insurance for all income brackets
Create skills-training classes for women
Build strong community networks
Prepare emergency kits and disaster plans
Spread hazard warnings to all at risk

SOURCE: See endnote 7.

Part of the population growth in cities today is due to migration, which can be a source of vulnerability in itself. When people move to cities, they often lose the traditional rural networks of family and neighbors that they could rely on during and after a disaster.[9]

Another significant factor is geographic location. Eight of the 10 most populous cities in the world sit on or near earthquake faults, and 6 of the 10 are vulnerable to storm surges. (See Table 6–1.) Several are located near major volcanoes. Populous cities line many coasts and will be exposed to sea level rise due to climate change. Scores of cities are located in hazardous areas, as a matter of historical significance or modern expansion. Some cities have colonial origins and were situated in their current locations for reasons of economic access rather than safety. These same cities often carry a legacy of imported colonial building codes and planning systems that are not appropriate for their locations. For example, some of the first buildings constructed in Mexico City were built to the same standards as in New York City—a distant city with crucially different soil.[10]

Even where building codes and planning systems exist, urban growth tends to be organic and proceed as necessary rather than as prescribed on paper. Slums are often filled with densely built shelters constructed of flammable materials, making the area vulnerable to runaway fires. In addition, many residents rely on open cooking fires, and they usually lack emergency fire access. Proximity to environmental contamination only increases the potential for harm. In almost every large developing-country city with industrial production, housing is perched precariously on or around heavy equipment, pipelines, effluent drains, and toxic disposal sites.[11]

After a disaster occurs, due to a lack of other job opportunities or housing options, urban dwellers may be forced to stay in dan-

the condition of our neighbors' homes—and these are among the factors that may put us at risk for disaster. With many day-to-day crises to deal with, poor families often give higher priority to more-immediate needs than reducing the risk of disaster.[8]

High population density concentrates risk in cities. Nearly 3 of every 100 people on Earth live in one of the world's 10 largest cities. "Slum" populations are growing by 25 million people a year, adding to the estimated 1 billion people living in informal settlements worldwide. As Mark Pelling of King's College London explains, "Urbanization affects disasters just as profoundly as disasters can affect urbanization." Rapid urbanization constantly changes disaster risk.

Table 6–1. Ten Most Populous Cities in 2005 and Associated Disaster Risk

City	Population (million)	Disaster Risk					
		Earthquake	Volcano	Storms	Tornado	Flood	Storm Surge
Tokyo	35.2	x		x	x	x	x
Mexico City	19.4	x	x	x			
New York	18.7	x		x			x
São Paulo	18.3			x		x	
Mumbai	18.2	x		x		x	x
Delhi	15.0	x		x		x	
Shanghai	14.5	x		x		x	x
Kolkata	14.3	x		x	x	x	x
Jakarta	13.2	x				x	
Buenos Aires	12.6			x		x	x

SOURCE: See endnote 10.

gerous areas of cities despite the risk of future disasters. Many residents of Kampung Melayu, for instance, a slum outside Jakarta that regularly floods, own land in other parts of Indonesia but return to their difficult life near the city out of economic necessity, even with the knowledge that disaster will almost certainly happen again.[12]

Risky Environment

Cities are often characterized by their skylines—the quintessential celebration of sculpture and architecture perched on geographic contours that is Rio de Janeiro, for example, or the uniform division of city blocks celebrating nature conquered, as in New York. Because each represents the pulse and culture of a particular city, it is shocking to see a skyline crumble—the grand facade shaken loose within minutes by a natural occurrence. But this has happened before, and it will continue to happen: earthquakes devastated San Francisco in 1906; Tokyo in 1923; Valdivia, Chile, in 1960; Managua, Nicaragua, in 1972; Mexico City in 1985; and Bam, Iran in 2003. Peer behind the skyline's veneer, and it

becomes apparent why cities are particularly vulnerable to such disasters.[13]

Each natural hazard varies in its scope and effects. (See Table 6–2.) Earthquakes, for example, occur in widely varying strengths, strike at any time of the day, and create effects that still baffle geologists or urban planners. Even more predictable hazards, such as hurricanes, are not fully understood. They may rapidly gain and lose strength, veering off expected trajectories and narrowly hitting or missing city centers. In 1991 Tropical Cyclone 05B, one of the most deadly cyclones to hit India, stalled in its path over the city of Bhubaneswar in Orissa, dumping torrential rain on the area for 30 hours. And in 2005 Hurricane Katrina lost strength before it made landfall to the east of New Orleans, although it still wreaked havoc on that low-lying city.[14]

What is clear is that natural hazards, combined with high levels of vulnerability, routinely turn into major urban disasters. The tsunami that struck Indonesia in December 2004 reached three kilometers inland in the province of Aceh, devastating the city of Banda Aceh. (See Box 6–3). Lava sluiced

Table 6–2. Selected Urban Disasters, 1906–2006

Date	City	Disaster	Deaths	Economic Losses
			(estimated number)	(billion 2005 dollars)
2005	New Orleans	hurricane	1,800	125.0
2005	Mumbai	flood	400	0.4
2003	Bam (Iran)	earthquake	26,300	1.1
2003	Paris	heat wave	14,800	4.7
2001	Bhuj (India)	earthquake	19,700	5.5
2000	Johannesburg	flood	100	0.2
1999	Istanbul/Izmit	earthquake	15,000	14.1
1995	Kobe, Japan	earthquake	6,400	128.2
1985	Mexico City	earthquake	9,500	7.3
1976	Tangshan (China)	earthquake	242,000	19.2
1970	Dhaka (Bangladesh)	flood	1,400	10.1
1923	Tokyo	earthquake	143,000	31.8
1906	San Francisco	earthquake	3,000	10.9

SOURCE: See endnote 14.

directly through the city of Goma, in the Democratic Republic of Congo, when a volcano erupted there in 2002, leaving 300,000 homeless. And landslides provoked by heavy summer rains routinely devastate neighborhoods flanking the steep hills of Rio de Janeiro.[15]

Unintentionally, many cities actually increase disaster risk. The concentration of heat and pollutants from power plants, industrial processes, and vehicles in cities contributes to a well-documented phenomenon called the "heat island effect," which can exacerbate heat waves and other warming trends. Most pronounced over megacities— those with at least 10 million people—this effect causes a city to be up to 10 degrees Celsius (18 degrees Fahrenheit) hotter than surrounding areas. In New York City, where the heat island effect has been observed for a century, average night-time temperatures hover just over 7 degrees Fahrenheit above the temperature in surrounding suburban and rural areas. Beyond making temperatures uncomfortable, heat islands can—especially during the summer months—increase air conditioning demand, lead to more air pollution, and add to injuries and deaths related to heat stress. (See Chapter 7.)[16]

Urban buildings and the layout of cities can contribute to vulnerability as well. Antennas and electrical equipment atop buildings can attract lightning; the effects of a lightning strike can extend for kilometers if sensitive equipment is hit. Indeed, every major city is at risk for lightning strikes, according to reinsurance agency Munich Re. The layout of the streets at the foot of these buildings can also encourage extreme weather events. Straight streets lined with tall buildings create canyon-like environments that whip up strong turbulence and wind gusts. They can even lead to hailstorms and heavy localized rainfall. Looking forward, these are vulnerabilities that can be avoided in areas of new growth or reconstruction, given smart planning and dedicated leadership from a city government keen on keeping its residents safe.[17]

Cities do not stop relying on their immediate environmental resources as they develop and expand; in fact, with more inhabitants living in a finite area, cities need to enhance ecosystem services to prevent severe natural disasters. Without trees to purify the air and stabilize soil, without open space to absorb storm water and provide wildlife habitats, and without natural coastal habitats to protect them from storm surges, cities become much more unpleasant and dangerous places to live.

A prime example of the latter point comes from Sri Lanka, where coral reef mining,

---

Box 6–3. Banda Aceh and the Tsunami

When an undersea earthquake in the Indian Ocean triggered a massive tsunami on December 26, 2004, the Indonesian province of Aceh was hit hardest. Of the 230,000 people believed to have died in a dozen countries along the rim of the Indian Ocean, close to 170,000 (74 percent) perished in Aceh alone. The provincial capital, Banda Aceh, suffered greatly: 61,000 died, nearly a quarter of its population of 265,000.

Tsunami waves carried debris inland at least three kilometers in Banda Aceh, and more in some places. Two years later, the affected ribbon of coast still evokes memories of Hiroshima in 1945—lifeless communities, twisted skeletons of buildings, rutted and cratered roads. Just a bit further inland, however, life seems perfectly normal—coffeehouses are bustling, teenagers are busy text-messaging, and motorcycles, *becaks* (motorcycle taxis), minibuses, and cars create a daily cacophony of traffic.

Meuraxa, a harbor area noted for its dense warren of roads and houses, was nearly obliterated. As far as the eye could see, almost every building was washed off its foundation or reduced to rubble. Beside a short bridge connecting this area to the mainland, however, a mosque withstood the waves. Within a year after the tsunami, damage to the mosque had been repaired: its exterior repainted in white and green, the Koranic verses on its facade highlighted in gold.

Why was this mosque, among others elsewhere in Aceh, able to parry the onslaught of the waves? The answer is sturdy construction—a reflection of the mosque as an anchor of daily life in this religious land. Islam came to Aceh in the eighth century, spreading later to the rest of what is today Indonesia.

Although a handful of private houses—undoubtedly those of wealthier people—also withstood the waves, the vast majority of homes along Banda Aceh's coast did not. In many cases, profiteering contractors used bricks of inferior quality and other substandard materials, predisposing the structures to collapse.

Shockingly, post-tsunami reconstruction suffers from some of the same problems, perpetuating Aceh's vulnerability to future disasters. In a number of cases, unscrupulous reconstruction contractors were found to have built flimsy schools and homes. Instead of laying proper foundations, they simply propped wooden stilts on stones. The timber and bricks used were substandard and warped.

Reconstruction has been slow but is now picking up some speed—by April 2006, some 47,000 new houses had been built across Aceh (141,000 had been destroyed), and reportedly between 3,500 and 5,000 additional homes are now being built each month. Banda Aceh still faces years of rehabilitation and serious challenges to ensure that reconstruction does not fuel future vulnerability to disaster.

—*Michael Renner*

SOURCE: See endnote 15.

---

sand dune grading, and mangrove cutting increased wave energy, building damage, and loss of life during the Indian Ocean tsunami. A study of land use around Galle, a major Sri Lankan city whose vibrant markets and cricket stadium were devastated by the tsunami, showed that areas protected by intact mangrove forests and coral reefs suffered significantly less damage than those with degraded natural defenses.[18]

Dealing with Losses

When hazards combine with vulnerability in urban areas, the results can be extremely costly, both in their impacts on human lives and in economic terms. Economic costs set urban disasters apart from rural ones: with a concentration of people and infrastructure, the economic impacts of a disaster are bound to be high. The Kobe earthquake that hit

Japan in 1995 caused $128.2 billion (2005 dollars) in damage, one of the costliest natural disasters ever recorded. Hurricane Katrina in 2005 led to $125 billion in damages to the Gulf Coast.[19]

Windstorms usually bring along substantial amounts of precipitation—often leading to flooding and landslides—and secondary hazards can be the main cause of losses. When Typhoon Nari hit Taipei in September 2001, heavy rains caused flooding in the city's underground railroad stations, closing off an important traffic artery for weeks. Even with relatively slow wind speeds, this transportation shutdown was largely to blame for insured damages of $500 million. The same was true for Tropical Storm Allison, which caused $1.5 billion in damage in Houston in 2001, mostly through flood damage to hospitals, many of which had storage rooms below ground level.[20]

With massive investments in urban infrastructure, the private sector has vested interests in reducing losses wherever possible. The insurance industry, in particular, is keenly interested in disaster prevention and modeling. Extraordinary losses from active hurricane seasons over the past few years have driven insurance companies to pass on their expenses to urban homeowners and businesses alike. Berkshire Hathaway's insurance subsidiaries, for example, lost $3.4 billion from the 2005 U.S. hurricane season.[21]

Several corporate groups are now focusing on risk management, including disaster mitigation. The American Insurance Group, Inc. released a statement saying that it would develop projects to keep greenhouse gases out of the atmosphere—following $2.1 billion in insured losses from hurricanes in 2005. The Corporate Network for Disaster Response, formed in 1990 in the Philippines, is a network of 29 businesses groups and corporate foundations that works toward disaster mitigation and prevention, conducts disaster assessments, delivers relief supplies, and mobilizes post-disaster donations.[22]

But business investments in disaster prevention are often predicated on the government's abilities to keep crucial infrastructure (such as electricity and water) online. Losses from power disruptions following a natural disaster can account for as much as 40 percent of the total insured loss for businesses, according to a report by the Lawrence Berkeley Laboratory. If a business suspects that there is a good chance that the electric utilities and water supply will be nonfunctional for a considerable time after a hurricane hits its area, it will be discouraged from taking preventative and protective measures because it will be unable to operate without vital services.[23]

Climate Change in the City

Urban disaster risk management and planning require a discussion of climate change for two reasons: first, cities produce large amounts of greenhouse gases and, second, they will be considerably affected by climate change.

The vast majority of the world's carbon dioxide emissions can be traced to cities, even though cities cover only 0.4 percent of Earth's surface. We are already seeing hints of the ways that climate change will affect cities by amplifying natural hazards: since 1880, the duration of heat waves in Western Europe has doubled and the number of unusually hot days in the region has nearly tripled, according to the Swiss Federal Office of Meteorology and Climatology. And the U.N. Environment Programme estimates that the devastating heat wave that hit Europe in 2003 meant that climate change effects cost the world $60 billion in that year alone—up 10 percent from the previous year.[24]

Several models show the potential impacts of sea level rise on specific cities, some of which are already occurring. New Orleans is losing coastal wetlands to rising seas at a rate of one and a half football fields each hour. Boston is slated to be significantly affected by climate change by the end of this century, according to a 1997 U.S. Environmental Protection Agency estimate, given that sea level is already rising by 28 centimeters (11 inches) per century and is likely to rise another 56 centimeters by 2100. In New York City, sea level rise could affect the city's water supply by increasing the salinity of the water drawn in by a pump station on the Hudson River. Worse, the pump station is most needed during drought periods, when the salinity problem will be at its worst. It could cost between $224 million and $328 million to fix this problem.[25]

In the absence of national leadership on climate change issues in the United States, some city governments are committed to reducing greenhouse gas emissions independently.

Some countries are already implementing projects to deal with climate change effects. People in the Netherlands have built amphibious houses that bob up and down on sturdy piles as water level changes. In Venice, the adjustable barriers of the controversial MOSE project (from Modulo Sperimentale Elettromeccanico) seek to prevent tidal flooding that currently engulfs the city's tourist center 50 times a year—a phenomenon that would worsen with sea level rise—despite concerns from environmentalists that the project will have a negative effect on water exchange with the surrounding Adriatic Sea.[26]

Yet most people who will be affected by sea level rise live far from Venice or the Netherlands. With a one-meter rise in sea level, for example, Bangladesh stands to lose 17.5 per-

cent of its country, and 13 million people there would be affected, while Egypt and Viet Nam would each need to tend to 8–10 million displaced residents. Of the 33 cities projected to have at least 8 million residents each by 2015, some 21 are coastal cities that will certainly have to contend with sea rise impacts, however severe they may be.[27]

There are also projects under way to draw down the impacts that cities have on global—and local—climates. Preserving and encouraging tree cover and green space are two important actions that can cool local climates and help absorb greenhouse gases. (See also Chapter 5.) Cities around the world vary widely in the amount of green space that they have per capita. Nearly 42 percent of Beijing is covered with greenery right now, for example, and efforts are under way to bring that figure to 45 percent by 2008.[28]

In Chicago, Mayor Richard Daley has taken a position of great leadership in strengthening the environmental attributes of the city. He oversees the planting of about 30,000 trees each year and has added 500,000 trees to the city since he took office in 1989. Despite efforts by the Casey Trees Foundation and other dedicated urban forestry groups, this is not the case in Washington, DC: whereas trees used to cover about one third of the city, they now cover just one tenth of the area. This is due primarily to the loss of 64 percent of the heavy tree cover between 1973 and 1997. During these years, there was an increase in storm water runoff, more frequent basement flooding, and more sewer backups.[29]

In the absence of national leadership on climate change issues in the United States, the popularity of two relatively new initiatives—the Clinton Climate Initiative and the US Mayors' Climate Protection Agreement—indicates that some city governments are committed to reducing greenhouse gas emissions independently. Riding on a "greater sense of urgency" about climate change, the Clinton Initiative is leading a coalition of 24 major cities that have joined together to share ideas on how to limit greenhouse gas emissions and bargain for cheaper energy-efficient products. The Initiative is targeting 40 cities—each with more than 3 million inhabitants—that are responsible for 15–20 percent of the world's total emissions.[30]

Seattle Mayor Greg Nickels began the US Mayors' Climate Protection Agreement the day that the Kyoto Protocol on climate change came into force—without the United States as a signatory. Buoyed by the conviction that cities and towns across the country could lead the way in tackling this global problem, despite the federal government's decision not to engage with the issue on an international level, Nickels asked mayors to pledge that they would take local action to meet or exceed the targets set out by the Kyoto Protocol. By early October 2006, less than two years after the program launched, 313 mayors—representing constituencies of over 51 million Americans—had signed up.[31]

Government's Key Role

Even with modeling systems and weather stations available in some parts of the world, it can be difficult to tell exactly when an abnormal weather-related hazard becomes a life-threatening emergency. Heavy downpours are characteristic of the monsoon season in Mumbai, India's seaside financial capital. But at some point on July 26, 2005, a day when well more than 60 centimeters (two feet) of rainwater inundated the city, the summer storm gave way to disaster. The commuter train lines, which crisscross Mumbai like pulsing veins, stopped. Office workers drowned in their cars on their way home. Eighteen slum dwellers made homeless by the floods were trampled to death the next day, caught in a panicked mob after rumors of a tsunami or flash flood circulated among those seeking higher ground.[32]

Two key mistakes by the Mumbai government left the city, and especially the 92 percent of its residents who live in slums, vulnerable to floods during heavy monsoon rains: First, garbage overwhelmed the archaic sewer system, plugging pipes designed to funnel water out of the city. Second, this bathtub effect was exacerbated by development over mangrove swamps at the city's edge, which demolished the natural flood protection the forests had once provided. Mumbai's experience serves as a warning: we need to protect and enhance existing environmental assets around cities, while also planning infrastructure to meet the needs of growing populations.[33]

Many cities expand in size and geographic extent in ways not planned or expected by government officials. They have populations that far exceed the capacity of existing infrastructure networks, which creates dangerous conditions. Densely packed settlements can prove deadly, as their layout often inhibits effective evacuation in an emergency. Jane Pruess, a planner who visited Sri Lanka after the 2004 tsunami, observed that there seemed to be higher fatality rates in high-density neighborhoods that had narrow, interrupted lanes and alleys. Some 10,000 people died in one community in Sri Lanka alone, in part because of densely packed, poorly constructed residential and commercial buildings and very narrow paths for movement. As density increases, an individual's feeling of control over risk can fade, because there are so many unprotected shelters surrounding any one particular dwelling that damage could spread easily from one shelter to the next.[34]

Risk can also increase as housing and critical infrastructure extend away from dense areas toward more risky open land. In London, a new business and commercial center is being built in the former docklands area, which is at greater risk to storm surges than other parts of the city. In Santa Tecla, El Salvador, a national court order overruling municipal law allowed a developer to build new housing on a risky slope, which later slid during an earthquake, killing those living below the development.[35]

Indeed, many people would blame the most tragic "natural" disasters that have occurred in cities on lax government policies or oversight. After the devastating floods in Mumbai, Gerson D'Cunha, founder of civic group Agni (Fire), said: "The past has caught up with us, about which little can be done. It is bad weather that has caused part of the tragedy, but it is bad government policy that has compounded the bad weather." Mumbai is one case in which quick growth of the city led to the disaster management agency being unable to provide basic supplies and services for residents—this is common among rapidly expanding cities. In other cases, cities grow together and merge without effectively integrating their disaster agencies. This leads to confusion and an inability to coordinate aid.[36]

One group that aims to support such overstretched local authorities is CITYNET, a 20-year-old network for local authorities that promotes sustainable urban improvement initiatives in the Asia-Pacific region. It maintains a Web site where its 63 member cities and 40 member organizations can offer and ask for assistance after a disaster—such as the tsunami, which affected Colombo and Negombo in Sri Lanka, and the 2005 Pakistan/Kashmir earthquake that affected Islamabad. After the tsunami, CITYNET was the first international organization to dispatch local government officials to assist with planning and water purification in Banda Aceh, where a third of the municipal staff had died

during the disaster. And in Sri Lanka, CITYNET facilitated the construction of two community centers using donations from citizens in Yokohama, a member city in Japan.[37]

Another important way that local and national governments can encourage appropriate disaster risk management in cities is by writing legislation to mandate the establishment of preparedness and risk reduction activities. When Hurricane Mitch hit Central America in 1998, killing at least 11,000 people in 10 countries, it sparked the introduction of several new pieces of legislation on disaster management in the region.[38]

Box 6–4 spotlights a few of the notable examples of how disaster preparedness, prevention, and mitigation can save lives and protect property. Other important experiences specifically demonstrate the financial benefits of investing in preparedness: The Philippines government has put in place a variety of preventative measures against floods and volcanic mudflow, with benefits amounting to 3.5–30 times the projects' costs. Several studies have also calculated the savings that could have been realized if preventative measures had been taken. In Dominica and Jamaica, for instance, if projects to protect schools and ports had been put in place, the countries could have avoided hurricane losses in 1979 and 1988 that were two to four times the cost of the mitigation projects. The U.S. Geological Survey has estimated that economic losses worldwide from natural disasters in the 1990s could have been reduced by $280 billion if just $40 billion had been invested in preventative measures.[39]

Simple Solutions

Lovly Josaphat lives in Cité Soleil, the largest slum in Haiti's capital city, Port-au-Prince. Speaking to writer Beverly Bell, Josaphat described the confluence of problems in the

Box 6–4. Selected Examples of Disaster Prevention Projects

Medellin, Colombia: After a devastating landslide engulfed the city in 1987, killing 500, local people and government workers used the Colombian National System for Disaster Prevention and Response to educate and collect financial commitments, creating a safer living environment and integrating risk management strategies into development plans. The number of landslides in Medellin decreased from 533 in 1993 to 191 in 1995.

China: The government invested $3.15 billion on flood control measures over four decades in the late twentieth century, averting possible flood-related losses of about $12 billion. Despite a near-tripling in the country's population, from 555 million in 1950 to 1.3 billion in 2005, deaths from flooding dropped from 4.4 million people in the 1930s and 1940s to 2 million in the 1950s and 1960s and finally down to 14,000 in the 1970s and 1980s.

Seattle: Seattle Project Impact, a public-private partnership formed by the city government in 1998 and funded by the Federal Emergency Management Agency until 2001, retrofitted schools, taught residents about earthquake risks, and mapped earthquake and landslide hazards in the metro area. When a magnitude 6.8 earthquake hit Seattle in February 2001, no students were injured at the retrofitted schools nor was there damage to the 300 homes that residents had retrofitted with guidance from the initiative.

SOURCE: See endnote 39.

bronchitis, malaria, and even typhoid now.... The doctor said to give him boiled water, not to give him food with grease, and not to let him walk in the water. But the water's everywhere; he can't set foot outside the house without walking in it. The doctor said that if I don't take care of him, I'll lose him."[40]

Economic losses worldwide from natural disasters in the 1990s could have been reduced by $280 billion if just $40 billion had been invested in preventative measures.

Millions of people live today in conditions similar to those described by Josaphat. Economic poverty prevents them from buying nutritious foods or medicine and from recovering belongings lost during disasters. And so they must cope with the risk they are exposed to any way they can. (See Box 6–5.)[41]

At the same time, there are plenty of inspirational examples of people helping themselves, their neighbors, and even strangers to better prepare for everyday urban disasters and major disasters to come. Around the world, neighborhoods and communities routinely come together to learn about the risks they face and to forge social networks that will protect them in a time of disaster. On Tuti Island, in the center of Sudan's capital, Khartoum, residents set up communication, health, and food committees each flood season. River patrol volunteers warn residents about rising water levels and organize search and rescue teams when necessary. In Santa Domingo, capital of the Dominican Republic, residents are often exposed to earthquakes and hurricanes, as well as routine environmental health risks from uncollected trash. Six organizations formed a household community waste collection business, addressing a sig-

slums: "When it rains, the part of the Cité I live in floods and the water comes in my house. There's always water on the ground, green smelly water, and there are no paths. The mosquitoes bite us. My four year old has

Box 6–5. Coping Strategies in Urban Slums

"We are always trying to improve, little by little, step by step, in order to become more secure." This statement by a slum dweller living in San Salvador characterizes the coping strategies used to deal with disasters and risk. Coping strategies have mainly been studied in rural areas, especially in relation to droughts, where disaster and development specialists have learned to appreciate their value. However, there seems to be comparatively little interest in coping strategies in urban areas. This is unfortunate, as poor households' views and actions regarding disasters and risk can yield important insights for restructuring development aid.

Interviews with 331 people living in 15 disaster-prone slums in El Salvador showed a grand diversity of coping strategies to reduce risk: slum dwellers build retaining walls with old car tires; they remove blockages from rivers and water channels; they take jobs outside their own settlement to avoid the effects of local disasters; they temporarily move their families to the highest room in their dwelling; or they create communication structures for early warning. They also adopt emotionally oriented strategies, such as relying on their faith or simply accepting their high risk.

But coping strategies go beyond reducing risk: they also include self-insurance—security systems created during "normal" times to help gain access to financing sources or other recovery help in the event of a disaster. An example of self-insurance is to buy and maintain physical assets, such as construction materials, that can easily be sold if needed. Other examples of Salvadorans' self-insurance include having many children, putting money "under the mattress," encouraging family members to move to the United States, joining religious institutions (which offer help after disasters), and contributing to community emergency funds.

In the slum communities visited, self-insurance appears to play a more significant role than formal insurance. Only 26 of the 331 people interviewed had access to the Salvadoran social security system. Property insurance is not available. However, while there is a general conception that slum dwellers do not have a culture of insurance, some cases showed the contrary. Through deals with entrepreneurs, residents can obtain certificates of employment that enable them to pay into the social security system, even though they are not formally employed.

Most intriguingly, the research revealed that households spend an average of 9.2 percent (ranging from 0 to 75 percent) of their income on reducing disaster risk (about $26 out of an average monthly income of $284). This is in addition to construction materials that are obtained for free, family members' free labor, the opportunity costs of the considerable time spent on risk reduction, and the negative impacts of some coping strategies (such as high interest paid to informal money lenders).

Post-disaster expenses add up as well: replacement of belongings washed away during floods and landslides, recovery efforts, temporary income losses, and the gradual loss of investments put into the incremental building of housing and community infrastructure. If the "little-by-little, step-by-step" development process of slums cannot keep pace with the frequency of disaster impacts, then the outcome can be increased insecurity and "poverty traps."

Current urban coping strategies are important for sustainable settlement development, but they appear to be less deliberate, not as effective, and more individually oriented than in rural areas, with a stronger focus on housing construction and land issues. It is crucial that development efforts encourage—and eventually scale up—strategies that increase capacities to manage urban disasters and disaster risk in the short and the long term. Where these do not exist, alternative mechanisms could be offered.

—*Christine Wamsler*
Lund University, Sweden

SOURCE: See endnote 41.

nificant risk and uniting diverse groups of residents around a common cause.[42]

Innovative visual techniques have been developed to learn about risk in community meetings. Risk-mapping activities, for example, can be used as a simple tool to calculate how far community members must travel to reach safety. Traditional stories can be turned into plays and songs about how to reduce risk and can provide memorable messages about disaster preparedness. One such project uses a Vietnamese story about a mountain genie and a storm genie, in which the mountain genie emerges victorious despite the storm.[43]

Simple technological applications of risk reduction are available as well. Dr. Ahmet Turer has invented a way of decreasing excess waste while also reducing the number of dangerous buildings in Turkey, where 95 percent of the population lives in an earthquake-prone area and half of the buildings in the four largest cities are masonry houses. He has designed a way to convert old tires into wall reinforcements, with only a sharp knife needed to install the improvement. He plans to use media connections to broadcast videos of the technology at work in an earthquake simulator. Encouraging the construction of safer housing and shelter by using locally relevant disaster-resistant construction techniques can have the dual effect of making such shelters safer for the inhabitants while also providing an example of construction methods that others in the city will see and hopefully adopt.[44]

Three celebrated techniques for supporting self-help are microfinance, microcredit, and microinsurance. These financial tools enable those with limited income or assets to safely save money or gain access to capital. (See Chapter 8.) Under microcredit schemes, a group of community members will pay dues to a shared account at regularly scheduled meetings and then take turns withdrawing

from the account to begin a business or cover unexpected expenses. Likewise, microinsurance works by pooling risk either at the community level or through existing insurance or microfinance companies.

Microinsurance is a promising—and relatively new—way for poor people to cover losses of livestock, crops, and belongings in a natural disaster.

With only 1–3 percent of families in low- and middle-income countries carrying insurance against natural disasters compared with 30 percent in high-income countries, and with only 2 percent of natural disaster losses in developing countries covered by insurance compared with half the costs in the United States, microinsurance is a promising—and relatively new—way for poor people to cover losses of livestock, crops, and belongings in a natural disaster. When implemented before disaster strikes, microcredit and microinsurance schemes can effectively reduce financial risk—which in turn can lead to reduced risk and quicker recovery.[45]

While this combination is often associated with rural areas in developing countries, it is applicable to urban regions and industrial countries as well. Grameen Bank founder Muhammad Yunus, writing just after Katrina struck New Orleans, pointed out that "economic apartheid had created deep inequalities and resentment," in the area. He noted that before the hurricane, groups in New Orleans were exploring starting a microcredit program and planning to introduce entrepreunership training programs into local high schools.[46]

In Bhuj, India, a city severely affected by an earthquake in 2001, around 12 percent of the population is now covered, directly or indirectly, through a microinsurance scheme run by the All India Disaster Mitigation Institute. Most policyholders are microenterprise owners with annual incomes of $520–650, paying premiums of less than $2 per year. There are concerns, however, about the viability of this and other microinsurance schemes, since many members of a single community may be affected by a disaster, meaning they will simultaneously require capital to pay for health services and rebuild their livelihoods.[47]

From Information to Action

The images of disaster victims that we see are invariably taken in the aftermath of a catastrophe: houses and roads in disarray, people hurt and traumatized. It is easy to forget that those affected often had no warning. They were at work, at school, cooking dinner, socializing, sleeping—going about their everyday lives. Sometimes, if the hazard is a weather event, warnings can be issued with time to prepare. Less frequently, the warnings are carried beyond the reach of technology— by mouth, on motorbike—to those in the most precarious situations. But often, there is simply no warning.

The range of technology available to give urban warnings varies drastically by country. China recently established its first permanent digital earthquake monitoring network in Shanghai, hoping to avoid a repeat of the devastation caused by the 1976 Tangshan earthquake, which measured 7.8 on the Richter scale and claimed at least 242,000 lives. But many officials in countries in Africa, Asia, and Latin America struggle with little more than out-of-date and incompatible paper maps.[48]

Recognizing the discrepancies in the capabilities of various warning systems, several innovative programs are maximizing the assistance of technologies that can be used globally. The International Charter on Space and Major Disasters, initiated in 1999, allows qualified rescue and security personnel to request images taken from space of regions experiencing a disaster. The charter has been activated 85 times, including during urban floods that displaced 60,000 people in Dakar, Senegal, as well as in Prague in 2006, when— thanks to government preparations—floods created less devastation than in earlier years. While the International Charter cannot be invoked for disaster risk reduction needs, planners can anticipate having its valuable resources available to speed disaster response.[49]

In some cases, information is available but is not disseminated in time for action to be taken. An example that planners hope will never be repeated comes from the Democratic Republic of Congo, where a vulcanologist named Dieudonne Wafula sensed in 2002 that an eruption of the Nyiragongo volcano overlooking the city of Goma was imminent. He sent e-mails to international experts a week before the eruption, giving them pertinent information, and then also contacted local authorities, all of whom failed to act on the information. More than 100 people were killed and 300,000 displaced as lava filled the streets, cutting the city in half.[50]

While many cities still lack disaster management plans, including for temporary shelter and evacuation, some are proactively addressing their vulnerability to natural hazards. Shanghai, for example, has begun to build a system of public earthquake shelters. These are located above ground, away from tall buildings, and will have water and power supply systems as well as wireless communication capabilities.[51]

Implementation can be as crucial as planning, however, and complex operations such as evacuations must be carefully considered. When Hurricane Rita threatened Houston less than a month after thousands of people in New Orleans suffered through botched emergency management, officials in Texas pledged that they had improved evacuation systems. But with the city depending on most residents to provide their own means of evacuation, Houston's highways proved to be hazardous and frustrating: swarmed with traffic jams, drivers struggled against serious fuel shortages. The evacuation turned fatal when a bus caught fire in the traffic jam and 23 elderly patients died.[52]

Compare this with the standard experience of a Cuban awaiting the approach of a hurricane. Three days before the storm is predicted to make landfall, Cuban forecasters issue national warnings and begin to check state-run shelters. Localized warnings are targeted to high-risk areas, and residents—who learn hurricane preparedness in school and practice evacuation drills each hurricane season—evacuate to shelters within 12 hours of the storm's predicted arrival. Neighborhoods are cleared of debris and houses boarded up, per national legal requirements. Similarly, in Jamaica volunteers forge the crucial link between national warnings and local action by calling residents on cell phones, checking water levels, and borrowing vehicles to evacuate incapacitated neighbors.[53]

Government preparedness alone will not easily reduce vulnerability and risk, nor will purely grassroots efforts. The best instances of sustainable risk reduction come through effective partnerships and open communication. As American journalist Ted Koppel proclaimed, "we need to begin setting up a network that reaches from the federal government to the state level, from the states to the cities and townships and from every police, fire and sheriff's department into each and every neighborhood." And in the absence of such a network, we need a grassroots initiative that will unite neighbors, involve the private sector, and translate needs upward to the government.[54]

Until that happens, we are at risk of repeating the past. Just hours after the 1906 earthquake struck the San Francisco Bay area, the Oakland Tribune published this statement: "It may be a thousand years before there is such another disturbance in this locality, but the consequences of this one is an admonition not to repeat the errors of the past." Nearly 100 years—and countless disasters—later, speaking just days after the 2004 Indian Ocean tsunami, U.N. Under Secretary-General Jan Egeland reiterated the need to learn from past calamities: "The best way to honor the dead is to protect the living. Good intentions must be turned into concrete action."[55]

Shanghai has begun to build a system of public earthquake shelters above ground, away from tall buildings.

Both statements are testaments to the immense amount of work that remains to be done in order to protect cities from future disasters. As we witness each year, natural disasters continue to occur, in both expected and unexpected locations. Though our knowledge of and ability to predict natural hazards grows constantly, there is much still to learn. Our ability to monitor and communicate these hazards and their effects only provides us with a clearer picture of the devastation wrought by natural forces around the globe, and that information must be communicated in such a way that it inspires action. Maxx Dilley, a geographer with the Disaster Reduction Unit at the U.N. Development Programme, points out that hazard information must be targeted to planners and disaster managers: "What you need are planner-friendly common views, as opposed to highly specialized scientific papers."[56]

As the population of many cities around the world continues to soar, the vulnerability of their infrastructure and inhabitants will grow as well unless measures are consciously taken to avoid a "business-as-usual" approach to disaster risk management.

Disasters can hit cities twice—incapacitating people and the city systems that serve them and stalling or reversing work toward a better future.

Urban disaster risk reduction goes hand-in-hand with the aims of poverty reduction, and it can easily be linked to international efforts to achieve a better standard of living for the growing number of urban dwellers struggling to make ends meet. A central tenet of slum improvements, for example, should be the locally appropriate provision of better sanitation and shelter, improvements that will prevent small-scale natural hazards from sparking urban disasters.

Among the myriad other actions that governments, organizations, businesses, and individuals should set as priorities to reduce disaster risk are the following:

• Foster a "culture of prevention" where disaster prevention, preparedness, and mitigation are streamlined into all planning and budgeting processes.

• Direct special attention to the urban poor in developing countries—those who are most at risk. Focus on efforts that will work simultaneously toward poverty alleviation and sustainable development.

• Protect and enhance environmental assets as a means of risk reduction. Take climate change and its effects on cities seriously, as millions of people are already at

risk because of it.

• If a city, neighborhood, business, or family has not already done so, plan ahead. Any disaster management and planning done prior to the onset of a natural hazard will give those affected a crucial head start on recovery.

In January 2005, more than 160 governments and 40,000 individuals met in Kobe in the Hyogo section of Japan for the World Conference on Disaster Reduction, less than a month after the tsunami that devastated Thailand, Indonesia, and other nations in the region. At the end on the meeting they signed the *Hyogo Framework for Action 2005–2015: Building the Resilience of Nations and Communities to Disasters*. The national governments that signed on are expected to actively work toward disaster risk reduction and to publish the results of their efforts through the convening body, the secretariat of the International Strategy for Disaster Reduction.[57]

The Hyogo Framework clearly links disaster risk reduction to sustainable development, stating that disasters "pose a major obstacle" to development goals, especially in Africa. Many cities have suffered significant setbacks to development projects and social programs when disasters destroyed infrastructure or when resources had to be transferred to fund recovery efforts. In this way, disasters can hit cities twice as hard—incapacitating people and the city systems that serve them and stalling or reversing work toward a better future.[58]

When a natural hazard hits an urban area, disaster does not have to automatically follow. In many cases, the knowledge and information needed to understand hazards and decrease social and environmental vulnerability already exist. For example, a new operational framework guides aid agencies on how to best incorporate disaster risk reduction into housing and settlement development projects. It recommends ways to build safe housing, generate income through local risk reduction, make housing affordable to the most vulnerable families, and develop financial tools to sustain the aid agency itself.[59]

As cities swell in size, converting such information into mitigation, prevention, and preparedness projects will have significant benefits for city officials and for the residents they serve. Governments, the private sector, communities, and individuals have the opportunity—and indeed, the responsibility—to genuinely commit to reducing urban disaster risk. With such a commitment, lives can be saved and property protected.

Efforts to increase disaster resilience should begin at home, but they cannot end there. Effective urban disaster risk management hinges on advocacy for risk awareness, good governance, proper technical and communication infrastructure, and the empowerment of all those who are at risk.

CITYSCAPE:
JAKARTA

River Management

When water levels in the Ciliwung River begin to rise, residents in flood-prone districts in Indonesia's capital, Jakarta, brace for the coming disaster. As water gates in Depok and Manggarai reach capacity, those who have been warned gather their belongings to higher ground, where they wait out the flood on second stories and rooftops. Trash from polluted rivers and diseases, including diarrhea and fever, spread rapidly with the rising floodwaters. People living in riverbank slums suffer the worst consequences: floods could spell the loss of their homes or even their lives.

This scenario is all too common in Jakarta's 78 zones regularly affected by flooding. With 40 percent of its land below sea level and 13 rivers flowing through it, this city of 8.4 million people on the marshy northern coast of Java has long struggled with floods. Neighborhoods such as Kampung Melayu, Jatinegara, and Cawang experience yearly inundation—often just a few feet, but sometimes as deep as several meters. According to one riverbank resident, who works in a small kiosk in front of his house, "many people prefer to live here even though they must evacuate at least twice a year."[1]

The problems have intensified with rapid urbanization and development. Construction along the riverbanks has destroyed native vegetation in many areas, leading to erosion and silting of the riverbeds. As a result, sections of the Ciliwung, the city's largest river, have been reduced to a fraction of its original width and depth. And as more area is covered by cement and asphalt, less rainwater is able to seep into the ground, so it runs into the river. In 2002 Jakarta experienced the worst floods in recent history; they affected over half the city, forcing some 300,000 people to flee their homes and leading to at least 30 deaths.[2]

While deterioration of the watersheds that feed the rivers has increased flooding pressures, water quality has also become a serious problem. With 14,000 cubic meters of household garbage and 900,000 cubic meters of industrial waste emptied into Jakarta's rivers each year, and with riverfront inhabitants using it for laundry, bathing, and their own human waste, the Ciliwung has been called the "longest toilet on Earth."[3]

The Ciliwung crosses several jurisdictional borders, beginning its 117-kilometer journey in the highlands of Bogor before winding its way down to the ocean. Much of the increased runoff is due to development in the high altitudes of Bogor, Puncak, and Cianjur. In some areas, flooding is most pronounced when heavy rainfall occurs in these upstream areas, even when skies are clear in the capital. In 2005 the Jakarta government offered to help neighboring districts improve river management and protect water catchment areas that soak up rainfall and prevent erosion.[4]

Several major initiatives to improve urban river management are now under way. In 2004, the Department of Public Works began construction of a 24-kilometer canal known as Banjir Kanal Timur to act as a floodplain for the city's northern and eastern sections. The Ministry of Forestry has begun rehabilitating 17 river catchment areas across the islands of Java and Sumatra as part of its $1.6-billion reforestation program, including key areas surrounding the Ciliwung. An improved early flood warning system and emergency response plan is being developed for the Ciliwung and Cisadane Rivers. Community groups are involved in reforestation and preservation of undeveloped land in areas such as the historic Condet district of Jakarta; others have volunteered to scour sections of the Ciliwung for trash, to demonstrate the importance of keeping the river clean.[5]

Accion Contra la Faime

Flooding in Kampung Pulo, Jakarta, in January 2006

The Indonesian government established a partnership with Dutch counterparts in 2002 to devise an "integrated" river management plan to address flood control, water quality, and the social issues related to riverbank encroachment. The Dutch government has helped finance projects to map the borders of the Ciliwung, reconfigure human settlements and public spaces near the rivers, and improve water quality through better waste management.[6]

The Bidara Cina neighborhood, which had benefited from the internationally recognized Kampung Improvement Program to improve living conditions in low-income urban neighborhoods, is now the site of a pilot project to test the new river management scheme. This has introduced dikes, roads, sanitation facilities, and improved drainage systems. Efforts to move residences away from the banks of the river have met with less success, however.[7]

Low-income settlements along the rivers—often singled out by officials as a source of riverbank encroachment and pollution—are commonly most affected by floods. Although the government budgeted $45 million in 2005 to develop low-income housing, this number falls far short of what is needed. Relocation has proved to be highly contentious, as residents question compensation levels and the housing alternatives offered. Many urban poor are not eligible for compensation because they lack official land rights and legal residency status.[8]

In some instances people receive eviction orders at short notice, despite having lived in a location for decades. Community groups charge that all too often informal settlements are simply destroyed and that residents are frequently abused by security forces in the process, displacing thousands of people without providing viable alternatives. In Muara Angker, local fishers organized to oppose their proposed resettlement, claiming that officials turned a blind eye to the real culprits—upscale apartment complexes and golf clubs built in adjacent areas.[9]

Having only just emerged from three decades of authoritarian rule and a rocky transition period, Indonesian citizens still mistrust the government, and corruption remains a serious problem. Many community organizations question the sincerity of the government's commitment to achieving an integrated solution. They claim that the government inconsistently enforces and implements the river management plans by focusing on informal settlements and domestic sources of pollution rather than on commercial construction and industry polluters, which provide government revenue. Groups also maintain that participatory processes intended to foster trust and bottom-up approaches are dominated by elite members of the community and that the large-scale, high-cost projects bring opportunities for corruption. Given the slow rate of progress, it is no surprise that many Jakartans remain skeptical.[10]

—*Biko Nagara*
International Policy Studies,
Stanford University, Stanford, CA

Policing by the People

Mumbai has the distinction of having almost half its population of more than 12 million living in slums or informal settlements. Today, these very slums are pioneering a new concept: policing with people's involvement.[1]

Commissioner of Police A. N. Roy says that slums do not get the attention they deserve from any part of the government, including the police. "Do we assign 50 percent of our resources to policing slums even though they make up half the population?" he asks. He also points out that the majority of crimes do not take place in slums. Most of the complaints registered with the police are not offences that would come before a court—minor scuffles and disagreements, often over things like the common water tap, or a fight between children, or a drunken husband beating up his wife.[2]

Roy believes that policing needs to be customized, which led him to the idea of slum *panchayats*—a Hindi word for council of five respected elders chosen and accepted by a village to take decisions on key issues. As slum dwellers needed ways to handle minor offences, and as the police do not have the resources to fully meet this need, he thought the best solution was to get slum dwellers to devise their own policing.

As Roy considered what should be done, he met with A. Jockin, president of the National Slum Dwellers' Federation (NSDF). Together they came up with ideas on how to involve the community. The police, Jockin says, were generally unresponsive to the issues that concerned slum dwellers. They were also particularly indifferent to women's concerns and often refused to entertain complaints brought to the police station by women.[3]

"We felt we should find a way to involve women in this attempt to change the policing system in slums," says Jockin. With this as a central thought, in 2003 Roy and the NSDF launched the slum *panchayats* in Pune, 170 kilometers southeast of Mumbai, where Roy was Police Commissioner at the time. The first slum *panchayats* in Mumbai were established in October 2004.[4]

A slum *panchayat* has 10 members: 7 women and 3 men. In addition, the local "beat cop"—the police officer assigned to the area by the local police station—also becomes a member. Roy believes that women are better equipped to deal with disputes than men, which explains why women are in the majority in the *panchayat*. Jockin concurs: "We felt if women were involved, this would not just empower them but also bring about an attitudinal change in the community." He adds that NSDF and its alliance partner, Mahila Milan (Women Together), chose the members of the *panchayat* to ensure that they were viewed as credible and approachable.[5]

Each *panchayat* covers a population of roughly 10,000. Every member of the *panchayat* is given an identification card. The *panchayats* have the authority of law and moral authority in the community. Roy notes that they really are in a position to resolve disputes, many of which are not crimes but could become so if they were not resolved.[6]

Thirty-four-year-old Malti Ambre is a member of the slum *panchayat* in Mankhurd, a northeastern suburb of Mumbai. This mother of three, who has studied only up to grade seven, says that the *panchayat* in her area has reduced petty crimes "hundred per cent." Proudly displaying her identification card, Ambre says that the presence of women has made a difference because "women are at home for 24 hours and they know what is going on in the slum." One of their main achievements, she says, is closing down the illicit liquor dens despite resistance from the

REUTERS/Punit Paranjpe

Mumbai slum dwellers involved in daily chores

men and even from some of the local police. "Our men would drink away all they earned at these dens and then come home and beat up their wives," she says.[7]

Explaining how the *panchayat* works, Ambre says that they summon both sides involved in a dispute, listen to both points of view, and then work out a compromise. They also explain to those involved in the dispute the benefits of reaching a settlement in the *panchayat* instead of registering an official complaint with the police. The latter would involve a great deal of expense, as the matter would go to court and would take much more time to settle. Most of the complaints that come to the *panchayat* have to do with minor disagreements between people or within families.[8]

In 2005, the police and the nongovernmental groups involved in this experiment took stock. "All the *panchayats* were not functioning effectively. In fact, out of 131, only 81 were working well," admits Roy. In the more efficient ones, however, the number of offences registered with the police had declined noticeably. "There is general harmony in the community and the difference is palpable. This is more important than crime statistics," says the Police Commissioner.[9]

The other important "collateral benefit," according to Roy, is the change in the attitude of the slum dwellers toward the police. This has happened because both sit at the same table, unlike in the past, when the slum dwellers would have to plead with the police to register their complaints. The *panchayat* has also made the police realize that more effective policing was possible through such collaboration. Roy adds that without the active participation of groups like NSDF, the slum *panchayat* would not work. In addition, notes Jockin, the *panchayats* have shown how community women can change the policing system.[10]

"One test of success," says Roy, "is that some of the slum women call me directly. The confidence they now have is one of the biggest successes. This is democracy working at the grassroots. The *panchayats* should function like people's committees. The police should just facilitate."[11]

—*Kalpana Sharma*
    Deputy Editor, The Hindu, *India*

CHAPTER 7

Charting a New Course for Urban Public Health

Carolyn Stephens and Peter Stair

Unoma is nine years old. She lives in Asaba, Nigeria. Her home, next to the River Niger, is one small room in a low-income settlement shared with her unmarried aunt and five other girls. Four other families share the house. Unoma and her neighbors have no access to clean water and no toilets. She does not go to school. Unoma begins her 15-hour day at 6:00 a.m. cleaning the house. She then spends the day selling food she has prepared—carrying up to five kilos of *fufu* on her head all day. After selling for nine hours, she returns to collect water for the house and cook. She collects 300 liters of water from a borehole and finally goes to sleep at 9:00 p.m. She is always very tired and often gets ill. But Unoma still hopes to go to school one day.[1]

For millions of young people just like Unoma all over the world, cities are places of hope and growth, but also despair and death.

For a tiny minority, cities and towns are places of long life, health, and, for some, luxury. But for the majority, cities are places that they hope will give them and their children better opportunities—yet often they find only pollution, disease, and insecurity.

The future of our planet now seems irrevocably urban, and we need to be sure that this urban life is healthy, equitable, and sustainable. Charting a healthier course for cities will require planning with a people-oriented view. It will require creative local governance to harness the human and environmental resources of a city, and urban communities bold enough to resist the momentum of many currently destructive trends. This chapter discusses the public health challenge of urbanization. It looks at existing health problems in cities and reviews some innovative solutions to them.

Carolyn Stephens is a senior lecturer in environment and health at the London School of Hygiene & Tropical Medicine and a visiting professor in the Federal University of Paraná in Brazil. Peter Stair, a former MAP Fellow at Worldwatch Institute, is a Master's candidate in the Department of City and Regional Planning, University of California–Berkeley.

The Health Challenge of Urbanization

The earliest city dwellers had difficulty with their new urban home. Tuberculosis, typhus, bubonic plague, and smallpox arose in many early cities. People often lived shorter lives on average than their rural ancestors. For example, evidence suggests that people during the Stone Age lived longer than those in ancient Rome centuries later. Likewise, the rapid growth of European cities during the mid-1800s was hard on people's health. There is some evidence that life expectancy in London then was lower than in the rest of England.[2]

Today, as people move into low-income settlements vaster and denser than ever before, they are struggling to prosper in environments at least as challenging as the cities of Victorian Europe. These are places unable to draw in enough fresh water or to channel excrement away safely. People live in dilapidated, intensely crowded homes. They have little access to health services, and few are able to get the education or jobs that could take them out of their situation. Although cities have gained a reputation as healthy places to live, the urban poor very often have higher rates of infant deaths and under-five mortality than their rural counterparts. (See Box 7–1 on the myth of the healthy city.)[3]

In poorer countries, only an urban minority lives in healthy living conditions and has access to good health services, education, and employment. In some towns and cities in Africa, Asia, and Latin America, up to 80 percent of the population lives in extreme poverty—with all that this implies in terms of

---

Box 7–1. The Struggle to Collect Good Data on Health in Cities

The myth of the healthy city has been linked for decades to a problem of data aggregation—where total health statistics are sometimes presented for cities with populations greater than those of nation states. Megacities suffer particularly from a problem of super-aggregation, in which health data on the whole city tell only part of the story. Each city's health will depend on a range of contextual factors, just as a national health profile does.

A key predictor is often the overall state of "development" of the city—measured in proportions of people with clean water, sanitation, adequate housing, and access to health measures such as vaccination and primary care. Thus a megacity with a large proportion of people living in poverty will have a health profile that reflects the profile of these people. However, there is another problem: megacities account for only about 9 percent of total urban population of approximately 3.2 billion citizens. Just over half of the world's city dwellers live in settlements with fewer than 500,000 inhabitants, and we still know little of the health situation in smaller towns and cities internationally.

In every town and city, data that have been broken down by income group tell a different story of urban health—it is the tale of urban inequality. Within cities and towns, disaggregated data show that there are inequalities in living conditions and in access to services such as health, water, and sanitation. There are also inequalities in access to education and work. Thus it is hard to generalize about urban health profiles, as each city and town may have a distinctive pattern of development and distribution of resources. Even a rich town in a rich country may have sharp inequalities that affect health drastically—for example, a classic study in New York found that black men in Harlem were less likely than men in Bangladesh to reach the age of 65.

SOURCE: See endnote 3.

living conditions and health. (See Box 7–2.) In wealthier countries, where most people have adequate living conditions, urban inequalities show themselves in social outcomes such as educational attainment and employment and in health outcomes related to social inequality, such as violence. But in poorer countries, urban areas often have the worst of all worlds—contaminated air, land, and water; deep poverty; and a health profile that includes both the infectious diseases of deep poverty and the so-called diseases of modernity (obesity, cancers, and heart disease). In these cases people carry a "double burden" of disease that poses an even more daunting challenge for human health on an urban planet.[4]

The poorest people in these countries sometimes include previously isolated indigenous rural peoples who have been forced off their land into the worst areas of towns and cities. These communities have lost their way of life and often are forced into the most marginalized roles in cities. Their health indicators can be the poorest among the poor.

For example, indigenous San communities in Botswana who have been moved off their traditional lands into resettlement camps in isolated towns face multiple challenges to their health and well-being. Their environment plays a vital role in their identity and way of life as an indigenous community; radical changes in that environment destroy the social structure of communities at the same time that members are forced into mainstream society at the bottom of the social and economic hierarchy, vulnerable to alcoholism and HIV/AIDS. The San are the only language group in Botswana whose Human Development Index score, as calculated by the U.N. Development Programme, fell during the 1990s.[5]

Both the physical and the social environment of cities and towns today affect urban

---

Box 7–2. Cities Out of Balance

Both historically and today, the biggest toll of urbanization has been concentrated on the poorest residents of a city. In the absence of amenities such as spacious housing, flushing toilets, running water, and regular garbage collection, shantytown dwellers around the world spend their days maintaining cramped homes, walking long distances to a menial job, accepting contaminated water when they have to, and living day-to-day in some of the planet's least secure and dirtiest neighborhoods.

Often they must also tolerate sharp inequalities of wealth and influence that seem to keep the whole city out of balance. Residents of *villas de emergencia* in Buenos Aires, for example, are compelled out of desperation to inhabit the notoriously polluted banks of the Reconquista and Matanza Rivers, which smell nauseating and are "overrun with rats, mosquitoes, flies, and other insects." Less than a mile away, right on the edge of another shantytown, some of the wealthiest residents of the city live in gated communities of lawns, swimming pools, and security guards.

One effect of living in such unequal societies is that the poorest people must live in environments that concentrate hazards and multiply health problems. In Dhaka, Bangladesh, for example, the wealthy can afford to buy land at prices inconceivable to most of the city's population. As a result, 40 percent of the people are crammed onto just 5 percent of the land area. Some of the most desperate of these settlers, victims of eviction or floods elsewhere, live on a ledge precariously situated between a toxic factory and a poisoned lake.

SOURCE: See endnote 4.

health. The overall quality of the urban environment is important for health, and so is the extent of inequality within an urban environment, as just described. Some problems of the urban physical environment, such as ambient air pollution, may affect almost everyone in a city. Other problems, such as contaminated water, indoor air pollution, or lack of sanitation, may disproportionately affect some groups more than others. (See Chapter 9.) Urban violence may also affect some urban dwellers more than others. Rapid urbanization in most cases exaggerates these problems, as cities are unable to build enough infrastructure and provide enough jobs for an influx of migrants, many of whom may be fleeing war or drought.

Today, the lack of facilities to provide clean water and adequate sanitation are major problems—with huge health repercussions. (See Chapter 2.) Largely due to contaminated water and poor sanitation, digestive-tract diseases are a leading cause of death in the world and a major urban health problem.[6]

Crowding is another major health hazard for the urban poor. People in low-income settlements often live in highly crowded homes, with four or more people per room, often shift-sleeping and with many children per bed. Contact-related diseases such as measles, tuberculosis (TB), and diarrhea are all linked to living in crowded environments. Notably, some richer cities have such deep pockets of disadvantage that diseases related to poverty and crowding have re-emerged during periods of increased inequality within the city. A study looking at TB in New York from 1970 to 1990, for example, found an increase in childhood TB in the period and that children living in areas of the Bronx where over 12 percent of homes are severely overcrowded were 5.6-fold more likely than children in other New York neighborhoods to develop active TB.[7]

Although undernutrition is generally considered to be a rural problem, the urban poor—forced to pay high prices for food shipped into the city, and often unable to grow their own food—can often have the most difficulty obtaining enough nutritious food. This has major impacts on urban health—particularly for children. A recent analysis of child health in 15 countries in sub-Saharan Africa found that differences in child malnutrition within cities were greater than urban-rural differences and concluded that these were a major problem for the region.[8]

In poorer countries, urban areas often have the worst of all worlds— the infectious diseases of deep poverty and the so-called diseases of modernity, a "double burden" of disease.

Deficiencies of calories or protein are not the only food-related problem faced by the urban poor. The cost of importing fresh raw fruits and vegetables from rural outskirts can be high, and for working women and men, street foods are more available. These food sources create their own health risk, with biological contamination a frequent cause of parasite infestation and diarrhea. Also, dense fatty, sugar-rich, and salty foods can often be a more economical—or more readily available—source of calories. In urban health terms, however, this has impacts on malnutrition too: obesity now ranks alongside undernutrition in many urban areas, with risks of diabetes and cardiovascular disease becoming more common in many towns and cities.[9]

Cooking food properly can decrease the risks of biological contamination. But some cooking methods can also lead to another urban health hazard. Both rural and urban poor families often cook with biomass or coal-burning stoves without proper ventilation. This creates a major threat to health, particularly for mothers and infants. In urban areas, however, crowding combines with use of these stoves in poor settlements. The health impacts are huge. The World Health Organization (WHO) estimates that indoor air pollution created by stoves that burn solid fuels is responsible for 1.6 million deaths and 2.7 percent of the global burden of disease—many of them affecting the urban poor.[10]

The industries that power urbanization also create quantities of air pollution in contained population centers. (See Table 7–1.) And the combustion of solid fuels (biomass and coal) in millions of homes can contribute to ambient air pollution. In developing countries, this domestic pollution combines with the greater cocktail of coal-burning industries, diesel trucks or cars, and small, two-stroke motorcycles. Globally, urban air pollution is estimated to kill 800,000 people each year. Notably, about half of these deaths may occur in China.[11]

Urban industries do not simply pollute the air—they often contaminate the land and water of the city. This can create a paradoxical kind of urban development: an immediate economic benefit, but at great expense for the health of current and future residents. As with other urban health problems, pollution

Table 7–1. Air Pollution Types and Effects and Urban Pollution Hotspots

Air Pollutant	Description	Health Effects	Hotspots
Large suspended particulates (PM 10)	Soot from combustion; soil from erosion	Can enter lungs and irritate bronchi	Northern cities in China, which rely on high-sulfur coal for heating and industrial processes
Small particulates (PM 2.5, especially sulfur dioxide)	By-product of burning sulfurous coal and diesel fuels	Smaller particles penetrate more deeply into lungs, causing asthma and bronchitis; tied to some cancers; can contribute to smog formation	Northern cities in China
Carbon monoxide	Incompletely combusted fuel	If concentrated in contained area, can cause asphyxiation	Along roads during rush hour in car-reliant, high-altitude cities, during winter
Nitrogen oxides	Atmospheric nitrogen oxidized during high-temperature combustion	Reacts with VOCs (and carbon monoxide) to form ozone, which irritates eyes and respiratory tracts and increases prevalence of asthma	Car-reliant cities during hot summers
Volatile organic compounds (VOCs)	Evaporated or incompletely combusted fuel; evaporated solvents, paints, adhesives	Often the limiting component of ozone formation; benzene causes leukemia	Urban areas crowded with older car models
Lead	Combustion of gasoline with added lead	Debilitates cognitive functioning	African cities where lead is still commonly used to increase octane

SOURCE: See endnote 11.

impacts may be worse for the poorer people who both live around and work in these industries—in both rich and poor countries. The urban poor may get work—but at what risk? Bhopal in India was perhaps the most notorious example of this kind of urban hazard and inequality: in 1984 more than 40 tons of methyl isocyanate gas leaked from a pesticide plant, immediately killing at least 3,800 of the city's poorest people and causing significant morbidity and premature death for many thousands more. Not only does this kind of urban industrial development often fail to move people out of poverty, it can harm thousands of lives over the long term.[12]

In many areas, poorly planned urban transport systems are another paradoxical urban development. They can increase commerce and the accessibility of jobs, but if they rely on motorized movement they can also produce other health hazards. Motorized forms of transport cause ambient air pollution and contribute to a massive toll in road traffic crashes. WHO estimates that each year about 1.2 million people around the world die as a result of road traffic crashes, and perhaps as many as 50 million more are injured. A study in Kumasi, Ghana, found that pedestrian road traffic crashes were the single largest cause of injury in the town. Even in cities in conflict, such incidents are an important cause of death by injury. In Maputo, Mozambique, road traffic crashes were the biggest cause of fatal injuries reported in

2000. In some cities, these are the main cause of death for young people, followed by urban violence.[13]

In many cities and towns of Africa, Asia, and Latin America, "modern" diseases such as asthma, heart disease, and cancer are arriving in places that have yet to fully get a handle on "old" diseases such as tuberculosis, cholera, and diarrhea. Figure 7–1 illustrates the double burden of disease brought by dirty industrial development and polluting motor vehicles in Kolkata, India. Up to 40 percent of people still live in poverty, but the route to development is sought through highly polluting industries, and providing basic services to millions of people remains a dream. The Figure shows the cause of death by age—demonstrating, albeit crudely, that if people in Kolkata survive the insults of contaminated water, they live on to experience the severe risks of highly contaminated air. This "double burden" is particularly exaggerated in unequal cities in Africa, Asia, and Latin America, where affluent people are

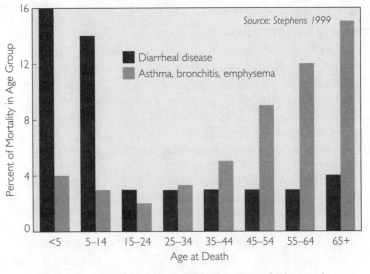

Figure 7–1. The Urban Double Burden of Disease in Kolkata, India

increasingly dealing with expensive modern maladies such as diabetes and heart disease while lower-income people are still plagued by undernutrition and infectious disease.[14]

There is another risk in cities that is not easily addressed with simple interventions such as improved living conditions or infrastructure: urban violence, which is in epidemic proportions in many cities both North and South. In some cities violence has actually started to reverse overall trends of health improvement, particularly for young people. For example, a longitudinal study looking at trends in mortality for 15–24 year olds in São Paulo and Rio de Janeiro from 1930 to 1991 showed a steady improvement in adolescent health until the 1980s, when death rates started to rise again due to violence. The death rate due to violence can be up to 11 times higher for young people from the poorest communities than for youngsters from a wealthier community.[15]

In some cities violence has actually started to reverse overall trends of health improvement, particularly for young people.

In Brazil, urban violence takes an enormous toll on poor young men in terms of both mortality and morbidity: death rates from homicide in Brazilian state capitals in 2003 for young men aged 20–24 was 133 per 100,000. This means that 1 child in every 1,000 will be a victim of homicide in these cities, a rate higher than that of any childhood cancer. The United States has a similar problem, with violence affecting principally economically disadvantaged young urban men from ethnic minorities, who are disproportionately represented in jails and in homicides.[16]

Injuries related to traffic accidents and violence are now respectively the second and

fourth causes of hospitalization in Brazilian cities. This makes trauma treatment a top priority for urban health services in this country and challenges health systems that are generally better prepared for infectious or chronic diseases.[17]

The fear of violence can also be a significance hindrance to mental well-being. Long-term anxiety, stressful life events, lack of control over resources, and lack of social support are all key preconditions for depression. Poor women in cities are often the most vulnerable group.[18]

Of all the urban challenges to public health in the future, perhaps the one that will affect more people than any other is global climate change. More frequent and extreme droughts are likely to test cities' ability to provide adequate food and water. More-intense floods are likely to destroy the precarious housing ubiquitous in slum areas and to increase outbreaks of waterborne diseases. Fatally strong heat waves are likely to become more frequent, while tropical diseases are becoming viable further from the equator and higher up mountains. Cities such as Nairobi and Harare, once located strategically beyond malaria's reach, are likely to lose their favorable position.[19]

Higher temperatures that many scientists associate with climate change have already had an impact on mortality rates in cities. In Chicago, 700 people died in a matter of days during a heat wave in 1995. In the summer of 2003, the toll was far higher: the deaths of more than 52,000 people in Europe were tied to the record-breaking heat in many countries there. In cities, the stress on people's health from the unprecedented heat was no doubt aided by the "heat island effect"—the higher temperatures experienced in urban areas everywhere due to the concentration of buildings, paved areas, and other infrastructure and the waste heat from cars and factories. (See Chapter 6.)[20]

Why does it matter if there are all these health problems in cities when urbanites supposedly have the best access to medicine? There is one pervasive health myth about towns and cities that decades of research does not seem to shift—that people might get ill in urban areas but at least they have access to health services. Technically, this is true: there are more health services concentrated in towns and cities than in more geographically dispersed rural areas. Hospitals are also usually found in cities—which should mean that urbanites have access to emergency and specialist care when they need it.

Yet the evidence suggests that the urban poor have little access to even basic health care at public facilities when they need it and that preventive programs such as vaccination are not reaching these people. Studies in India, for example, suggest that while "60% of the children aged 12–23 months in urban India are fully immunized; coverage among urban poor children is a dismal 43%." The urban poor also are more likely to use private health services than public ones—having no choice but to use their hard-earned resources to pay for illnesses created by the dense, polluted environments they are forced to live in.[21]

The Move Toward Healthier Cities

While urban living has the potential to immerse people in a dense array of hazards, it is also true that wealthier cities like London, Berlin, and Tokyo—once home to the world's least healthy living conditions—have achieved historically unprecedented qualities of life. A combination of improved housing conditions, better nutrition, cleaner food, and new medical treatments brought a significant and constant decline in death rates across England and the United States, and then France, Germany, Italy, and Japan. Better scientific understanding of infectious disease and improved sanitation systems were linked to a gradual end to outbreaks of cholera, while improvements in London's housing conditions between the 1830s and 1920s brought a substantial decline in the number of deaths from tuberculosis.[22]

By improving the road system and rail connections, introducing more extensive refrigeration, and improving agriculture, clean and healthy food became more readily available in cities like London. Death rates from typhoid, which spread via contaminated food, fell dramatically between 1840 and 1900. Over the same time, cities developed more professional and standard systems of garbage collection, as London's municipal government finally identified city-wide garbage collection as a fundamental task for a healthy city.[23]

Today, children born in urban areas of Europe, the United States, New Zealand, and Japan will see a doctor as soon as they need to if they fall ill. They are more likely to be vaccinated against diseases such as measles, polio, mumps, and rubella, and as they grow up they are more likely to live in adequate housing, go to school, and find work that is not hazardous. They have a better chance of surviving past the age of five. And as they become adults they will have better access to medical facilities in the case of illness.[24]

During the twentieth century, people living in cities have not had to repeat the entire experience of dirty industrialization. There is some evidence that Latin America accomplished an urban epidemiological transition over a much shorter period of time through a combination of preventive health interventions such as vaccination, some public health infrastructure, and macroeconomic improvements that provided jobs and put food on poor people's tables. The more urbanized the country, the faster mortality rates fell.[25]

But facilitating a similar transition for all urban dwellers, even as it is under way today, is still a daunting challenge. Moreover, macroeconomic changes can affect urban peoples' lives very quickly and can be beneficial for the urban poor or not. Fundamentally, city dwellers everywhere rely on the macroeconomy for their well-being, and shifts in national prosperity can be felt more sharply by the urban poor due to their profound dependence on the economy for everything from access to food to education and work. The movement of a major industry from one location to another can change conditions profoundly: it can provide jobs and better living conditions for people in one city, while those left behind in another city can see a drop in their living standards and well-being.

Asking the most disenfranchised people in any city about their priorities can stimulate change in almost all urban health problems.

It may be that trends in macroeconomic inequality will have severe effects on the urban poor. There is some evidence of this from two very different contexts: Studies in the United States looking at trends in mortality between the 1970s and 1990s found that African Americans living in cities experienced extremely high and growing rates of excess mortality compared with rural and wealthier communities. A similar study in five African countries of rising death rates for children under the age of five between the 1980s and 1990s found that in Zimbabwe the increase was largest among urban children.[26]

In this context, what can we do to encourage the move toward healthier cities? There is evidence that urban inequalities decrease when local governments listen to low-income residents and when such individuals are included in urban plans and actions. Inequalities may also lessen when health is used as a criterion to establish priorities for urban policy. In Kolkata, India, for example, an environment and development plan was based on health priorities and on a consultation weighted toward the views and needs of the poorest citizens. This did not change the city overnight, but it gradually put the needs of the urban poor on the policy agenda. There are also many successful stories of self-help in water and sanitation—where health and city equality have both improved when local governments work with local people. (See Chapter 2.)[27]

Donors, too, who are seeking to help the urban poor can better support health by listening to them. Looking for a set of best practices, the World Bank embarked on an institute-wide analysis of 45 participatory urban development programs in the 1990s. The final report concluded that community participation was absolutely essential for any slum upgrading projects. The authors noted that the people who must move their homes to make way for roads, public spaces, and sewage lines "must be involved in the decision making process if they are to cooperate with it." The report further concluded that community participation was "the single most important factor in overall quality of project implementation—efficiency, effectiveness, timeliness, responsiveness, and accountability."[28]

In some countries, such as Brazil, recent local and national government strategies have systematically supported governance and budget setting that gives priority to the needs of the urban poor and that puts public services high on the overall agenda. (See Chapter 9.) Local government support to the efforts of low-income city dwellers can have effects on a surprising range of health challenges. In Ilo, a small port town in Peru, for example, air pollution policy finally got on the agenda when communities were included in debates with the local mining company, and access to public services increased dramatically as community concerns were included in local planning and budgeting. Green spaces increased, water contamination decreased, and sustainable development gradually rose in importance on the agenda. The mayor during this period noted that Ilo's people and their mobilization put the environment, poverty, and equity high on the agenda—all themes that affect urban health.[29]

It is not just inequalities in the physical environment of cities that can be improved with more involvement of citizens. It is notable that some of the most intractable problems of urban inequality—such as urban violence—are often only tackled in this way. It is young people who are most affected by urban violence, and the urban poor are often in constant threat of violence—either from young people in other poor areas or from the police. Interestingly, young people themselves often have the solutions. For example, young men from one of the most difficult *favelas* of Rio de Janeiro started a musical alternative to drug gangs, eventually spreading their combination of music and education to other *favelas* and starting a health education program for young women and men.[30]

Film and music can often engage young people in these towns and cities in ways that more conventional approaches cannot. And stories can be told from the perspective of young people who feel that their lives are ignored. Health lessons can come through this, while shared visions are exchanged.

Inequalities exist in even one of the world's most famous new cities: Curitiba, Brazil. Touted as a healthy, social, and sustainable city, Curitiba has achieved world fame for its vision. But it is also home to thousands of people living in *favelas*, and like other cities it is affected by violence and drug wars. Yet

here, too, young people have come up with an answer. In 2006 Alex Roberto and his friends from a poor neighborhood in the city made a film about "the ignored society," conveying to the city government their vision for a healthier city. In this film, the young people asked their city to listen to the people of the poorer communities—not to ignore their part of the urban world, but to include them in the vision of this beautiful city.[31]

Asking the most disenfranchised people in any city about their priorities can stimulate change in almost all urban health problems. Their solutions are often the cheapest and most appropriate ones in particular situations. Involving the poor can also reveal health problems—and solutions—that most planners have never thought of. A 1997 study that asked homeless women in Kolkata about their greatest health threat, for instance, revealed an enormous burden of urban sexual violence. One intervention that the women really valued was some light around the water posts and toilet blocks to illuminate the unsafe areas at night. Both the issue and the solution would not have occurred to city planners had they not asked the advice of the women who were the least heard in this vast city.[32]

Old Stories and New Solutions

We know that many of the solutions applied by urban planners in the past would alleviate a good deal of the health problems of today's towns and cities. But now, perhaps for the first time, we are also aware that urban systems and structures need to be environmentally sustainable if we are to secure the health of people living in cities in the future. Sustainability, as we now know it, is something that the urban planners of the past did not think of as they pursued industrial development. Fortunately, there are some ways cities can be both healthier and less environmentally destructive.

Perhaps the first sustainable challenge is to make a city that breathes life and does not consume more than it needs. Plants are vital to this—whether they exist within urban boundaries or not, they are a vital part of a city's metabolism. They provide the food that replenishes cities. They produce the oxygen people breathe and they filter the air and water fouled by cities. In many cases, they absorb much of the organic waste that cities dispose of.

There are health benefits to ensuring that urban areas contain more of the plants they need within their own borders. In cities, plants can absorb air pollution before it does harm to people, reduce the heat island effect, filter sullied water before it enters a stream or river, help dispose of sewage before it poisons other ecosystems, and even help ensure a continuous supply of fresh food—all at a potentially low cost. Greening a city can thus reduce the ecological footprint of its residents while also improving their health.

Unfortunately, most cities around the world are reducing the amount of green spaces within their borders. Cities in the eastern United States, for example, have lost 30 percent of their tree cover over the last 20 years. Buildings and roads take priority, making some areas into "pavement ecosystems" that are even more lifeless than deserts.[33]

Different species of plants can bring diverse benefits to urban areas, but crops such as tomatoes, carrots, and lettuce have the most obvious value for people. Fresh vegetables and fruits are essential for providing people with enough vitamins and fiber, yet the cost of maintaining their freshness en route from rural areas is often more than poor residents can afford.

Urban agriculture can be one of the most effective ways to improve health in a variety of urban areas. Food harvested in cities can provide cheap, reliable nutrition, as well as an extra source of income for the poorest residents. Urban agricultural systems are also often the cheapest way to handle sewage wastes, which can provide a nutrient-rich source of irrigation water as long as care is taken to eliminate pathogens that pose a risk to health. In wealthier countries, community gardens are a cheap way to improve ruined properties and to foster community cooperation. (See Chapter 3.)

Plants, and large trees in particular, can also improve urban air quality. Plants naturally absorb pollutants such as nitrogen oxides, carbon monoxide, and ozone, and their leaves can filter dust and particulates from the air. Moreover, because trees can shade the asphalt and pavement that contributes to the urban heat island effect, they can help lower the temperature of a city and, with it, reduce the rate of ozone formation.

Although trees can only filter out a small portion of the pollutants released into large cities, the effect of an entire municipal forest can add up. The trees in New York City, for example, remove about 1,821 tons of air pollution a year, providing a value to society of about $9.5 million. A computer simulation in Atlanta, Georgia, found that a 20-percent loss of the city's urban forest cover would lead to a 14-percent rise in ozone concentrations. The U.S. Environmental Protection Agency has suggested urban tree planting as a way to reduce ozone formation in cities that are out of compliance with ozone standards.[34]

Some evidence suggests that urban trees and green space can also help decrease violence in a neighborhood. A study of police records in Chicago found that after randomly assigning people to live in 98 different apartment buildings with varying degrees of vegetation surrounding them, the greener the apartment block, the fewer crimes were reported—both property and violent.[35]

Availability of green space is also linked to urban equity. In Bogotá, Enrique Peñalosa made the provision of green public spaces one of the centerpieces of his administration when he was mayor. "With economic development lower-income groups get goods which once seemed inaccessible to them," he noted. "But they will never have access to green spaces unless governments act judiciously." Bogotá's program was remarkably decentralized: citizens participated in the siting, design, and creation of new city parks.[36]

A novel example of links between medicine, health, and urban green spaces is Mumbai's new 37-acre Mahim Nature Park. Previously the area was a treeless garbage dump, with poor communities living on one side and a polluted creek on the other. The city government is using the park as a way to encourage residents of Mumbai to plant their own green spaces closer to home. But this is not just about health as an abstract concept: at the center of the park there are more than 100 species of ayurvedic plants for medical use.[37]

As with community gardens, community efforts to plant green spaces can bring mental health benefits. One study of a community effort to preserve the Truganina Explosives Reserve in Altona in Australia found that participants visited doctors less frequently, perceived their health as better, felt safer in their community, felt they had more opportunity to share their skills, and had a greater sense of belonging. One participant who had suffered from depression said the work was "like a dose of medicine." Another said, "You become part of what is around you; you see people enjoying themselves and you benefit from that vicariously."[38]

In many urban areas, transport is one of the central causes of the "double burden" of diseases. Former Mayor Peñalosa of Bogotá has noted, "As developing country cities become more economically developed, the automobile becomes the main source of deterioration of the quality of life." Cars become the least controllable air polluters. Wide, fast, and dangerous roads are "like fences…making the city less humane." And cars demand, "unlimited investments in road infrastructure, which devour scarce public funds [for] water and sewage supply, schools, parks, and meeting the other basic needs of the poor."[39]

Food harvested in cities can provide cheap, reliable nutrition, as well as an extra source of income for the poorest residents.

Air pollution from cars remains one of the biggest problems in a number of big cities around the world. In Mexico City, three quarters of air pollution comes from cars. Contained by a basin and an old volcano crater and at 2,000 meters above sea level, a gray-brown noxious haze often hangs stagnant over the city.[40]

Compounding these chronic health problems, collisions between cars and pedestrians are the single most common type of accidental death and injury in the world. The problems can be especially bad in cities where people are rapidly starting to use cars. Vehicle fatality rates nearly tripled in Beijing during the 1990s, making car accidents the leading cause of death among people under 45 there and the leading cause of working-life years lost. WHO has identified traffic as one of the greatest dangers to poor urbanites. Not surprisingly, people who will never own a car in their lives—many of whom are forced to walk for miles each day along dangerous roads—are at the greatest risk.[41]

If cities and town were instead designed without preference for automobile travel, they could be the places where people walk and cycle to better health. (See Chapter 4.) WHO has estimated that physical inactivity already causes 1.9 million deaths a year as the result of heart ailments, cancer, and diabetes, and this number is rising as people increasingly travel by car instead of foot. In the United States, where childhood obesity and diabetes are growing problems, the proportion of children or adolescents walking or bicycling to school dropped from 48 percent to less than 15 percent between 1969 and 2001. At the same time, childhood obesity and diabetes are on the rise.[42]

There are two urban worlds emerging: the urban world of the wealthy, imprisoned in gated communities, and the world of the economically and socially disenfranchised, who work desperately to escape poverty.

Cities that are oriented toward walking can encourage physical activity where people are dying for lack of it. In the United States, residents of highly walkable communities got an hour more exercise each week and were 2.4 times more likely to meet the physical activity requirements for a healthy life. Other studies have found that residents of walkable communities are at a lower risk for being overweight or obese.[43]

Our Holistic Urban Future?

Improved transport or greener environments alone cannot bring good health to city dwellers. Community involvement in urban decisionmaking can improve cities immensely by putting people's views on the urban planning agenda, but again that alone is not enough. We also need a holistic vision of the urban space—we need to put intersectoral

collaboration and sustainability on the agenda. Cities and towns all over the world are now the nub of overconsumption, as well as being the heart of darkness for millions of people with little chance of escaping urban poverty.

There are two urban worlds emerging. There is the urban world of the wealthy, who are imprisoned in their gated communities and live long lives whether they are born into Asian, African, or American cities. And there is the world of the economically and socially disenfranchised, who work desperately to escape their poverty while risking their lives every day in their homes, on the roads, and in their workplaces. There are many differences between these two worlds, but perhaps the main one lies simply in the numbers: the wealthy urban world is for a tiny minority; the disenfranchised urban world is for the vast majority.

Perhaps the most startling thing about these urban worlds is that they are not separated in space or time. In every city the peoples of these worlds live alongside each other. They breathe the same air and may travel along the same streets. They also have, in essence, the same set of aspirations: to live well and long and to give their children a better future. Our urban future sets us perhaps the biggest challenge as human beings that we have ever had to face. We have to bring these worlds together and to move them both toward a sustainable future or there will be no chance of healthy lives for those living in tomorrow's cities.

There is no easy solution to the challenges of urban health internationally. The majority urban world demands, rightly, that their health involves their basic rights to clean water, sanitation, adequate housing, education, access to health services, and work that will be safe and remunerative. Interventions toward this would guarantee survival for the urban millions who die unnecessarily today.

But alongside this, our urban future needs creative new solutions too: there are no simple interventions that will address urban violence, for instance. Technical interventions will not address an urban health problem that is rooted in complex social injustice. Obesity and mental health problems also need a new lens. Cities that isolate us from nature and from each other cannot be described as healthy cities for the twenty-first century.

Equity is perhaps the key to the more complex social problems of cities—and it also can lead toward sustainability. (See Chapter 9.) A city where all peoples live together in peace, sharing the same spaces and the same resources, is far from today's urban reality. A city where people think of the next generation and the planet as a whole is also far from this reality. But neither vision is impossible—either to imagine or to achieve.

Perhaps we can only achieve both urban health and urban sustainability with a new vision of the planet and of each other—a vision based on the reality of our urban future.

Maybe we need a real renaissance of vision—a new urban ethic that is taught and shared between all of us, one that builds on the ethic of the ancient worlds we have lived through. An ethic that is anti-materialist, resource-sharing, peaceful. An urban ethic that connects us all and essentially argues that "No man is an island, entire of itself...any man's death diminishes me, because I am involved in mankind; and therefore never send to know for whom the bell tolls; it tolls for thee."[44]

What will Unoma and all her contemporaries think of our urban world? Will they ever live in a healthy urban setting—with adequate water, food, and shelter and free of risks of ill health and early death? Urban health has no easy solutions. But for many problems it is not technical knowledge that we lack but political will, social solidarity, and a commitment to sustainability and equity. In the cities that have political and citizen commitment to these values, we see a healthier, more sustainable, and more equitable urban space.

CITYSCAPE:
NAIROBI

Life in Kibera

From a distance, Kibera—the largest informal settlement in Kenya's capital city, Nairobi—is visually stunning. Seen from the air, its corrugated iron sheets twinkle like stars scattered on the ground. Step a little closer, though, and the first thing that will hit you is the stench of human waste.

On a typical day in Kibera, the smell of chips, *mandazi* (a local doughnut), and roast meat mingle with the odor of raw sewage. Plastic bags, some used as "flying toilets" (bags that people defecate in and throw on rooftops or just in the streets), litter the lanes that separate each shack. Occasionally you may stumble upon someone lying on the ground intoxicated with *changaa*, a lethal local brew. But for the most part, it is a bustling settlement, where restaurants and businesses thrive.

Some seven kilometers southwest of Nairobi's central business district, Kibera covers approximately 225 hectares split by the Nairobi-Kisumu railway line and surrounded by middle-income housing estates, a dam, vacant land, and a lush golf course. Kibera, a corruption of the Nubian term "kibra," means wilderness or bush. What was once a sprawling, sparsely populated settlement inhabited by retired Nubian soldiers who were allocated the land by the British Army in 1912 has today become one of the most crowded places in this city of nearly 3 million people.[1]

Kibera—one of nearly 200 informal settlements or "slums" in Nairobi—is estimated to have more than 1,200 people per hectare, most of whom live in tiny wattle, daub, and tin shacks with no electricity, no running water, no breathing space. Population estimates for the area range from 400,000 to over 600,000, making it the densest, most populous slum in the city. A recent survey found that over 80 percent of inhabitants live in single rooms that measure, on average, 9.4 square meters and are shared by five persons. Water sold by vendors is affordable only in small quantities. One pit latrine is usually shared by 75 people. Fewer than half the residents have access to a bathroom, so many bathe in their one-room hovel. Over 80 percent of residents are tenants in illegal structures, paying an average monthly rent of $12.[2]

Although most people reported that they did not live in Kibera out of choice, they cited affordability as the most critical determinant of whether they would stay. In Nairobi's skewed housing market, in which more than 80 percent of residents are tenants, rental options for low-income families range from shacks in slums to single rooms in multistory tenements. A one-bedroom apartment in Nairobi's Umoja estate, intended for the lower end of the housing market, costs a little less than $100 a month, out of reach for most Kiberans, nearly half of whom earn $70–140 a month and a third of whom earn less than $70. In response to the housing squeeze, unauthorized multistory tenements with shared toilet and washing facilities are springing up within and on the outskirts of slums; most are not built to housing standards or building codes.[3]

Patrick Obwaya, 53, who earns 5,500 Kenya shillings ($77) a month as a security guard, arrived in Kibera over 20 years ago and lives there still with his three teenage sons in a dark, mud-walled shack with no running water or toilet. His bed and those of his sons are separated by sheets, an extraordinary feat, as the shack is no more than 12 square meters. One of his sons sleeps on Patrick's bed when he is on night duty, a kind of slumberland musical chairs. On his walls, next to newspaper cutouts of Mother Teresa

Hiroshi Sato

A vegetable vendor in Kibera

and Jomo Kenyatta, Kenya's first president, is a photo of him and his wife at their wedding. It's been 20 years since he lived in the western Kenyan village where his wife lives.[4]

The scale of deprivation in Kibera is so huge that nongovernmental interventions have only managed to marginally improve access to basic services. Schools, health facilities, and water points remain inadequate. A study found that 14 public primary schools were situated within walking distance of Kibera, but they could only accommodate 20,000 of the 100,000 primary school–age children in the area. To have significant impact, the public sector must get more involved.[5]

Government efforts remain mired in conflict and confusion, however. The Kenya Slum Upgrading Programme, initiated in 2000 in an agreement between the government of Kenya and UN-HABITAT, has remained largely unimplemented, mainly because no one can agree on the best way forward.[6]

Suspicions arise due to Kenya's corrupt land allocation system. Even the most well intentioned housing projects have been marred by political patronage, nepotism, and profit extraction. Kibera residents fear that any attempts to "redevelop" or "upgrade" their homes will leave them homeless, as upgrading may lead to land speculation that prices them out of the upgraded settlements. All units in a recent National Housing Corporation project were allocated or traded to the middle classes. The housing units were built to middle-class standards, which increased their market value, encouraging beneficiaries to sell or trade in their units.[7]

The answer to Nairobi's slum problem lies in stronger and more integrated intervention by government ministries and agencies. The government needs to step into the distorted rental market with effective regulation and to invest in low-cost public housing. Ownership solutions, such as communal titling, could prevent titles from being traded to non-slum dwellers. This does not require huge foreign loans. Most countries that have reduced slum populations and slum growth used domestic resources to do so.[8]

A common view among policymakers in Kenya and other countries is that once rural areas are made more enticing, rural-to-urban migration—and slum growth—will be curbed. But history has shown that urbanization is irreversible and closely linked to development. People move to Kibera because it means better access to employment opportunities. Slums like Kibera are sites of immense opportunity and enterprise, places of transition where dreams of escaping poverty are first nurtured. But they are also sites of immense misery and poor health and environmental conditions.[9]

—Rasna Warah
Freelance Writer, Kenya

149

Managing Tourism

Petra, in southern Jordan, has been a center of human activity for more than 9,500 years and was declared a UNESCO World Heritage Site in 1985. The city flourished for three centuries under the Nabateans, from 200 BC to 100 AD, with a population of 20,000–30,000 at the crossroads of trade routes. After the last Nabatean king died in 106 AD, Petra began a slow decline. For centuries, only a few Bedouin tribes lived in the area. Seeds of change were planted in 1812, when a Swiss explorer rediscovered the city, paving the way for archeological missions.[1]

The modern fortunes of Petra are tied to its archeological significance and related tourism. The region's population surged from 4,610 in 1979 to approximately 20,000 in 1994 and an estimated 26,000 in 2006. The largest urban center in the area is Wadi Musa, built on the hillsides surrounding the wadis, or gorges, of Musa and Sadr. The traditional village lies on the lower and midrange slopes, while new, functional three- and four-story residential buildings dot the higher mountainsides, dispersed among derelict grazing lands.[2]

The town center comes to life during the tourist season, with Internet cafes and a large number of hotels—both international chains and small, locally owned inns. The ruins of the original stone dwellings are either abandoned or used as animal shelters. The dusty white sprawl of Wadi Musa contrasts starkly with Petra's carefully planned buildings, which strike visitors with their size, colorful facades, and the sheer magnitude of effort it must have taken to build them.

The rise of small urban centers and the pressures on Petra's archeological site prompted the government in 1993 to delineate the "Petra Region"—with the Petra Archeological Park at its center, surrounded by Wadi Musa and five other small settlements. The

area covered approximately 3,000 households.[3]

The Petra Region Planning Council (PRPC) was set up by the government in September 1995 to manage the 1,000-square-kilometer region, including the park. PRPC representatives were selected from among the local community, relevant governmental agencies, activists, and nongovernmental organizations (NGOs). The council's responsibilities included building roads, streets, sidewalks, and utility infrastructure; running environmental and tourism sites; managing and protecting the environment; and ensuring people's welfare.[4]

Although the council provided a forum to engage the community in debate and to develop policies, it struggled to influence urban development and the preservation of cultural and natural heritage. It was formed to centralize development efforts, but instead it splintered them, as the PRPC's mandate to manage the park overlapped with those of relevant ministries and government departments and of nongovernmental agencies.

Unplanned development was causing problems. In response, the government commissioned a study in 1996, which noted that in Wadi Musa "recent rapid and largely uncontrolled growth has resulted in the emergence of urban characteristics previously unknown in the region." Employment depends on the growth of tourism, the single largest source of income. Buildings have encroached on agricultural areas, changing the terraced topography into sporadic built-up areas between patches of grazing land. Pollution is damaging the ancient buildings and local residents' health. Wadi Musa's cesspits have discharged into the wadis, degrading the water and at times causing foul odors.[5]

To streamline policy planning for Petra, the government of Jordan established the

Petra National Trust

Al Hussein bin Talal University

Petra Region Authority (PRA) in 2001 headed by the Minister of Tourism. This office developed a zoning plan with provisions for land use, ownership, landscaping, transportation, roads, and waste management. While the plan marked an important change, critics argued that the government failed to discuss it in advance with local residents and did not account for sensitive areas with high impact on the national park.[6]

Again, internal conflicts ensued due to overlapping mandates and ineffective policies. In 2005, the government passed a law linking the PRA directly to the Prime Minister's Office, giving it higher-level executive authority.[7]

Conflict has limited the effectiveness of the PRA and its predecessor the PRPC, which has succeeded in zoning some land and issuing building licenses but has yet to produce building regulations. New buildings in Wadi Musa, which many consider eyesores, continue to lack local character. Even the PRA has built large new offices overlooking the Archeological Park, in a stark pink hue. Also overlooking the park is the Al Hussein bin Talal University, a series of monotonous beige stone buildings that dominate the landscape.[8]

The PRA has had little success encouraging business development or promoting income-generating projects in the poorer villages. Some land surrounding Petra has not been zoned, including the area from the Crowne Plaza Hotel to Um Seyhoun and the scenic road to Beidha. Although the government claimed these lands for their ecological and cultural importance, they are incrementally being sold to private developers.[9]

The institutional framework has produced policy gridlock. According to one defender of Petra's archeological and cultural heritage, Aysar Akrawi of the Petra National Trust, regulations and zoning laws have not been revised to reflect the increasingly vulnerable environment or the impact on ancient buildings. "Wadi Musa," she said, "is a buffer to Petra, and they [the PRA] have not taken into consideration the high impact it has on the park. The six towns surrounding the park must be considered a buffer zone and not an urban center in the development plans."[10]

The Petra region, and Wadi Musa in particular, continues to develop rapidly. But the fortunes of the region are unpredictable, as they depend on the fluctuating number of visitors to Petra, which in turn depends on the perceived political environment in countries bordering Jordan. The continuing challenge for Petra, as for many other cities, is to forge a balance among economic, environmental, and sociocultural development goals. This challenge requires effective institutions. The PRA, as currently structured, is not up to the task.[11]

Fortunately, the management of some national park areas in Jordan has been "subcontracted" to NGOs. The Dana Nature Reserve is one successful example. The Petra region would probably benefit from a similar arrangement that streamlines policymaking for the park and the buffer zone while regulating the urban areas to promote sustainable tourism and economic development.[12]

—Dana Firas
Jordanian author on sustainable development

CHAPTER 8

Strengthening Local Economies

Mark Roseland with Lena Soots

The wealth of a nation depends in large measure on the economic health of its cities. Cities make countries rich. Highly urbanized countries have higher incomes than other nations, more stable economies, stronger institutions, and more ability to withstand the volatility of the global economy. Cities around the world are playing a growing role in creating wealth, enhancing social development, attracting investment, and harnessing both human and technical resources for achieving unprecedented gains in productivity and competitiveness. Cities are also engines of rural development. For example, improved infrastructure between rural areas and the cities that rely on them increases rural productivity and enhances rural residents' access to education, health care, markets, credit, information, and other services.

For the first time in history, more than half the world's people will soon live in urban areas. Cities are both surrounded by and made up of communities. Geographic or territorial communities have a shared destiny, represented by a municipal or local or, in indigenous communities, band or tribal form of government. They may include the built-up or densely populated area containing the city proper, suburbs, and continuously settled commuter areas. They may be larger or smaller than a metropolitan area.[1]

Strong local economies are the foundation of strong communities that can grow and withstand the pressures created by an increasingly urbanized world. And strong communities require a holistic approach that not only provides the traditional deliverables of economic development—jobs, income, wealth, security—but also protects the environment, improves community infrastructure, increases and develops local skills and capacity, strengthens the social fabric,

Mark Roseland is Director of the Centre for Sustainable Community Development (CSCD) and a professor in the Department of Geography at Simon Fraser University in Vancouver, Canada. Lena Soots is a researcher at the CSCD.

and respects heritage and cultural identity. In this way, strong local economies also provide a foundation for strong national economies. Cities and towns provide enormous untapped opportunities to strengthen local economies by pioneering new approaches to sustainable development and community management.[2]

This chapter explores various approaches to strengthening local economies, some of the tools and strategies available, the actors involved in local economic development, and a development framework that addresses this broad range of concerns.

Economic Benefits— At What Cost?

Whether human settlements are villages, towns, suburbs, or megacities, local economic development has a critical impact on the sustainability of urban areas. Whenever agricultural or forest land is cleared for other purposes, whenever roads are built or expanded, whenever a new shopping center or subdivision is created, whenever an urban area is "redeveloped"—in short, whenever the natural or built environment is changed through human action—the health of communities and the planet is affected.

The economic development that generates these changes in the natural and built environment should benefit urban residents by improving their economic lives. Weak local economies are expensive for residents who suffer from poverty and its associated consequences, such as malnutrition and disease. Yet despite the enormous potential of cities to reduce poverty, recent evidence shows that the wealth they generate does not automatically lead to poverty reduction. On the contrary, inequalities within cities are on the rise, particularly in Africa and Latin America. Poverty is a severe, per-

vasive, and largely unacknowledged feature of urban life.

Even "successful" national economies can be expensive in terms of local human, social, health, and ecological costs. China's recent gains in economic growth and productivity provide a vivid illustration, as China is now home to 16 of the 20 most polluted cities on the planet. Economic gains there have in many cases exacerbated environmental problems in the cities.[3]

In 2001, for example, people living in the Chinese city of Huashui welcomed the first chemical factories to open in that area as a source of jobs and economic growth. However, their perspective changed as stillbirths increased and as more children were born with deformed limbs, unable to cry or blink, or with learning disabilities. As more chemical factories moved in, residents saw a "death zone" expand around the industrial area, killing trees and crops as far as 10 kilometers away. After four years of having their concerns dismissed by government and factory officials, on March 20, 2005, residents blocked the main road leading to the factories with homemade bamboo tents and mounted slogans on the factory walls that read: "Give us back our land" and "We want to survive." Three weeks later, an estimated 10,000 police officers and desperate residents faced off one night in a pitched battle that lasted for hours. This rare case of citizen outrage about deeply vested interests resulted, several months later, in the closure of the last of the area's 13 poison-spewing factories.[4]

China Daily reports that there were 50,000 environment-related riots, protests, and disputes in the country in 2005, an increase of nearly 30 percent. Many were linked closely to other divisive and sensitive social issues, including the nation's increasing wealth gap. Analysts blame these environmental flash points on a top-down single-party system obsessed with economic growth. Governments worry about instability when economic growth is not very fast, but the Huashui example shows how such growth can also lead to social and environmental instability.[5]

Though an extreme example, what happened in Huashui demonstrates that strong local economies must do more than contribute to national revenues—they need to also generate local improvements in social and environmental conditions. Many governments continue to assume that poverty is mainly a rural phenomenon and that people who live in or move to cities escape its worst consequences, including hunger, illiteracy, and disease. This view is reflected in most national poverty reduction strategies, which remain rural-focused, and in international donor assistance to cities, which continues to be modest in both scale and impact. As a result, development projects during the last two decades have had the net effect of increasing poverty, exclusion, and inequality in cities. The notion that cities are islands of privilege and opportunity is supported by statistics on health, education, and income, which generally reflect better outcomes in urban areas. But they fail to reveal the severe inequality within cities and the various dimensions of urban poverty that are not captured by income-based indicators, such as political exclusion and poor-quality, hazardous, insecure housing. (See Chapters 1 and 9.)[6]

Cities around the world can be thought of as lying on a continuum of economic development—from wealthy cities such as New York, Los Angeles, Vancouver, London, and Stockholm to poorer cities such as Lima, Harare, and Mumbai. Yet within each wealthy city there is also a poor city (as evidenced, for example, by the 2005 riots in the outskirts of Paris), while seasoned travelers can attest that within most poor cities

there are also wealthy enclaves. Sustainable development looks very different in each of these contexts. It means economic development that, on the one hand, encompasses multiple bottom-line objectives to enable continued prosperity without compromising the planet's natural support systems or, on the other hand, alleviates poverty and creates sustainable livelihoods so that people can live secure, healthy, and dignified lives.

From Global to Local Economies

Urban economic challenges are often ignored in part because of our flawed understanding of the economic system itself. Most conventional economic development efforts are rooted in the belief that economic benefits trickle down—from the rich to the poor, from the state to the city, from the market to the consumer. But such trickle-down strategies may actually exacerbate the very problems they are designed to ameliorate. Rather than trickling—or even pouring—economic development down, weaknesses in the trickle-down effect illustrate that there is a strong case for economic development to also be "from the bottom up."

Yet the past several decades have witnessed unprecedented economic expansion and the emergence of the "global economy." International development efforts of the last 20 years have focused on facilitating the integration of developing countries into this global economy through mechanisms such as the structural adjustment programs and policies of the International Monetary Fund and the World Bank, which were designed, among other objectives, to alleviate poverty through macroeconomic restructuring.

Recognizing the failures and shortcomings of past approaches to economic development, the United Nations Millennium

Declaration, adopted in September 2000, committed nations to a new global partnership to reduce poverty. Eight Millennium Development Goals were set to be achieved by 2015, with the aim of not only addressing issues of extreme hunger and poverty but also promoting gender equity and the basic human right to health, education, shelter, and security.[7]

There is a consensus that there have been more failures than successes. In light of this, in 2005 the UN Millennium Project presented its final recommendations, *Investing in Development: A Practical Plan to Achieve the Millennium Development Goals*, calling for both an increase in aid from western countries and a reallocation of funding priorities in developing countries. More interesting, however, is that the report also called for more aid to be given at the local level. With an emphasis on local assistance, the Earth Institute at Columbia University in New York has also started the Millennium Villages Project, a "bottom-up approach to enabling villages in developing countries to lift themselves out of the poverty trap." The 12 underlying principles of this project include community empowerment through participation and leadership, local-level capacity building, and the strengthening of local institutions.[8]

Many international development efforts seek not only to integrate developing countries into the global economy but to globalize the economy itself. Capital mobility, increasing trade, and the decline of national state regulation have made individuals and the communities they live in more vulnerable than before. Conventional approaches to economic development at best only address the conditions under which local economies connect more or less favorably to external economic circuits. For example, "cluster strategies" that emphasize the competitive advantage of a local commodity are concerned with how to plug local economies into the high value-added parts of global commodity chains. While these strategies are important, they do not guarantee sustainable, equitable, local outcomes.

In North America and Europe, the global economic expansion of recent decades has been felt in local communities in a variety of ways. Perhaps the most noticeable and currently debated result has been the commercial retail "big box" boom—that is, the proliferation and increasing dominance of multinational superstores.

As an example, consider Wal-Mart, the world's largest retailer. If Wal-Mart were a country, it would be the twentieth largest in the world; if it were a city, it would be the fifth largest city in the United States. Its critics have long complained that Wal-Mart is a bad employer, neighbor, and corporate citizen. But the company might be changing. In 2004 Wal-Mart launched a long-term sustainability initiative with leaders and executives from virtually every branch of the company formed into entrepreneurial teams focusing on areas such as packaging, real estate, energy, raw materials, and electronics waste. They partnered with environmental consultants, nonprofit organizations, and other groups that helped them examine business practices through the lens of restoration and sustainability.[9]

In October 2005, Wal-Mart's CEO announced three new goals for the company: to rely 100 percent on renewable energy, create zero waste, and sell products that sustain resources and the environment. In April 2006 Wal-Mart was one of a handful of major retailers and energy companies urging the U.S. Congress to impose mandatory carbon caps on their businesses. It has also become the world's largest supplier of organic food, not only reducing its ecological footprint but making organics more accessible for everyone.

If global forces such as Wal-Mart are signing on to save the planet by supplying organic produce and lobbying for carbon caps, why bother trying to strengthen local economies?[10]

There are several critical reasons to pursue such a strategy. First, economic development rooted in local ownership and import substitution has clear benefits in terms of stopping economic leakage, a term that refers to community income that is spent outside the local economy. For example, each time a resident shops in a different town or larger center, the amount spent represents dollars and income lost to the home community. There are two basic forms of leakage: immediate and secondary. Immediate leakage occurs when members of a community travel to another center and use their locally generated incomes to make non-local purchases. Secondary leakages occur when a resident makes a purchase in the community but the product was purchased or manufactured outside the community. Money spent outside the community represents a loss to the local economy.[11]

A 2004 study of a Chicago neighborhood found that a dollar spent at a local restaurant had 25 percent more local economic impact than a dollar spent at a chain store. The local advantage was 63 percent more for local retail and 90 percent more for local services. Other studies have found that local businesses yield two to four times the multiplier benefit compared with non-local businesses. The underlying reason that local businesses have this positive impact is that they spend more locally, particularly on management, business services, and advertising, and they enjoy their profits locally. These four items can constitute one third or more of total expenditures.[12]

Second, locally owned businesses are more likely to be a stable generator of wealth for many years, often for generations, and they are more likely to adapt to rather than flee the introduction of reasonable labor and environmental standards. During business downturns, they are less likely to relocate production to lower-cost regions, and in boom times they are less likely to move for a slight increase in the rate of return on investment. This anchoring of locally owned businesses minimizes the incidence of sudden, calamitous, and costly departures, which are often followed by massive unemployment, shrinking property values, lower tax revenues, and deep cuts in schools, police, and other services, which in turn throws still more people out of work. This is far less likely in a regional economy made up of primarily locally owned businesses.[13]

Third, while big-box stores such as Wal-Mart expand commercial choice and offer good consumer value, some studies have found that these stores do little to contribute to local economies. As noted, compared with locally owned stores, multinational chain stores siphon revenues out of communities through economic leakage. By displacing local businesses, they also contribute to increasing unemployment and decreasing overall incomes. Studies have found that big-box retailing triggers a decline in total economic activity despite increasing overall sales. A study by the Department of Agricultural Economics and Rural Sociology at Pennsylvania State University found that the presence of Wal-Mart "unequivocally raised family poverty rates in the US counties during the 1990s relative to places that had no such stores."[14]

Fourth, in North America a prevailing myth is that in order to foster economic development a community must accept growth. The truth is that growth must be distinguished from development: growth means to get bigger, development means to get better—an increase in quality and diversity. Local governments will often subsidize superstore development through infrastructure expansion in the name of economic development, only to facilitate more sprawl and municipal debt. Two alternatives for development without growth are supporting existing businesses and increasing the number of times a dollar is spent in the community. Local purchasing is the primary means both of supporting existing businesses and of increasing the economic multiplier, resulting in a more efficient, self-reliant, economically resilient community.[15]

Finally, strong local economies reduce the negative ecological impacts of global trade, in particular fossil fuel emissions from long-distance transport. For example, "food miles" refers to the distance food travels from where it is grown to where it is ultimately purchased or consumed. Locally produced food is clearly more sustainable and more environmentally desirable than food with frequent flyer miles.

Reclaiming Local Economic Control

There are few limits to what can be achieved when people work together for their mutual benefit. Since the 1800s, people have been forming cooperatives in order to meet a wide range of local needs. (See Box 8–1.) From agricultural producer cooperatives and consumer co-ops to worker co-ops and social cooperatives aimed at delivering health care and social services, cooperative enterprises can be found in nearly every country. In the context of increasing urbanization, cooperatives offer a community-based strategy for reducing poverty.[16]

The International Labour Organization (ILO) defines a cooperative as "an autonomous association of persons united

---

Box 8–1. Emilia Romagna—A Cooperative Economy

Situated in the north of Italy, just below the foothills of the Italian Alps, Emilia Romagna is a region of 3.9 million people. Bologna, the region's most populous city, is the commercial and communications hub of the area and has re-emerged in recent decades as one of the most vital and livable cities in Italy, with a strong sense of history, art, and culture. But beyond its cultural pleasures, this region is a unique example of a living cooperative economy. The region has more than 15,000 cooperatives in both the commercial and civil sectors, which collectively contribute over one third of the region's gross domestic product.

In the 1970s, Emilia Romagna was near the bottom of Italy's 20 regions in terms of economic performance. Today, it ranks first. It also currently ranks tenth out of 122 economic regions in the European Union (EU) and has a per capita income 30 percent higher than the national average and 27.6 percent higher than

the EU average.

What makes this region particularly fascinating is that its economic success in the past several decades is a result of ideas and practices that should be considered the antithesis of mainstream economic ideology. Cooperatives are the foundation of the region's economic makeup—the strongest sectors include retail, construction, agricultural production, housing manufacture, and social services. Most public works, including large-scale engineering, construction, and heritage restoration projects, are carried out by building co-ops.

With one of the highest qualities of life in all of Europe and strong economic performance, the Emilia Romagna region of Italy is a working example of community-based alternatives to the global market economy and living proof that strong local economies are possible through cooperation.

SOURCE: See endnote 16.

voluntarily to meet their common economic, social and cultural needs and aspirations through a jointly owned and democratically-controlled enterprise." Cooperatives thus practice a unique form of economic participation that is based on membership rather than amount of investment. (See Box 8–2.)[17]

Improving access to financial resources is an important component in the fight against poverty.

Within cities, worker cooperatives—businesses owned and controlled by their employees—are the most common form of cooperative enterprise. Decisions for the operation of the business are made democratically on the basis of one member/one vote. Common forms of worker co-ops include manufacturing plants, retail stores, communications companies, technical firms, and various forms of service providers. Worker cooperatives typically form as a result of a group of people organizing to create employment for themselves and to overcome barriers to employment such as disabilities or racial, gender, or ethnic prejudices. Alternatively, worker co-ops form as a result of an existing

---

Box 8–2. International Co-operative Alliance Principles for Co-ops

Voluntary and Open Membership
Democratic Member Control
Member Economic Participation
Autonomy and Independence
Education, Training, and Information
Cooperation among Co-operatives
Concern for Community

SOURCE: See endnote 17.

---

company that has "mutualized" by selling shares to its employees, who then take over ownership and management of the business.[18]

Argentina provides a good example of the power of cooperation in saving the local economy. During the 2001 economic crisis, many Argentines decided to take collective action to save their jobs and their livelihoods. By occupying bankrupt factories and businesses, the workers turned their workplaces into cooperatives without upper management or unions. The National Movement of Recovered Companies spread across Argentina as a bottom-up approach to economic recovery with the motto: "Occupy, Resist, Produce." There are now roughly 200 worker-run factories and businesses in Argentina, employing over 15,000 people. Most of these cooperatives started during 2001.[19]

In 2002, the Ghelco Ice Cream Factory in southern Buenos Aires went into bankruptcy. The workers were owed thousands of dollars in back pay. So they formed a co-op and stopped the owners from removing the machinery and dismantling the factory by protesting in front of it. Three months of protest led to an offer for the co-op to rent the factory. Five months after that, the factory was seized by the Buenos Aires legislature and given to the co-op. Now the factory is run by 43 members of the co-op who earn more than they ever did, and they have better working conditions. The co-op does not have to pay high managerial salaries, unlike the previous owners; in addition, it gets to keep the large profits that used to end up in the owners' pockets.[20]

Historically, cooperatives have not only enabled people to lift themselves out of poverty, they have also become a way for low- and middle-income people to continue accumulating economic advantages they would not be able to achieve individually. Beyond this, co-ops can contribute to strengthening the social fabric of communities and can build social cohesion among community members as well as facilitate the equitable distribution of resources. In a press release on the occasion of the United Nations International Day of Cooperatives in 2001, Secretary-General Kofi Annan underscored the role of cooperatives in development, noting "the values of cooperation—equity, solidarity, self-help and mutual responsibility—are cornerstones of our shared endeavour to build a fairer world.... The cooperative movement will be an increasingly adaptable and valuable partner of the United Nations in pursuing economic and social development for the benefit of all people."[21]

Community-Based Finance

Finance plays a fundamental role in the process of economic development—capital investments and growth in financial assets are important factors in the development of economies of low-income countries. Yet access to capital is not easy for those who live in poverty. People are often blocked from climbing out of poverty not by a lack of skills or motivation but by their lack of access to capital.

Poor people are often forced to rely on informal financial relationships, which are usually erratic, insecure, and costly to borrowers. Improving access to financial resources is thus an important component in the fight against poverty, and microfinance has proved to be an important tool in this regard. "Microfinance" refers to the provision of financial resources and services to people who are generally excluded from traditional financial systems based on their low socioeconomic status. Generally it includes the provision of loans, savings, and other basic financial services that poor people need to protect, diversify, and increase their sources of income.[22]

In 1997, delegates from 137 countries gathered in Washington, DC, for a Microcredit Summit and launched a campaign to reach 100 million of the world's poorest families, especially women, with credit for self-employment and other financial services by the end of 2005. According to the *State of the Microcredit Summit Report 2005*, by the end of 2004 a total of 3,164 microcredit institutions had reached 92,270,289 clients, 72 percent of whom were among the poorest of the poor when they took their first loan.[23]

Nearly 84 percent of these poorest clients were women. Assuming five persons per family, the loans given to the 66.6 million poorest clients affected some 333 million people. Just over half of the 3,164 institutions reporting were in Asia, 31 percent were in Africa, 12 percent were in Latin America and the Caribbean, and just under 5 percent were in North America, Europe, the newly independent states, or the Middle East.[24]

Contrary to some opinions, poor people are very good at saving their money. In fact, their savings represent a higher proportion of their net assets than the savings of their higher-income counterparts do. With access to well-designed savings products and services, low-income people can accumulate wealth and begin to climb the ladder out of poverty.[25]

The Grameen Bank in Bangladesh has become the international model for microcredit programs. Its central feature is its credit program, which provides small loans to the poor for self-employment activities. The project was initiated in 1976 when Muhammad Yunus, head of the Rural Economics Program at the University of Chittagong, launched a research project on the possibility of designing a credit delivery system to provide banking services to the rural poor. The Grameen Bank Project began with the following objectives:

- to extend banking facilities and services to poor men and women;
- to eliminate the exploitation of the poor by money lenders;
- to create opportunities for self-employment;
- to bring the disadvantaged, mostly women, within an organized format in which they can understand and manage themselves; and
- to reverse the vicious cycle of "low income, low savings and low investment."

After initial success in the village of Jobra, the project soon expanded to other villages. In 1983, government legislation transformed the project into an independent bank.[26]

The unique and innovative lending scheme of the Grameen Bank requires borrowers to voluntarily form small groups of five people to provide mutual, morally binding group guarantees in lieu of the collateral required by conventional banks. Initially, only two of the five group members are able to apply for a loan. Access to credit for the other members depends on the successful repayment of these initial loans. Although the Bank monitors borrowers fairly closely, organizing the borrowers into groups provides incentives for peer monitoring.

The success of the Grameen Bank is astounding, as the Nobel Committee noted when it awarded the 2006 Nobel Peace Prize to Muhammad Yunus for his pioneering work in this field. As of April 2006, it had a total of 6 million borrowers, 96 percent of whom are women. The bank has 2,014 branches working in 65,847 villages with a total staff of 17,816. Since it started, the total amount of loans dispersed is 271.94 billion taka ($5.46 billion). Of this, 241.63 billion taka ($4.83 billion) has been repaid. The loan recovery rate is 98.4 percent.[27]

Because of the interconnection between financial power, poverty, and women, microfinance has an important role in empowering women and improving economic equality. (See Box 8–3.) In many countries, particularly in low-income regions, women have few ownership rights. Cultural norms and expectations place additional constraints on women's access to assets and income-generating opportunities.[28]

Studies have shown that income earned by female borrowers has more beneficial effects on the well-being of children and household members generally than income earned by male borrowers. And microcredit lending by the Badan Kredit Kecamatan in Indonesia has been found to increase women's participation in decisionmaking, reduce fertility, and improve household nutrition.[29]

Particularly in low-income regions of the world where poverty rates are high, microfinance can be an important tool for improving the socioeconomic conditions of communities. It helps to foster financially self-sufficient private sectors and create wealth for low-income people. As it does this, microfinance creates new consumers and markets for existing businesses as well, thus contributing to the overall integrity of local economies.

Community-based financial systems are not only found in low-income countries struggling with widespread poverty. In countries such as the United States and Canada, there are community-oriented financial institutions that have a vision of social and environmental as well as economic benefits.

In a financial climate dominated by big banks, community credit unions demonstrate that financial benefits can coincide with broader community values and objectives. A credit union is a nonprofit financial institution that is cooperatively owned and controlled by its members and managed through the election of a volunteer Board of Directors. Credit unions offer the same financial services as

---

Box 8–3. The Power of Microcredit—A Personal Story

Unitus is a nonprofit organization based in Redmond, Washington, with a mission to fight global poverty by increasing access to microfinance. Using the example of the Grameen Bank and under the guidance of Muhammad Yunus, the Unitus Acceleration Model was developed combining best practices from venture capital and investment banking to create large-scale, poverty-focused microfinance institutions.

Unitus shares the following success story of one of their clients:

"Susan grew up in a poor rural area of Kenya. She was…forced to drop out [of school] after fourth grade when her family could no longer afford the school fees. Her parents kicked her out when she became pregnant at 17. Hoping to find work, Susan and her infant son moved to Nairobi, where she married and had a daughter. Her husband left her when they learned she was HIV-positive. Unable to find work and with no means to support her two

small children, Susan ended up in prostitution.

"Susan learned about Jamii Bora, a Nairobi-based microfinance institution, from neighbors in her slum. She completed their business training, which improved her business skills and gave her the confidence to begin her clothes mending and sales business. Jamii Bora's microfinance services enabled her to quit prostitution and move her family into a safer house….

"With each increasing loan, Susan buys more raw materials in bulk at lower costs, thus increasing her business's profitability. She is convinced she would not be alive without Jamii Bora's medical insurance and access to HIV medication, and can't imagine what would [have] become of her children…. Susan has savings for the first time and is striving to earn enough to ensure her children's education so they can break free from the chains of poverty."

SOURCE: See endnote 28.

---

banks (savings, investments, loans, and so on), but they generally market themselves as providing superior member services because of their community orientation. As cooperative institutions, their policies are set up to benefit the interests of their memberships as a whole. Credit unions typically pay higher interest rates on shares and charge lower interest rates on loans than traditional financial institutions.

Worldwide, there are over 157 million credit union members in 92 countries. Canada has the highest per capita use of credit unions, with over one third of the population enrolled as members. As community-based financial institutions, credit unions ensure that financial investments are not only economically successful but also in line with the broader social objectives and values of the communities in which they operate. As such,

credit unions strengthen local economies by providing positive social and economic returns to communities.[30]

The Downtown Eastside of Vancouver, Canada, is home to many people battling drug addictions, mental illness, and homelessness. Most local residents are low-income earners with little to no access to traditional financial services. Without identification to open a bank account or enough money to keep one open, people find themselves unable to attain any level of financial security, and many rely on costly private check-cashing facilities. With the help of VanCity Savings Credit Union (see Box 8–4), Pigeon Park Savings opened its doors in 2004 as a provider of low-cost, reliable financial services in a supportive environment. People can now cash or deposit checks and use a variety of financial services otherwise unavailable to

The Vancouver City Savings Credit Union (Vancity) in British Columbia is an example of a community-based "triple bottom line" financial institution. Established in 1945 on the premise of "banking with the unbankable," Vancity is a cooperatively owned credit union with over $10 billion in assets and more than 340,000 members. Member ownership ensures that the institution offers favorable social and environmental benefits to the community without compromising financial returns.

In the 1960s, Vancity was the first financial institution to offer loans to women without requiring a male cosigner. Beyond financial considerations, Vancity supports community members in socially and environmentally conscious choices. It was the first financial institution in Canada to offer special financing for hybrid vehicles and energy-efficient home renovations. Every year it awards hundreds of grants to community groups on environmental and social justice projects. Other initiatives include an EnviroFund Visa card that donates 5 percent of its annual profits to local environmental projects. Since 1990, more than $1.45 million has gone to such initiatives.

Vancity also offers its employees some of the best benefits packages in the industry. In 2005, it was named "Best Place to Work in Canada" by *Maclean's* magazine and "Best Work Place in Canada" by *Canadian Business* magazine.

SOURCE: See endnote 31.

Downtown Eastside residents. Pigeon Park's banking system runs entirely on VanCity's network, which also provides the operating infrastructure, technical support, administrative services, and security.[31]

Cooperatively owned credit unions are not the only form of community-based financial structures. Shorebank Pacific is a chartered, commercial bank in Washington State with a commitment to environmentally sustainable community development. It is the result of an innovative partnership between Ecotrust, a nonprofit environmental organization dedicated to fostering a conservation-based economy, and Shorebank Corporation of Chicago, a pioneer in developing inner-city community projects. To track financial progress, clients and loans are reviewed by a bank scientist who uses a scoring system of sustainability principles.[32]

Buy Local, Trade Fair

Although a significant proportion of goods used in cities are also produced there, cities are by no means self-sustaining entities. In terms of local economic activity, urban centers rely heavily on rural areas for resources such as fuel and food. The trading systems that facilitate the production, distribution, and consumption of goods in urban areas thus affect more than just the economies within cities.

In the context of an increasingly globalized economy, strengthening local economies means having strong local networks and trading systems that support economic activity within and among communities and that contribute to the overall health and well-being of these areas. Promoting trading systems that contribute to strong local economies allows cities and communities to participate in and contribute to the larger, global economic system in ways that are sustainable, equitable, and just.

Fair trade is a rapidly growing movement (see Table 8–1) that seeks to challenge unequal international trade relations and that makes trade beneficial to disadvantaged and vulnerable producers. It does

Table 8–1. General Sales of Fair Trade Labeled Products

Year	Amount	Growth
	(tons)	(percent)
1997	25,972	
1998	28,913	11.3
1999	33,495	15.8
2000	39,750	18.7
2001	48,506	22.0
2002	58,813	21.2
2003	80,633	42.0
2004	125,596	56.0

SOURCE: See endnote 33.

this by establishing direct and positive links between producers of the South and consumers of the North. Fair trade attempts to level the global playing field by enabling poor producers to be part of a trading system that ensures a fair and stable price for their products. It also offers producers and their organizations support and services and promotes the use of ecologically sustainable production practices.[33]

Similar to certification in organics or labor standards, fair trade involves the implementation of voluntary global production standards. Fairtrade Labelling Organizations International (FLO) is the primary body that sets the labeling standards to be met by producer groups, traders, processors, wholesalers, and retailers. Although most commonly associated with rural production, FLO sets standards for hired labor situations as well, including standards for working conditions, employee returns, discrimination policies, and child labor. In the context of manufacturing and the production of goods within cities, the concept of fair trade can play a significant role in creating strong local economies.[34]

For many years, coffee was the main product sold with the fair trade label. Now at least 20 different products—from tea and chocolate to sporting goods, textiles, and handicrafts—are exported from developing countries to more than 20 countries in Europe and North America plus Australia, New Zealand, and Japan. There are now 531 producer organizations certified with FLO, representing some 1 million farmers and workers from 58 countries in Africa, Asia, and Latin America. There are 667 registered traders, consisting of exporters, importers, processors, and manufacturers from 50 countries all over the world.[35]

Strongly linked to the cooperative movement (see Box 8–5), fair trade is helping communities build the financial resources to sustain livelihoods and alleviate poverty. It is also building community capacity, fostering strong relationships within communities, and promoting sustainable and equitable systems of production and trade.[36]

As recent mass protests in Seattle, Quebec City, Prague, and Genoa demonstrated, concerns about corporate dominance in the global economy cover more than just the issue of fair trade. A localism movement has emerged promoting alternative approaches to global economic development, particularly with respect to energy, materials, and food. Campaigns to promote the localization (or "de-globalization") of business and trade are sprouting up all over North America and Europe, espousing the social, economic, and ecological benefits of more localized economies.

Although the definition of "local" varies— from bioregions to geopolitical boundaries— the idea is consistent: local is better. The benefits of localized economies are many: they support local businesses and keep money and profits within communities, they re-establish producer-consumer relationships and build social cohesion, and they reduce the negative ecological impacts of global trade, namely fossil fuel emissions from long-distance transport.

Box 8–5. A Women's Fair Trade Sewing Cooperative in Nicaragua

After the devastation of Hurricane Mitch in 1998, the Center for Development in Central America began to search for ways to tackle the massive unemployment problems that had emerged in one of Nicaragua's relocated cities, Nueva Vida (New Life) in Cuidad Sandino.

Through the establishment of a marketing partnership with Maggie's Organics, an organic clothing company in Michigan in the United States, a sewing cooperative with the women of Nueva Vida was formed. The idea behind the co-op was to create fair trade employment in a region dominated by free trade zones and sweatshops with substandard labor conditions.

In 2001, the women's worker co-op became incorporated as Cooperativa Maquiladora Mujeres de Nueva Vida Internacional—Women's International Sewing Cooperative of Nueva Vida—and in 2004 it was certified as the world's first worker-owned free trade zone. This designation not only allowed the co-op to provide fair wages, good working conditions, and employee control, it also enabled them to compete on a level playing field with traditional sweatshops in the free trade zone. With the successful establishment of the cooperative, the workers' goal now is to put a percentage of the co-op's profits toward social projects in the community.

SOURCE: See endnote 36.

The Business Alliance for Local Living Economies (BALLE) is a growing alliance of businesspeople around the United States and Canada who join networks dedicated to building "local living economies" with a goal to green and strengthen their local economies. Businesspeople organize themselves into local business networks—each fully autonomous—and share a commitment to living economy principles. BALLE provides support and tools to catalyze, strengthen, and connect local business networks. Members of local networks join together to:

- support the development of community-based businesses;
- encourage local purchasing by consumers and businesses;
- create opportunities for business leaders to share best practices; and
- advocate public policies that strengthen independent local businesses and farms, promote economic equity, and protect the environment.[37]

Countless cities all over North America are working to promote the importance of strengthening local economies. The San Francisco Locally Owned Merchants Alliance, for example, has over 50 members and works to promote locally owned independent retailers in the area. The Buy Local Philly campaign is sponsored by the Sustainable Business Network of Greater Philadelphia and is a network of more than 200 local and independently owned businesses. The city of Portland, Oregon, has a Think Local First Campaign to raise public awareness of the benefits of a more localized economy. Vancouver, Canada, has started a Buy Local, Support Yourself campaign to encourage local purchasing and support local fashion, food, art, and more.[38]

As the world faces the emergence of peak oil and related energy challenges, cities need to reduce their dependence on outside markets and fossil-fuel-dependent transportation systems. What is needed are more localized systems of production, distribution, and consumption. Local trading systems not only allow cities to increase self-sufficiency, they also contribute to strengthening local economies by giving communities the ability and resources to meet their own needs.

Local Economy Actors

Who are the local economic actors and what are their roles in strengthening local economies? Local authorities are clearly important players. They build and maintain infrastructure that is essential for economic activity, and they set standards, regulations, taxes, and fees that determine the parameters for economic development. Local authorities procure large numbers of services and products and can influence markets for goods and services.

Like private enterprises, local authorities serve as public enterprises to produce "products" that are sold on the market. These products include environmental services (such as water, waste management, and land use control), economic services (transportation infrastructure, for example), and social services (such as health and education). Local authorities also operate under the considerable constraints of most public agencies: limited resources, jurisdiction, imagination, courage, time, and so on. For them to fulfill their potential in strengthening local economies, they need community organizations as partners.[39]

Local authorities are an influential employer and consumer in most communities. All community members have a legitimate interest in knowing what measures their local authority is, could, or should be taking to strengthen the local economy. These might include local studies of indicators, assets, imports, or subsidies; local training via entrepreneurship programs linked to incubators for locally owned businesses; help for those trying to purchase locally with a directory, a buy-local campaign, time dollars, or a local currency; local investing of municipal funds; and local public policy such as smart growth zoning or a living wage bylaw.[40]

In many low-income countries, local authorities have neither the resources nor the power to provide support of any kind to local business, let alone dedicate time and resources to overall localization initiatives. Cities with developed formal economies and stronger local authorities fortunately have more options.

Local trading systems contribute to strengthening local economies by giving communities the ability and resources to meet their own needs.

Local tax shifting is one local authority strategy with enormous potential to strengthen local economies. Taxes generate revenue for governments, but they can also serve as an effective tool of governance, supporting community values and goals. The main principle of tax shifting is simple: tax "bads," not "goods," so that markets work to direct the economy where we want it to go. Tax pollution, waste, urban sprawl, and resource depletion, for example—not jobs, income, investment, good urban development, and resource conservation. There are many successful examples of tax shifting in Scandinavian countries, which have been using this tool for at least a decade. Local authorities in Canada and elsewhere are recognizing the benefits of tax shifting and are beginning to examine its potential.[41]

For example, a 2005 study at Simon Fraser University in Vancouver examined six potential areas where the city could shift taxes in one municipal precinct: carbon emissions, drinking water, parking, solid waste, stormwater runoff, and sewage. The results were staggering: tax shifting on average could lead to a 23-percent reduction in environmentally harmful activities and could generate an extra $21 million in tax-based revenue. The surplus funds could in turn provide tax grants, subsidies, and rebates for environmentally friendly and sustainable development programs.[42]

The private sector also has a significant role to play in strengthening local economies. Business culture in general is starting to take on a different shape in light of public pressures and today's social and environmental realities. Over the last 10–15 years the socially responsible business movement has taken great strides in raising awareness that businesses need to serve the common good rather than simply maximize profits. There is growing recognition among companies and organizations that they have a social obligation to operate in ethically, socially, and environmentally responsible ways. Using "triple-bottom-line" accounting, some companies are measuring their performance in terms of social and environmental as well as economic results. Many business leaders are discovering the economic advantages to understanding and aligning business strategies with the values of stakeholders.[43]

Corporate social responsibility (CSR) is now a recognizable catchphrase in business circles, with standards that serve as a guide to conducting business in socially and environmentally responsible ways. General principles of CSR fall into the categories of ethics, accountability, governance, financial returns, employment practices, business relationships, products and services, community involvement, and environmental protection.[44]

Most examples of CSR are found in high-income countries, although, as noted earlier, the fair trade movement is changing business practices in the developing world as companies realize the social and environmental costs of doing business. Fair trade standards embrace the principles of CSR, including environmental protection, community benefits, and fair wages.

Emerging business trends such as CSR represent a significant shift in thinking about business. Indeed, socially and environmentally responsible businesses play an important role in sustainable economic development, particularly at the local level. Given that most business transactions take place in cities, it is essential that sustainable economic development within urban centers include the commitment of businesses to adopt practices that consider people and the planet as well as profit.

In addition to local authorities and the private sector, people in their daily lives are of course important actors in local economic development in their multiple roles as workers, consumers, voters, volunteers, and advocates. People who are disengaged limit their constructive interaction and participation in their communities. Some may express their disengagement by looking elsewhere to do business or by sending their children away to learn and work; others may give up completely and relocate to another community, another city, or another country. It is through participating in local communities that people can take the necessary measures to create sustainable economies, so citizen engagement is an essential component of strengthening local economies.

Community Capital: Using All Our Resources

Alternative economic theories and ideas are not new: in 1973 E. F. Schumacher proposed the idea of "new economics" in his influential book *Small is Beautiful: Economics As If People Mattered*, promoting small-scale development based on meeting people's local needs. Since then, alternative local economic approaches have been put forward in both industrial and developing countries— approaches that are rooted in community and designed to meet local needs and objectives. These have emerged in response to the negative effects of globalization and as a policy approach to sustainable development at the community level. Community economic development (CED) and sustainable livelihoods are two examples of these alternative strategies.[45]

Community economic development provides a conceptual means of addressing sustainable economic development at the community level. Its core principles include a community-based approach to development; direct and meaningful community participation; integration of economic, ecological, and social aspects of local development; asset-based development based on community strengths and resources rather than deficiencies; and support for community self-reliance. Its distinguishing features are captured in this definition: CED "is a process by which communities can initiate and generate their own solutions to their common economic problems and thereby build long-term community capacity and foster the integration of economic, social, and environmental objectives."[46]

Just as sustainability has prompted a shift in transportation and energy planning away from traditional concerns with supply to a new focus on managing demand, the focus of economic development needs to shift from traditional concerns with increasing growth to one of reducing social dependence on economic growth—or what could be called EDM, economic demand management. This has distinct implications for sustainable community development, particularly regarding employment and community economic development.[47]

Community economic development not only promotes initiatives that contribute to the economic health and viability of communities, it also emphasizes environmental considerations and the importance of social considerations in broader economic thinking. Examples of sustainable CED initiatives include:

- car cooperatives to reduce the cost and necessity of car ownership (Bremen, Germany);
- sustainable employment plans to create jobs, spur private spending, and reduce pollution through public investment in energy conservation and audits (San Jose, California);
- new product development to encourage manufacturers to develop environmentally friendly products through municipal research and development assistance (Gothenberg, Sweden);
- increases in affordable housing supply through zoning codes that promote a variety of housing types, including smaller and multifamily homes (Portland, Oregon);
- experiments with local self-reliance through establishment of closed-loop, self-sustaining economic networks (St. Paul, Minnesota);
- community-supported agriculture projects to preserve farmland and help farmers while making fresh fruits and vegetables available in city neighborhoods (Vancouver; London, Ontario; New York City);
- creation of local currencies such as LETS, Local Exchange Trading Systems, which seek to recirculate local resources and strengthen social ties (Toronto; Ithaca, New York; United Kingdom);
- a local ownership development project with a revolving loan fund to encourage employee-owned businesses, which are considered more stable over the long term and more likely to hire, train, and promote local residents (Burlington, Vermont); and
- a community beverage container recycling depot that employs street people—"dumpster divers"—and provides them with skills, training, and self-esteem (Vancouver).[48]

Closely related to CED is the sustainable livelihoods approach to poverty alleviation, which aims to address the immediate as well as the long-term needs of individuals and households and which takes into consideration the social and environmental as well as the economic sustainability of livelihood activities and strategies. The idea of sustainable livelihoods provides a framework to understand the practical realities and priorities of those struggling in poverty—that is, what they actually do to make a living, the assets they are able to draw on, and the everyday problems they face.[49]

How does the overall approach to economic development need to change in order to facilitate the development of strong local economies?

Beyond income generation, successful strategies under a sustainable livelihoods approach should serve to improve access to and control over local assets and help to make individuals less vulnerable to shocks and stresses (such as illness, natural disasters, or job loss) that could otherwise exacerbate situations of debt and poverty.[50]

Historically, the sustainable livelihoods framework has been used primarily in the context of rural poverty alleviation. But the same framework can easily be applied to situations of urban poverty and livelihood generation. In fact, a sustainable livelihoods approach is necessary in order to tackle issues of urban poverty over the long term. According to the ILO, 184 million people in the world do not have jobs, although this figure reaches at least 1 billion if underemployment is also taken into account.[51]

The concepts of community economic development and sustainable livelihoods together provide a useful framework for an alternative approach to economic development that emphasizes the development of strong local economies.

Will the cumulative effects of the small initiatives described here be enough to create communities and economies with enough resilience and strength to withstand the pressures and problems of an increasingly urbanized world? As noted in the beginning of the chapter, conventional approaches to economic development leave little room for strengthening local economies. Strong local economies are those that not only generate revenue but also take into consideration the equitable distribution of wealth within communities and the environmental implications of economic activities. How does the overall approach to economic development need to change in order to facilitate the development of strong local economies?

Cities, communities, and local economies are multidimensional, with a complex interaction of social, economic, ecological, and cultural factors. Some analysts think of local economies in terms of assets or capital. The term community capital, conventionally used to refer just to economic or financial capital, has more recently been used to include natural, physical, economic, human, social, and cultural forms of capital. Strengthening local economies means focusing attention on these six forms of capital:

- *Minimizing the consumption of essential natural capital* means living within ecological limits, conserving and enhancing natural resources, using resources sustainably (soil, air, water, energy, and so on), using cleaner production methods, and minimizing waste (solid, liquid, air pollution, and so on).
- *Improving physical capital* includes focusing on community assets such as public facilities (hospitals and schools, for instance), water and sanitation provision, efficient transport, safe and high-quality housing, adequate infrastructure, and telecommunications.

- *Strengthening economic capital* means focusing on maximizing the use of existing resources (using waste as a resource, for example), circulating dollars within a community, making things locally to replace imports, creating a new product, trading fairly with others, and developing community financial institutions.
- *Increasing human capital* requires a focus on areas such as health, education, nutrition, literacy, and family and community cohesion, as well as on increased training and improved workplace dynamics to generate more productive and innovative workers; basic determinants of health such as peace and safety, food, shelter, education, income, and employment are necessary prerequisites.
- *Multiplying social capital* requires attention to effective and representative local governance, strong organizations, capacity-building, participatory planning, and access to information as well as collaboration and partnerships.
- *Enhancing cultural capital* implies attention to traditions and values, heritage and place, the arts, diversity, and social history.[52]

Strengthening local economies requires mobilizing people and their governments to shore up all these forms of community capital. Community mobilization is necessary to coordinate, balance, and catalyze community capital. This approach to stronger local economies requires some relatively new thinking about broad questions of community sustainability and self-reliance, as well as more-specific innovations concerning community ownership, management, finance, organization, capacity, and learning. This approach is increasingly referred to as sustainable community development and includes both community economic development and sustainable livelihoods strategies.

While individual actions and lifestyle choices, such as buying organic produce, are important personal contributions, strengthening local economies requires a collective shift in individual actions and political choices. Community mobilization has been effective in some contexts and some regions. The cooperative economy of Emilia Romagna in northern Italy, the Grameen Bank in Bangladesh, Vancity Credit Union in Vancouver, the Women's International Sewing Cooperative of Nueva Vida, and the campaigns for local trade across North America are all examples of the potential of community mobilization to help strengthen local economies.

Strong local economies are a fundamental part of sustainable communities. They give communities the capacity and resources to address specific and immediate problems such as the provision of health care, adequate housing, clean water and sanitation, and disaster prevention and response. Human settlements—large and small, rich and poor—need strong local economies to withstand the pressures created by an increasingly urbanized world.

CITYSCAPE:
BRNO

Brownfield Redevelopment

Vaňkovka, the former engineering works of Friedrich Wannieck, became a Czech industrial architectural heritage site in 1992. When it was built in 1864, Vaňkovka stood outside the historic city center. But thanks to construction all around it, the site eventually found itself at the heart of a larger city downtown, on a busy street between the main bus and railway stations. During the industrial period, Brno was a center of technical invention, and Vaňkovka at its height represented the city's innovative and technological progress.[1]

Following the Velvet Revolution that brought an end to communist rule in 1989, traditional markets were lost. Central planners in Eastern Europe had built industrial plants in the city's core and high-rise housing estates along tram and bus lines on the outskirts. When production declined, nearly one third of the industrial zones—mostly situated in the southeast part of the city—became "brownfields": obsolete and permanently underused sites such as former factories and military bases.[2]

Brno now wants to hark back to its roots. In the early 1990s, the city began preparing the South Centre project to develop a new city quarter on the extensive underused land between the core of the historic city and the proposed new railway station. In the eastern section of this site stood the former Vaňkovka engineering works and iron foundry. The derelict buildings were a blight on the city's frequently traveled transportation corridors.

The idea to revitalize Vaňkovka was first put forward in 1993 during Brno's 750th anniversary celebrations. The following year a group of activists from various bodies set up a foundation, which later became the Vaňkovka Civic Association, with the aim of aiding the reconstruction and revitalization of the site for business, cultural, and educational purposes. The Project for Public Spaces, a nongovernmental organization (NGO) from New York, supported this work, recognizing that Vaňkovka could be transformed into a charming place that could spur the revitalization of the surrounding neglected land. Similar projects had been undertaken in the United States and Europe. And in many cases action was initiated by small groups of passionate and tireless people.[3]

Various community organizations, arts and educational institutions, private-sector donors, and the authorities—including the Brno City Council—formed partnerships to bring life back to Vaňkovka. The Chief Architect's Office conducted a survey to determine the conditions required for monument protection, as well as feasibility studies for new use of the area. The city government supported a privatization process to remove the area as a protected zone from the "package" of the residual property of the state-owned ZETOR company. The city agreed to pay 51.6 million Czech crowns ($1.34 million in 2000 dollars) to resolve the company's financial debts and get ownership of Vaňkovka.[4]

The Brno city environment department paid for research on cleanup to gauge the risk for future investors. The area was loaded with heavy metals, cyanides, polycyclic aromatic hydrocarbons, and oil residue, with decontamination estimated at 7–8 million Czech crowns ($181,000–207,000 in 2000 dollars).[5]

Exhibitions, theater performances, concerts, workshops, festivals, and children's activities were organized on the Vaňkovka premises, drawing many thousands of people. Most took place in the Pattern Shop and the Core Room of the old factory buildings. By the end of the 1990s, Vaňkovka had hosted more than 170 events and had become an

Libor Teplý, www.volny.cz/fotep

After renovation, the old foundry wall became the front of the NGO center

item on the cultural map of the region, with attention and support from abroad.[6]

Shortly after the city of Brno became the owner of the premises in September 2000, the German firm ECE presented plans to invest $120 million in the construction of a 37,000-square-meter shopping center. The firm accepted the Vaňkovka Civic Association's invitation to open discussions, bringing together Czech and foreign experts and members of the public to consult and contribute to the plan. While most of the block was transformed into the Galerie Vaňkovka shopping center, two buildings remained from the original monument-protected area—the Machine Works hall and the "office building"—while the Core Room and the facade of part of the Foundry were incorporated into the new building.[7]

When the shopping arcade opened in March 2005, a special section that is wheelchair-accessible from outside was made available for a cultural and community center for NGO activities. The name of this center, Slévárna Vaňkovka, calls to mind the former foundry. It consists of Cafe Práh ("halfway" cafe), an information center for young people, and an all-purpose hall. To renovate the two original buildings, the Brno South Centre city development company invested the funds that ECE had paid for its part of the site. The large space of the former Machine Works became an exhibition hall for contemporary Czech paintings. It is named after the old factory's founder—the Wannieck Gallery.

The planning process also initiated discussions concerning public spaces in and around the Vaňkovka complex—the "Vaňkovka streets"—and upgrades for the surrounding streetworks and public transport.

After more than 15 years, this piece of land at the center of Brno is once again alive, bringing the city benefits and income. The redevelopment of Vaňkovka, with its central location and public transport accessibility, has increased prospects for new developments on surrounding sites. Its success has made Brno a leader in forging a systematic policy approach to brownfields, while refocusing strategic priorities on redevelopment of land within the city. Vaňkovka has given life to the hopes of revitalizing other brownfield sites in the region.

The experience also demonstrates the importance of the creative and visionary role played by civic initiative and the nonprofit sector in regenerating brownfields. The Vaňkovka Civic Association not only documented Vaňkovka and made it more visible, it also helped solve a complicated ownership problem. These activities increased the site's social potential and its commercial attractiveness while reducing the pressure for greenfield construction. The association's innovative work to support cultural activities is now being replicated in brownfield redevelopment campaigns in other Czech cities.[8]

—*Eva Staňková*
Vaňkovka Civic Association

CHAPTER 9

Fighting Poverty and Environmental Injustice in Cities

Janice E. Perlman with Molly O'Meara Sheehan

"Cities, like dreams, are made of desires and fears," wrote Italo Calvino in *Invisible Cities*. If so, our cities range from daydreams to nightmares, depending on the city, the moment in time, and a person's position within the social and physical landscape. Cities, like regions and countries, have experienced uneven development exacerbated by government inability to offset the inequities produced by globalization. Within many cities, inequalities have deepened—between rich and poor, between included and excluded, and between the "formal" and the "informal" city.[1]

The informal city consists of squatter settlements, clandestine subdivisions, invaded residential and commercial buildings, provisional housing for refugees or migrant workers, and often degraded "social housing" complexes. These communities account for some 40 percent of the total urban population of the South, including 41 percent in Mumbai and 47 percent in Nairobi. Residents typically lack basic urban services (water and sanitation, electricity, paved roadways) and security of tenure, including official title to homes or land and freedom from eviction. Even where informal communities have urban infrastructure and de facto rights to use the land, they remain stigmatized spaces, while the low-cost labor of their residents helps sustain life's daydream for the privileged in the formal city.[2]

As housing prices in the formal city are prohibitive for the poor, they have no choice but to live in the most dangerous areas: on the streets, as is common in India; in alleyways outside wealthy homes, as in many Asian cities; on hillsides too steep for conventional construction, as in the *favelas* of Rio de Janeiro; on stilts in marshes, as in Bahia's *alagados*; on floodplains, as in many

---

Janice E. Perlman, a Guggenheim Award recipient, is the founder and President of the Mega-Cities Project, an international nonprofit, and a former professor of city and regional planning who consults widely on urban poverty and environmental justice issues.

of Jakarta's *kampungs*; atop garbage dumps, as in Manila; or even in cemeteries, as in Cairo. Families often remain through several generations and upgrade their homes and communities over time. Even young people who manage to enter university often have no place to live outside these "marginalized" spaces.[3]

Poor urban neighborhoods face the worst of two worlds: the environmental health hazards of underdevelopment, such as lack of clean drinking water, and of industrialization, such as toxic wastes. Yet their residents tread lightly on the planet, using few resources and generating low levels of waste in comparison with their wealthier neighbors. The gap between rich and poor in cities from Nairobi to New York means that those with the fewest resources suffer most from pollution generated by the wealthiest.

Those who advocate "sustainable development"—meeting the needs of people today without despoiling the planet for future generations—too often overlook the striking environmental injustice in our cities. The logical sequence linking global sustainability to urban poverty is synthesized in what have become known as the Perlman Principles:

- There can be no global environmental sustainability without urban environmental sustainability: Economies of scale in cities create energy and resource efficiencies. Transforming the urban metabolism through circular rather than linear systems is the key to reversing global environmental deterioration.
- There can be no urban environmental solution without alleviating urban poverty: The urban poor tend to occupy the most ecologically fragile areas of cities and often lack adequate water, sewage, or solid waste management systems.
- There can be no solutions to poverty or environmental degradation without build-

ing on bottom-up, community-based innovations, which are small in scale relative to the magnitude of the problems.
- There can be no impact at the macro level without sharing what works among local leaders and scaling these programs up into public policy where circumstances permit.
- There can be no urban transformation without changing the old incentive systems, the "rules of the game," and the players at the table.
- There can be no sustainable city in the twenty-first century without social justice and political participation as well as economic vitality and ecological regeneration.

A casual look at book titles throughout the 1960s and 1970s shows that the call for attention to the urbanizing world has been sounded for decades: *The Urban Explosion in Latin America, The Exploding Cities, The Wretched of the Earth, Uncontrolled Urban Settlement.* There may have been slightly more interest in urban poverty during the cold war due to fear that migrants and squatters might lead to leftist regimes. But once it was recognized that squatters were more interested in better opportunities for their children than in social protest, this interest fell off. It has only recently resurfaced in light of urban violence and security issues.[4]

Gradually, international agencies have begun to acknowledge the significance of cities and urban poverty. In 1999 the World Bank and UN-HABITAT formed the Cities Alliance to coordinate slum upgrading, in 2001 UN-HABITAT became a full-fledged United Nations program, in 2003 the international community agreed on a single definition of "slums," and in 2004 United Cities and Local Governments was formed, giving formerly competing local authority networks a unified voice.

These recent milestones are important, but the pace of change remains too slow.

One of the U.N. Millennium Development Goals explicitly focuses on urban poverty: improve the living conditions of 100 million slum dwellers by 2020 (Target 11, which is part of Goal 7 on environmental sustainability). Even if achieved, this would make but a small dent, as it aims at just 10 percent of the existing slum population, and an additional 1 billion people will be living in urban areas of developing countries by 2020. Why are we always playing catch-up?[5]

Barriers to Equitable Cities

Among the obstacles to reducing urban poverty and promoting environmental justice are inept and corrupt governance, violence, anti-urban bias, skewed development assistance, counterproductive incentives and resistance to change, and a lack of reliable city-level data necessary to benchmark progress.

Weak governance. The urban poor, generally excluded from decisionmaking, are the greatest untapped source of ideas about improving their cities and lives. Over the last several decades, mayors have been elected for the first time in many countries, arguably bringing the government closer to the people, but most poor people still do not have a voice in governance. Since decisionmakers tend to come from elite sections of society, they often have a vested interest in maintaining the status quo. Lawmakers in Kenya, for instance, who saw bicycles as toys for children rather than an important transportation mode, levied a luxury tax on bikes for years, keeping the cost too high for many low-income residents.[6]

Poor governance is reflected in uneven service delivery. World Bank Institute researchers have documented that people have less access to water, sewerage, and school-based Internet in cities that have a record of bribery in their utility companies, private firms that dictate local laws and regulations, and high levels of corruption nationwide. Ronald MacLean-Abaroa discovered this reality after becoming the first elected mayor of Bolivia's capital, La Paz, in the mid-1980s: "Whenever I found problems in service delivery or the prompt completion of public works or the collection of revenues, they happened not just to be associated with inefficient organization but almost always with corruption."[7]

Corruption, clientelism, and cronyism—the "three Cs" that undermine democracy—worsen urban poverty. Corruption skews public spending toward sectors where bribing is easier, such as large public works construction, and generally away from education, health, and maintenance of existing infrastructure. When people need to bribe officials to get needed services, those who can least afford the bribes suffer the most. A survey in Indonesia found that bribes required by police, schools, electricity companies, and garbage collectors ate into the already tight budgets of the urban poor.[8]

Violence and stigma. The increase in urban violence that has accompanied the rise in drug and arms trafficking has created particularly high mortality rates among urban youth. When dealers purchase complicity from the police for their illegal activities, they can hold entire low-income neighborhoods hostage. The consequent association of the urban poor and violence only serves to deepen the stigma that already constrains life opportunities for these people.[9]

A multigenerational study of families in Rio confirmed this trend. (See Box 9–1.) Fear of violence keeps people homebound, while job interviews end when the applicant's address is identified as in a *favela*. Among people of equivalent educational levels, those who lived in *favelas* had less success finding jobs.[10]

Box 9–1. Violence in Rio: Undermining the Urban Poor

A study of three favelas in Rio de Janeiro, Brazil, in 1968–69 and follow-up interviews with many of the original participants and their families in 2001–03 revealed that one of the most dramatic and devastating changes over the past 30 years has been the growth of lethal violence. In 1969, people were afraid their homes and communities would be removed by the military government. Today, they are afraid of dying in crossfire between drug dealers and police or between rival gangs.

Their fear is justified. In 2001, 20 percent of the original interviewees, 19 percent of their children, and 18 percent of their grandchildren reported that some member of their family had died through lethal violence—much higher rates than in cities in Colombia or Bolivia, which are drug-producing countries. Even when low-income families manage to move out of the favelas into public housing or peripheral but legitimate neighborhoods, the drug-related violence follows them. The death rates and life expectancies for young men in poor communities rivals those in countries in a civil war.

The poor feel trapped between the drug dealers and the police. In 2003, 81 percent said that neither helps them and that both commit violent acts with impunity. As the gangs are better financed and armed, it is easy to bribe the police. It is not uncommon for police to barge into low-income homes with the excuse of searching for a gang member and then tear the home apart and kill family members at random to demonstrate power and "instill respect." Seeking retribution from the courts is futile, and identifying the dealers is a death sentence, for the communities are entirely in the hands of the dealers once the police leave. People say

they must act like the three monkeys: they "see nothing, hear nothing, say nothing."

This "sphere of fear" has the devastating consequence of decreasing social capital, one of the few great assets for mutual assistance in getting ahead. Now people are afraid to leave their homes. As Nilton, a 60-year-old former favela resident who now lives in a housing project, put it: "To live here is to live in a place where daily you do not have the liberty to act freely, to come and go, to leave your house whenever you want to, to live as any other person who is not in jail. It is imprisoning to think: 'Can I leave now or is it too dangerous?' Why do I have to call someone and say that they shouldn't come here today? It is terrible, it is oppressive. Nobody wants to live like this."

Consequently, there is less sense of community unity, less use of public space, less socializing and trust among neighbors, and much less participation in neighborhood organizations. Almost all residents' associations are controlled by the drug and arms traffickers (except those controlled by the "death squads"), so that even the earlier low level of bargaining power of such groups is gone. And the link between favela residents and criminality has made it harder to get jobs or to hear about available odd jobs. The rental and sales values of homes—residents' greatest asset—has fallen, and service providers from teachers and health workers to nonprofits hoping to provide aid are afraid to enter the communities. Even commercial deliveries are no longer made. Those interviewed noted that marginalization and exclusion have only deepened since the end of the dictatorship.

SOURCE: See endnote 10.

Violent crime is a much greater threat in some places than others. Researchers have found the world's lowest levels of robbery and assault in Asia and the highest levels in Latin America and sub-Saharan Africa.[11]

Anti-urban bias. Environmentalists and development specialists have long portrayed cities as threats to nature. Many policymakers still adhere to this old mindset, pitting environmental concerns against economic growth and preventing further growth of cities, especially megacities. In 1986, a study commissioned by the U.N. Fund for Population Activities found that almost every nation had made some attempt to limit urban growth by investing in rural development, creating "growth poles" or "new towns," forcing people to relocate to undersettled regions, moving their national capitals, or introducing closed-city policies.[12]

Countries spent a good deal of money and political capital in these efforts, but none succeeded in stemming the tide of migration. Investment in rural roads, electrification, education, health, and industrialization, while important for improving rural living standards, in many cases just increased the rate of migration to urban areas. As people became more aware, they used their new roads and skills to seek wider opportunities in the city.[13]

Even where freedom of movement was highly restricted—such as colonial governments, command-and-control economies, and police states—people nonetheless found ways to sneak into cities. The most "successful" efforts to prevent urban growth were in apartheid South Africa, which required passports for non-whites to enter cities; China, which used rice ration cards and a household registration system; Russia, which used housing allocations; and Cuba, where a national policy of keeping people in the countryside was backed by the use of force. Yet "floating populations" of migrants entered Chinese cities, Moscow apartments became crowded with friends and relatives trying to move to the city, and governments everywhere kept inaccurate records of city size to maintain the fiction of non-growth.[14]

As recently as 2005, scores of nations were still attempting to curb urbanization. The United Nations Population Division recently analyzed migration policies of 164 countries and found that 70 percent aim to lower migration from rural to urban areas. This study confirmed that the impact on overall population distribution was "almost negligible."[15]

The myth that people will stop coming to cities if public housing is not built and squatter settlements are removed is unfounded and hurts the poor. Miloon Kothari, U.N. Special Rapporteur on Housing Rights, estimated in 2006 that the number of forced evictions had risen worldwide since 2000. "Without human rights safeguards," Kothari warns, "the commitment to the reduction of slums, including through the relevant MDGs, can easily become slum-eradication to the detriment of slum-dwellers."[16]

Skewed international assistance. Although virtually all of the world's population growth is expected in the cities of Africa, Asia, and Latin America, and most of this will be in low-income areas, development assistance has been reluctant to recognize the urbanization of poverty. From 1970 to 2000, all urban development assistance was estimated at $60 billion—just 4 percent of the total $1.5 trillion. Few bilateral aid agencies have any kind of urban housing program, or any serious urban program at all.[17]

The decisions of international development banks are important, even if aid is not the primary source of foreign investment in a given country. In recent years, aid has been roughly one tenth the level of private capital flows in developing countries. Yet international aid agencies leverage additional funding and influence the research agendas and spending priorities of governments, universities, and nongovernmental organizations (NGOs).[18]

Aid that does flow to urban areas often misses the poor. The London-based International Institute for Environment and Development found that urban projects accounted for 20–30 percent of all lending at several agencies from 1981 to 1998. But housing, water, sanitation, and other services that improve conditions for the urban poor received just 11 percent of total lending at the World Bank, 8 percent at the Asian Development Bank, and 5 percent at Japan's Overseas Economic Cooperation Fund.[19]

The World Bank reports that its lending and staff devoted to urban areas continues to lag behind the resources invested in the rural sector. All countrywide World Bank investments and most bilateral aid in developing countries are guided by Poverty Reduction Strategy Papers prepared by governments in consultation with the World Bank Group. These documents tend to neglect urban areas.[20]

The international assistance community cut its teeth on rural development and is professionally and structurally geared toward assisting rural peasants, not to continuing assistance for them when they move to cities. Development experts seem to prefer "missions" to attractive agricultural areas, fishing villages, and environmental preserves over those in the polluted, overcrowded, and often dangerous urban slums.

Pressure to bring more attention to urban poverty within the international development community has met with strong resistance. Every Executive Director of UN-HABITAT since its founding in 1978 has urged that we must act now, and yet none has succeeded in receiving funding parity with other U.N. agencies. When UN-HABITAT was headquartered in Nairobi along with the U.N. Environment Programme (UNEP) to assuage political pressures from African countries, it was no secret that these were considered the most expendable United Nations bodies. Although UNEP and UN-HABITAT have an impressive campus in Nairobi, both agencies remain politically and financially marginalized.

The voices for greater support for urban programs often face derision from those who believe that government spending is already pro-urban. But rigorous economic studies find that wealth generated in cities ends up subsidizing the countryside.[21]

Development assistance has been reluctant to recognize the urbanization of poverty.

On a more positive note, the World Bank is now considering subnational lending, which would allow loans to go directly to municipal governments, bypassing national finance ministries. This would help cities—particularly megacities—receive monies designated for them from international agencies that is now often held up by national governments that have imposed lending ceilings or that see political advantages in withholding funds. For instance, national governments in Brazil and Mexico do not want the mayors of their major cities to appear successful because they are potential competitors for the presidency and often from opposition parties.[22]

Counterproductive incentives and fear of change. The incentive systems of aid agencies are at odds with contextually specific, community-based, anti-poverty initiatives. Development professionals are promoted based on the size and rapidity of loans "pushed out the door," making one-size-fits-all approaches the best route to promotion, as opposed to smaller-scale projects where the priorities are set by local people. One analyst concluded that "the people whose needs justify the whole development industry are the people with the least power to influence development and to whom there is least accountability in terms of what is funded and who gets funded."[23]

The public sector is generally risk-averse. Elected and appointed officials often feel safer sticking with "the way things have always been done," even if results are suboptimal, rather than risk being fired or not re-elected for making a mistake. There is a high price to pay for an unsuccessful initiative—and little or no reward for innovation. As Alan Altshuler and Marc Zegans have noted, in the private sector the expectation that some ideas and initiatives will not be successful is built into the process, and funds are set aside for R&D where the entire point is to experiment and innovate.[24]

Federations of urban poor have pooled their money in savings groups to support community projects.

Fragmentation and competition among public agencies and academic disciplines further limits cooperation in solving urban problems. Each area—water and sanitation, transportation, housing, land use planning, private-sector involvement, poverty alleviation—is in a separate department, often run by people who may compete for resources, attention, or staff. Yet these issues are intertwined, so that an apparent solution to one may lead to new problems in another. Even when universities or international agencies have created interdisciplinary structures, individuals still have the strongest allegiances to their home department, where appointments, promotions, and payments are set.

Inadequate data for benchmarking. Lack of city- and neighborhood-level data makes it difficult to measure progress and hold governments accountable. Most statistics are only available at the national level, at best broken down between rural and urban, but not by specific cities. Where cities have managed to

mobilize resources to collect their own data, they often exclude informal settlements and are rarely comparable with earlier studies or with data from other cities. The Global Urban Observatory created by UN-HABITAT in 1998 to address this problem has had little success in finding indicators comparable across cities, despite its Web-accessible database of 237 cities that covers measures of poverty, environment, infrastructure, urban services, shelter, and land.[25]

Apparently neutral questions such as what to measure, what indicators to use, how to collect reliable data on these indicators, how to make sense of the results, and how to disseminate the results are in fact value-based issues that have political and social implications. Answers to these urban indicators questions will be a major topic of debate at the next World Urban Forum in Nanjing, China, in 2008.

Signs of Hope

Despite these obstacles, the energy and creativity in cities has generated scores of system-challenging innovations, many of which have spread from one place to another or—politics permitting—have been scaled up into public policy. Three examples described in this section are urban poor federations, which started in Mumbai, India; land-sharing, which started in Bangkok; and participatory budgeting, which started in Porto Alegre, Brazil.

Over the last two decades, a new kind of federation of urban poor has emerged from grassroots savings groups. The member groups, often led by women, learn from and support each other. The catalyst for this development was the National Slum Dwellers Federation in India. Its head, A. Jockin, was a community leader who had long fought to protect his own and other "slums" from being bulldozed.[26]

In the early 1980s, Jockin and other slum leaders in India demonstrated the capacity of their groups to lower costs and reach more people with their own form of housing and basic service programs. At the same time, women pavement dwellers in Mumbai supported by a local NGO called SPARC (the Society for Promotion of Area Resource Centers) created a savings program (Mahila Milan) that each woman contributed a few cents to every week. These three organizations formed an Alliance that since 1985 has been demonstrating to governments how to address the needs of slum dwellers in concrete projects from new housing to community mapping and public toilets. (See also Chapter 2.) The Alliance has strong links with the Asian Coalition for Housing Rights, and both work to support organizations of the urban poor in different Asian nations.[27]

Representatives from this Alliance were invited to South Africa in 1991 to help community leaders consider how that country's first democratically elected government should address housing issues. The visitors helped set up what is now the South African Federation of the Urban Poor, formed around indigenous savings groups, which began housing projects to demonstrate the capacity of their members. These groups, along with the Asian Coalition for Housing Rights, also connected with urban poor organizations and networks in Thailand. During the early 1990s, other national federations formed, drawing support from the already established federations. Most, like the Indian and South African federations, had a local NGO to support them.[28]

In 1996, with active federations in Cambodia, India, Namibia, Nepal, South Africa, Thailand, and Zimbabwe and with interest from community organizations in many other nations, the federations formed their own umbrella group: SDI (Slum/Shack

Dwellers International). Groups in Brazil, Ghana, Kenya, Malawi, Sri Lanka, and Swaziland have since joined SDI. Many have programs that have reached tens of thousands of members with better housing and basic services; some have reached hundreds of thousands.[29]

Federations of urban poor, and the NGOs that work with them, have devised numerous ways to tackle common obstacles. They have pooled their money in savings groups to support community projects. They have overcome the lack of data on informal settlements by doing their own censuses and community mapping. Their persistence has often prodded into action governments that were resisting change.[30]

Another important innovation—land sharing—began in Thailand. In the 1980s, NGOs and National Housing Authority architects began organizing residents of informal settlements facing eviction to negotiate with the owners of the land they occupied in Bangkok. Their goal was to enable squatters to get new apartments with full services and secure land tenure on a part of the disputed land in exchange for returning the rest of the land for commercial development. By agreeing to such land sharing arrangements, the landowners could avert years of conflict, lost profits, and loss of face, and they were able to develop market-rate projects on a portion of their property, the profits of which easily paid the costs of building the multi-family dwellings and infrastructure.[31]

The Thai government worked with one of the local leaders, Somsook Boonyabancha, to set up the Urban Community Development Office (UCDO) in 1992 and to bring participatory development to more urban poor neighborhoods. The board of this new agency included senior government officials, academics, and community representatives. Beginning with a $50-million capital fund,

UCDO was able to make low-interest loans to organized communities to buy land, construct or improve homes, or start small businesses. It encouraged communities to form savings and credit groups to hone their management skills and bring their own resources into development projects.[32]

In 2000, UCDO merged with the Rural Development Fund to become the Community Organizations Development Institute, which launched a huge campaign in 2003 to develop good housing, infrastructure, and secure land tenure with 300,000 households in 2,000 poor communities in 200 Thai cities within five years. By the time of the September 2006 coup d'état in Thailand, 450 community upgrading projects in 750 communities in 170 towns and cities were under way or completed, reaching more than 45,500 households.[33]

The ambitious scale of these Thai efforts has yielded results. In 2006, UN-HABITAT found that Thailand was among a handful of nations that are on track to meet the Millennium Development Goal of improving the lives of slum dwellers. The number of slum dwellers in Thailand has fallen by 18.8 percent a year, and UN-HABITAT credited community-driven upgrading supported by the government for this encouraging development.[34]

A third notable innovation is participatory budgeting, which was started in Porto Alegro, Brazil, when the city government used the 1988 post-dictatorship constitution to involve the urban poor in priority setting for community-level budget appropriations. The Workers' Party Mayor made public the existing neighborhood allocation of service delivery and investments and asked that participatory processes be used to set priorities on the use of municipal funds in each community. Once every penny of expenditure had been negotiated within and among the neighborhoods, corruption and clientelism became impossible to hide.[35]

Participatory budgeting has resulted in greater investment in poor neighborhoods, with greater targeting to issues identified by the poor as top priorities. A survey conducted after the first year revealed that most of Porto Alegre's poor people wanted clean water and toilets, whereas the government previously assumed that the top priority was transportation. After initial success in creating solidarity rather than conflict among neighborhoods of rich and poor, participatory budgeting spread to other cities in Brazil, the rest of Latin America, and elsewhere. (See Box 9–2.)[36]

While participatory budgeting has been adapted internationally, many cities lack the legal framework and strong local government needed to practice it. In 2002, researchers commissioned by the Inter-American Development Bank concluded that for participatory budgeting to happen, cities must have a competent local government committed to including the urban poor. This is generally not the case.[37]

Even when the political will exists to implement participatory budgeting, many obstacles remain. Yves Cabannes of University College London identifies the following as key challenges: How can municipalities maintain participatory budgeting after it is started by a political party that leaves office? How can it be sustained in conditions of scarce local budgetary resources? How can cities mobilize local resources to meet the expectations of those who participate so that false expectations do not backfire? How can cities expand participation to include the poorest and most excluded, especially young people, women, the elderly, and migrants? And how can cities best learn from the immense variety of experiences with participatory budgeting?[38]

Box 9–2. The Spread of Participatory Budgeting

In Brazil, where participatory budgeting was born, expansion is slowing but the process is reaching a greater range of cities. By 2006, some 200–250 municipalities in Brazil had participatory budgets. This process initially was used in Brazil's wealthier south and center, but now it has reached northeastern capitals, including cities with high poverty, such as Fortaleza, Recife, and Aracaju. Participatory budgeting is no longer unique to the Workers' Party, as parties from the center to the extreme left are adopting the process.

Peru has also passed and implemented comprehensive participatory budgeting legislation. Over the past five years, more than 800 local, regional, and provincial governments there have started to discuss budgets with residents. Although this "top-down" approach suffers from rigid procedures and a uniform system for civil society representation that does not always fit the local context, many initiatives are flourishing, supported by the Red Peru network and nurtured by Peru's strong participatory culture.

In Colombia, the U.S. Agency for Interna-

tional Development is supporting participatory budgeting through its Plan Colombia, and local governments, including in Pasto and Medellin, are separately experimenting in their own ways. Venezuela, Uruguay, Chile, and Bolivia all have the topic on their policy agendas.

Since 2003, the Municipality of Porto Alegre has coordinated a network, supported by the European Union, that includes over 350 cities and 100 institutions from Europe and Latin America. At the World Urban Forum in June 2006, participants from more than 30 countries attended a workshop on participatory budgeting; earlier that month, people from Africa, Asia, and Eastern Europe traveled to Brazil for a three-city workshop on the subject. Between 2000 and 2006, the number of municipalities with participatory budgets grew from 200 to roughly 1,200 worldwide. In Europe, more than 50 cities have adopted this approach, and initiatives are under way in Asia, Africa, and North America as well.

SOURCE: See endnote 36.

Sharing Approaches That Work

The experiences just described belie stereotypes about the incompetence of the urban poor, about the inability of community groups to collaborate with governments and international agencies, and about the narrow self-interest of urbanites seeking benefits for their own neighborhoods at the expense of the common good. Urban poor federations, land sharing, and participatory budgeting are only three of many innovations that should provoke policymakers to rethink whom the city is for and how it works—who sets the policy agenda, what groups are included in decisionmaking, how conflicts are handled, and which incen-

tive structures generate progressive change.

The Mega-Cities Project was founded in 1987 to address these issues and to stimulate transformation of policy and practice from the bottom up. Hundreds of successful innovative solutions are bubbling up in cities worldwide. The Mega-Cities strategy is to discover and uncover these, document and disseminate them, and scale them up into policy (where possible) or transfer them worldwide to cities seeking to adapt them and jump-start problem solving of their own. As a result, the lag time between ideas and implementation is shortened and the effects of approaches that work are multiplied.

The need to share workable solutions was highlighted by the results of a Global Leaders Survey conducted by Roper Starch International. Although 96 percent of leaders from all sectors agreed that their cities faced similar problems to other cities, and 96 percent said they would benefit from sharing information, only 11 percent felt very knowledgeable about what other cities were doing to solve problems. Despite the availability of instant messaging, Web sites, blogs, and data retrieval at the touch of a keypad, leaders have little time or propensity to go online for solutions to their problems. They need to hear a new idea in person from a peer or see it for themselves.[39]

Mega-Cities teams in the 21 largest cities of the world facilitate these face-to-face encounters. They include network leaders from NGOs, grassroots groups, government, business, academia, and the media. These teams detected, visited, and documented more than 400 successful innovations at the intersection of poverty and environment and brokered the transfer of 40 of them across national, city, and neighborhood boundaries. This section describes three such transfers— in Cairo, Rio de Janeiro, and Curitiba—as illustrations of both the power of mutual learning and the difficulties of sustaining new practices that threaten vested interests.[40]

Traditionally the Zabbaleen, Cairo's version of India's "untouchables," were trash collectors. They lived in an abandoned quarry that had become the city's garbage dump, and they earned a meager livelihood from selling waste products by the ton to brokers. In the 1970s, a consulting firm called EQI and an affiliated nonprofit began working with the Zabbaleen to convert trash into marketable products through micro-enterprises based in their homes. Scrap metal was worked into elaborately etched trays, fabrics were used to weave colorful placemats and quilts, plastics were recycled into shoes and games, and organic waste was composted.[41]

The success of these endeavors, which cap-

tured the value added of the final products, drew support from government and international institutions, facilitating the total upgrading of the community. People moved out of the dump into apartments with standard urban services, and the children left the donkey carts and entered school. Children who attended school could apprentice in the craft of their choice and earn income from their production. After centuries, the Zabbaleen began to overcome the stigma attached to trash workers.

In 1994, at a Mega-Cities Project meeting in Jakarta, Manila coordinator Me'An Ignacio realized that this approach might be adapted to the Payatas settlement in Manila and arranged to bring a delegation from there to the 1995 meeting in Cairo to study the Zabbaleen. The group returned inspired to study the contents of their local garbage stream and decided to start with a guild for paper recycling. They were successful in producing and selling all sorts of paper products and helped start other small enterprises, using plastic foam insulation and other waste products. The idea spread to other communities in Metro Manila and around the country and helped bring many families out of dire poverty. Today recycling continues on a small scale in Manila, but garbage dumping has become a source of large kickbacks to local officials, so they discourage these grassroots recycling efforts.[42]

In Cairo, the Zabbaleen have continued to generate income through recycling but have not been integrated into the city's waste collection system as hoped. In fact, globalization has entered their lives in recent years, as international waste collection businesses have been given contracts for garbage collection in various parts of the city and have pressured the Zabbaleen to move out. The international companies burn the garbage rather than reusing it; the Zabbaleen, in contrast, manage to recycle 85 percent of what they collect. At the moment one third of Cairo has no garbage collection at all. As Mona Serageldin, an architect from Cairo who teaches at Harvard, suggests, the waste problem there will not be solved until the traditional and modern waste collection systems are linked.[43]

Another innovation comes from Rio, where about one third of the residents live in *favelas*, many of which are too steep for conventional construction. Each year during the rainy season, floods and mudslides wash out hundreds of their homes, killing many people and polluting communities with sewerage overflow that then runs into Guanabara Bay. In 1986, the Municipal Secretariat of Social Development launched Projeto Mutirão (reforestation), which combined planting vegetables and fruit trees on upper slopes to prevent erosion and further settlement with installing underground sewerage pipes to separate human waste from open drainage canals. Community leaders managed the process, hiring qualified unemployed residents, and for the first time in history the city paid for the labor—from cultivating seedlings, planting trees, and maintaining new green areas to digging trenches and installing pipes.[44]

At the 1992 Mega-Cities Project meeting in Rio, Jakarta coordinator Darrundono was inspired to integrate elements of Rio's progam into the Kampung Improvement Project (KIP), an upgrading initiative that he directed. The core idea of environmental and nutritional awareness and of the importance of greenery to the quality of life in poor settlements was integrated into KIP, with the additional elements of installing communal washing and toilet facilities on sites of garbage dumps and encouraging each family to cultivate edible plants in the small spaces around their homes. But the Indonesian program did not adopt the practice of paying residents for their labor.

Back in Rio, the lessons from this experience evolved into Favela-Bairro, the most extensive squatter upgrading project ever undertaken. From 1995 to 2004 the reforestation program reached 87 communities and 250,000 people, and by 2006 Favela-Bairro had reached 556,000 people in 143 *favelas*. Workers are paid $150–500 a month, and some work at three nurseries the city supports, producing 120,000 trees a month. More than 1,500 hectares have been planted with some 3.5 million trees, which helps to generate income, provide food security, improve air quality, and prevent flood disasters.[45]

In Cairo, international companies burn the garbage rather than reusing it; the Zabbaleen, in contrast, manage to recycle 85 percent of what they collect.

The success the city government had in paying community residents through local leaders was transferred into several other projects, the most extensive being the garbage collection initiative, Favela Limpa (Clean Favela): the municipal waste collection company now hires *favela* residents to collect waste door-to-door and bring it to collection points for municipal trucks.[46]

A third transfer involves Curitiba, Brazil. In the 1970s, Mayor Jaime Lerner developed a bus system that was quick, comfortable, and convenient enough to draw riders out of their cars. (See Chapter 4 for further details.) Building and zoning regulations were adapted to allow for higher density construction at bus intersections, and transportation within and between neighborhoods was fully integrated. In addition, in poor communities inaccessible to garbage trucks, residents received free bus tickets for bringing their trash to the main road, and old buses were used as mobile classrooms for computer literacy.[47]

In 1991, New York City and the state government were seeking ways to reduce automobile emissions, as the city risked losing federal funding because it was out of compliance with the Clean Air Act. The Mega-Cities Project was asked to present some measures used successfully in other cities. A private foundation offered a challenge grant that would cover the costs of the innovation transfer if all the players could agree on one initiative. New York officials chose an environmental education program called Alert II, which linked car emissions with air pollution and public health risk in São Paulo, along with the Curitiba bus system.[48]

Grassroots groups have been the most fertile source of innovation, followed by local governments.

In early 1992 the Mega-Cities Project brought New York City Commissioners of Environment and Transportation to São Paulo and Curitiba to study these initiatives. The result was a public awareness event called "Green Alert" on Park Avenue, and, with financial help from Volvo, four of the Curitiba buses and loading tubes were given a test run in four key locations in lower Manhattan.[49]

The trial received good evaluations from those who used it, but the transfer lost momentum when the city administration changed in 1993 after an election. Critics could cite many reasons that Curitiba's system would not work in New York. The demonstration there had given people free trips, so one key element—prepaid boarding—was not tested. And the staffed turnstiles for payment at every bus station would be expensive to replicate in New York. Unlike

Curitiba, New York has a subway and narrow streets, which make it hard to dedicate full-time street space for buses. Also, the responsibility for New York streets is split between the state-run transit authority that operates buses and the city transportation department that manages street signs and signals, which are key to giving buses priority.[50]

The idea never died, however: it stimulated interest at the Federal Transit Administration, which then encouraged bus rapid transit in Los Angeles and other cities. In 2001, when New York Mayor Michael Bloomberg first ran for election, "subways on the surface" was one of his campaign ideas. And in 2006 officials considered a similar system on several corridors throughout the city. The pilot system anticipated for operation in 2008 will have drawn some of its inspiration from that Curitiba demonstration years ago. For the Mega-Cities Project, it was important to have one of its first transfers occur from South to North to show that learning could be the reverse of what many people expect.[51]

These cases and hundreds of others documented and disseminated by the Mega-Cities Project provide lessons about the process of social change. Grassroots groups have been the most fertile source of innovation, followed by local governments. To reach significant scale, innovators need to work in partnership with NGOs, business, and government. To overcome the myriad obstacles during the innovation life cycle, from origin to routine practice, a "product champion" needs to provide passion and continuity. Innovation transfer works best when the learning is peer-to-peer, when there is a face-to-face visit, and when continuity can be assured. (The rapid change of administrations makes transfers tricky between local governments, as they are prone to reject an idea they did not initiate.) But the biggest lesson is that social change is a Sisyphean struggle. While community groups and progressive governments struggle to push the boulder of innovation up the mountainside, the gravitation pull of "business as usual" is always working in the other direction.

Promising New Directions

Our global future will be urban, like it or not. People in developing countries will continue to vote with their feet, moving to cities or the urban fringe. It will be many decades before cities of the global South reach a stable population. Although birth rates decline with urbanization, this happens more slowly with the urban poor.

What can be done to make our urban future a desirable and sustainable one? What kinds of cities foster conviviality and creativity? How can poverty and environmental degradation be alleviated and a voice for the disenfranchised be ensured?

There is no magic bullet for creating sustainable, equitable, and peaceful cities. But there are some necessary if not sufficient conditions for such transformations: transparent governance, decent work or a basic income, innovative infrastructure to conserve the environment, intelligent land use with integrated community development, and social cohesion along with cultural diversity.

Foster transparent governance. Effective governance is essential to scale up promising innovations into public policies, to provide basic services equitably, and to forge partnerships with the private and voluntary sectors. Addressing the World Bank Board in Singapore in September 2006, Bank President Paul Wolfowitz emphasized this point: "Without [good] governance, all other reforms will have limited impact.... It is the view I have heard on sidewalks and in taxis—in the marbled halls of ministries and in rundown shacks of shantytowns." [52]

Tackling the corruption that weakens governance requires fostering competition so that governments do not have monopoly power, reducing bureaucratic leeway, and increasing accountability. For instance, La Paz began to reduce bribery in construction permits by simplifying and publicizing the rules, contracting the permitting out to architects, and reducing the city's job to overseeing the contracts—a job that could be done by fewer municipal employees who could be paid more. Some promising efforts to foster transparent governance make government rules, purchases, and investments public knowledge by posting them on the Internet. [53]

Ensure decent work or a basic income. Jobs are a top priority for the urban poor. In 2001, the majority of interviewees in the multigenerational study of *favela* residents in Rio said that "the most important factor for a successful life" was a good job with a good salary. They want a chance to earn their livelihoods whether as employees or informal workers—that is the key to their dignity. What is needed is both job creation and preparation of people within their communities for jobs that exist in market sectors that are growing. [54]

Job and skills training, mentoring, and help in finding a first job can play a major role if done correctly. There is no use building people's capacity for jobs that no longer exist.

Savings and credit are key to job creation. Without access to these financial tools, would-be entrepreneurs cannot start small businesses. Many forms of microfinance, including community savings and credit funds as well as small loans to make improvements to housing, have been successful in urban settings. (See Chapter 8.) [55]

Larger companies can also play a role in strengthening the economies of poor communities. In Guadalajara, Mexico, where a good share of the population lives in unplanned settlements, a large, multinational cement company, CEMEX, developed a savings-and-credit scheme to allow households that earn $5–15 a day (low-income families, but not the poorest of the poor) to buy materials to build and improve their housing. The program has since expanded to 23 cities in Mexico.[56]

In some cases, governments can help create local jobs by hiring the urban poor to help solve pressing environmental problems, as reforestation in Rio demonstrated. In low-income communities in Dar-es-Salaam in Tanzania and Kampala in Uganda that were damaged in flooding, local governments used "community contracts" to get local labor for the necessary rebuilding.[57]

There is also a role for a "negative income tax" to help in periods between jobs or to supplement incomes too low to live on. Mexico and Brazil started innovative "conditional cash transfer programs" in which the government directly deposits funds into personalized debit cards for low-income people as an incentive for desirable action. For example, a family gets a certain amount of cash for children as long as they attend school regularly, gets other funds upon proof of immunization against contagious diseases, and receives further aid if there are elderly or infirm people in the family. Such programs have led to higher school enrollment and better preventive health care, especially in Latin American cities, and New York's mayor now plans to try the approach.[58]

Develop innovative infrastructure to conserve the environment. Cities that do not yet have full infrastructure have the chance to "leapfrog" over outmoded and wasteful systems created during the Industrial Revolution. They can take advantage of resource-conserving technologies, both low- and high-tech, to revolutionize the built environment. Examples include installing water-conserv-

ing toilets, separating drinking and greywater into different systems, using passive solar energy or biogas for heating, and adapting the recycling technologies developed by NASA for life in outer space. Architect William McDonough is working on low-cost housing projects in China that use biogas from wastewater for cooking, local compressed earth as building materials, and passive solar design for heating and cooling. These sorts of technologies are also being used in Johannesburg. (See Box 9–3.)[59]

One challenge in adopting low-impact or "alternative" technologies is that they do not carry the prestige of "modernity" seen on televisions and in houses of the rich around

---

Box 9–3. Circular Technologies in Johannesburg, South Africa

In the low-income neighborhood of Ivory Park, Johannesburg, residents are discovering new job opportunities with technologies that turn wastes into resources. Local NGOs—EcoCity and GreenHouse People's Environmental Centre Project—have demonstrated the viability of locally available rammed earth building materials, solar water heaters and cookers, waterless sewage systems, recycling, and small-scale organic agriculture. Together with the Co-operatives and Policy Alternative Centre, EcoCity has helped set up more than a dozen "green" cooperative businesses. Clara Masonganye, from the women eco-builders cooperative Ubuhle Bemvelo, says the green homes "are warm in winter and cool in summer." Midrand Eco Savings and Credit Cooperative allows the local savings community to decide how to allocate loans.

—Annie Sugrue
EcoCity Trust, Johannesburg

SOURCE: See endnote 59.

the world. While cutting-edge neighborhoods in places like Stockholm show that eco-friendly design is compatible with a high standard of living, most people in developing countries do not see these models and continue to aspire to the worst mistakes of the U.S. and European systems.

Promote intelligent land use and integrated community development. Urban planning is making a comeback after decades of being dismissed as the province of useless colored maps. Creative urbanists have found new ways to involve communities in the trade-offs of physical planning decisions and to use old planning tools for progressive change. Zoning, building, and land use regulations have been adapted to foster mixed-use communities, with homes close to workplaces, commerce, and recreation. Incentives for development in areas with existing infrastructure have helped limit sprawl. "Areas of special interest" have been established to protect environmentally important areas, connect nature corridors, and allow the upgrading of informal settlements in flexible ways. Comprehensive transportation planning has now included investments in sidewalks, bicycle paths, and low-cost public transportation options with easy connections between local, regional, and longer-distance travel.

São Paulo has been among the leaders in advancing urban planning tools to create a more inclusive city. One of the biggest steps there was taxing building developers to create a fund for investments in the public interest, including public transportation, housing, and environmental improvements. (See Box 9–4.)[60]

Low-income housing options are so limited in most cities that new migrants end up living in the most dangerous places. A logical response would be to lay out small plots of land with connection to basic urban services, available for small sums or loans. This

---

Box 9–4. Planning for the Public Interest in São Paulo, Brazil

In São Paulo's plan, approved in 2002, the administration negotiated a change in the floor ratio, which dictates how large a building on a given lot can be. Throughout the city, the ratio that developers can build for no additional cost is now kept to 1, meaning the total area on all floors of a new building must be equal to the lot area. But in areas where zoning allows floor ratios of 2, 3, or 4, developers who wish to build more can pay for a permit. The proceeds go to a public fund for environmental, transport, and housing purposes.

The developers' association published four pages of ads against the proposed plan in all local newspapers. The administration fought back and the innovation was eventually agreed upon. But at the City Council several changes diminished the proceeds that could be expected from the new permit system. It is now the law, however, and developers have to abide, which they can do without any significant harm to their profits.

When a new administration has a different approach to social problems or another set of priorities, it puts less emphasis on some issues, even though they are enshrined in the law. During the Workers' Party administration (2001–04), for instance, São Paulo created an Urban Policies Council of 40 members, representing both government and NGOs. It is required by law. But the current administration has never convened the council.

Developers will always try to change the law in order to build more and more; it is in the logic of their business. But as Jean-Jacques Rousseau wrote, "the public interest is not the interest of all."

—*Jorge Wilheim*
Former head of São Paulo's
planning department

SOURCE: See endnote 60.

"sites-and-services" approach, first tested in Dakar, Senegal, in 1972, has been little used, as it does not offer politicians the chance for ribbon cuttings and photo opportunities. Yet it makes monetary and environmental sense: the costs of infrastructure are small compared with retrofitting a squatter community, and the new settlements help reduce the invasion of environmentally sensitive areas.

Cultivate social cohesion and cultural diversity. Diversity makes natural ecosystems and human economies more resilient, yet prejudice and misunderstanding between different groups of people often squanders the potential of cultural diversity to strengthen cities. Solving the complex problems that our cities face will require the greatest diversity of cultures and value sets possible.[61]

As violent crime rips apart the urban fabric and further isolates the poor, there is a need to focus on controlling the sale of arms and drugs, reducing the corruption that permits violence with impunity, and mobilizing the society at large to find solutions to the problem. Promising initiatives include efforts at community policing in low-income neighborhoods and all varieties of arts, culture, and sports programs for young people at risk. Some programs of arms amnesty and prevention of small arms sales have also been tried, with varying degrees of success in lowering urban violence.[62]

Our Urban Future

The directions in politics, economics, environmental policy, and society just described require at least three fundamental changes. The first is to revise the architecture for support to cities and the urban poor, giving cities their due and reversing the reward systems to promote innovation. As the global population shifts toward cities, the agendas of aid agencies, national governments, foundations,

research centers, and nonprofit groups need to reflect this reality. All too often in the discourse on the future of cities, the focus remains exclusively on global or world cities in their role as centers of capital and information flows and as corporate headquarters rather than on the more numerous and populous cities of the developing world.

To bridge the divide between official sources of assistance and the urban poor, governments and aid agencies might channel their support to a local fund or community foundation in each city. The fund would be earmarked for use by community organizations, have transparent decisionmaking, and make it easy for groups that receive assistance to exchange ideas with each other. The Thai government's Community Organizations Development Institute offers one example of how national governments could contribute to local funds. The Swedish International Development Cooperation Agency has set up local organizations to run urban poverty programs in Costa Rica and Nicaragua. In Ecuador, the government reached agreements between 1988 and 1993 with foreign governments to restructure commercial debt instruments and then channeled the funds through local NGOs to finance development projects throughout the country.[63]

Rethinking development assistance includes finding ways for federations of the urban poor and NGOs working with them to bypass official aid channels. Two former World Bank staff members have started a Web site, www.globalgiving.com, that allows individuals and institutions to support projects run by local people. Analyzing this initiative, former Bank economist William Easterly writes: "Think of the potential for creativity if thousands of potential donors, project proposers, technical advisers, and advocates for the poor were freed from the shackles of the large centralized bureaucracy and could find solutions that worked on the ground. This is not a panacea for redesigning all of foreign aid; it is just one promising experiment in how aid could reach the poor."[64]

A second change needed is to create systematic ways for benchmarking progress and measuring outcomes in cities. Without reliable, comparable indicators of poverty and environmental conditions, we will never know whether progress is being made or how to compare the impact of one set of practices and policies with another. As international and national efforts have not yielded city-level indicators, the need for local benchmarking is clear. One possibility is for local governments to hire residents to collect health, housing, income, and environmental data. Federations of the urban poor, from Mumbai to Nairobi, have shown how this can be done, organizing communities to perform their own censuses. Cities can hold information collection fairs that would motivate people to collect data on their own areas of interest and responsibility.

Exchange of information is especially essential among those who are most directly involved in fighting urban poverty and among the urban poor themselves. The journal *Environment and Urbanization*, produced by the International Institute for Environment and Development, offers one important forum for researchers, NGO staff, and others to exchange ideas. There is also a need for face-to-face discussions among community leaders.

Some new projects aim to collect and share information on urban poverty and the environment. The International Development Research Centre in Canada has launched a program to study interventions in urban agriculture, water and sanitation, solid waste management, vulnerability to natural disasters, and land tenure as a cross-cutting theme in a handful of "focus cities" in developing countries. A new Urban Sustainability Initiative, supported by the Moore Foundation in the United States, is planning city partnership projects in China, Mexico, South Africa, and Tanzania, an exchange of ideas among cities, and a set of scientific and social indicators to gauge urban progress.[65]

Despite apparent differences in politics, economics, and culture, cities in developing countries and the industrial world have many problems in common.

In these and other initiatives it will be important to realize that rather than "best practices" and "competitive cities," what is needed is "better practices" and "collaborative cities." The "best practice" model is flawed by its implication of one ideal way for all communities. As the Mega-Cities Project discovered, each new innovation gives rise to new problems and contradictions, which require yet further innovations and revised solutions. Another lesson was that a "best practice" in one place may be useless or detrimental in another. Each location needs to adapt solutions to its own history, culture, and local circumstances. The current system of nominating, judging, and rewarding "best practices" allows for self-nomination and self-promotion but leaves little room for neutral external evaluation.

Despite apparent differences in politics, economics, and culture, cities in developing countries and the industrial world have many problems in common, often more than they share with small towns or villages in their own countries. Nearly every wealthy city contains within it neighborhoods with high infant mortality, malnutrition, homelessness, joblessness, and low life expectancy. And nearly every city in the developing world contains within it a world of high finance, high technology, and high fashion. If cities are to be used as laboratories for urban innovation, they can harvest ideas to exchange from the South to the North, since low-income cities have a lighter ecological footprint and have more experience in reuse. It is time to move from NIMBY (not in my backyard) and NOPE (not on planet earth) to the recognition that all by-products of human activities end up in someone's back yard and in the atmosphere that surrounds us.

The last basic shift needed is for people in positions of power to listen to the most vulnerable portions of the population, particularly young people and women. The cities of the future belong to the children of today. Unfortunately, a review of municipal efforts to incorporate children's concerns in decisionmaking found that "there is generally more interest in showcase projects" than in broader changes.[66]

Cities could scale up programs that expose young people to arts and sports and develop areas in which they can excel and feel part of something worthwhile. In Rio, one such program—Affro-Reggae, started in the *favela* Vigário Geral—has used drumming, dance, and song lyrics that expressed the community's reality to attract youngsters, build solidarity, and develop a critical analysis of their situation. Its work, chronicled in the documentary *Favela Rising*, helped defuse a drug war with the adjacent *favela* and has spread to other communities.[67]

One of the most articulate explanations of the need to listen to the urban poor comes from Rose Molokoane of the South African Federation of the Urban Poor, or FED UP. She recently told an audience that included development professionals: "We are fed up of being the subject of the agenda. We are fed up with you not listening to us.... We are poor, but not hopeless. We have money, but no chance to come to the bank and open an account because we have no address. If you give me security of tenure, then I have an address, and I will open account. We will show you we can do it.... The only thing we are concentrating on is how to organize ourselves. If communities are organized, they are a tool to address issues that are giving you double stress."[68]

The gulf that Rose Molokoane identified between those who set "development goals" and those who are the target of that agenda is a subset of larger rifts between rich and poor, between the powerful and the powerless. Bridging these divides will require a new mindset. Unless and until we are ready to expand our conception of "we" from "me and my family" to my community, city, country, and planet, the gap will continue to grow. We may have come this far through competition and survival of the fittest, but if we are to make the leap to a sustainable world for the centuries ahead, we will need to be intelligent enough to do it through collaboration and inclusion. In the words of Australian aboriginal elder Lilla Watson: "If you've come to help me, you're wasting your time. But if you've come because your liberation is bound up with mine, then let us work together."[69]

Notes

State of the World: A Year in Review

October 2005. "Mexico First in Latin America to Set Aside Wilderness," *Environment News Service*, 4 October 2005; "Pakistan Quake Rocks South Asia; Over 18,000 Killed," *New York Times*, 10 October 2005; "Pakistan Quake Toll Tops 73,000," *The Guardian* (London), 3 November 2005; United Nations University, Institute for Environment and Human Security, "As Ranks of 'Environmental Refugees' Swell Worldwide, Calls Grow for Better Definition, Recognition, Support," press release (Bonn: 11 October 2005); "An Underclass Rebellion—France's Riots," *The Economist*, 12 November 2005; Gabriel D. Grimsditch and Rodney V. Salm, *Coral Reef Resilience and Resistance to Bleaching* (Gland, Switzerland: IUCN–World Conservation Union, October 2005).

November 2005. World Wildlife Fund, "New Species Found in Fiji's Great Sea Reef, WWF Survey Reveals," press release (Washington, DC: 2 November 2005); "Fiji Chiefs Create Marine Sanctuaries on World's Third Largest Reef," *Environment News Service*, 4 November 2005; "Chinese Petrochemical Explosion Spills Toxics in Songhua River," *Environment News Service*, 13 November 2005; U.N. Food and Agriculture Organization, *Global Forest Resources Assessment 2005* (Rome: 13 November 2005); U.N. Environment Programme (UNEP), "West Africa's 'Forgotten' Elephants Remembered at CMS Meeting," press release (Nairobi: 22 November 2005).

December 2005. National Center for Atmospheric Research, "Most of Arctic's Near-Surface Permafrost May Thaw by 2100," press release (Boulder, CO: 19 December 2005); World Trade Organization, "Day 6: Ministers Agree on Declaration That 'Puts Round Back on Track,'" news feature (Geneva: 18 December 2005); "Seven States Agree on a Regional Program to Reduce Emissions from Power Plants," *New York Times*, 21 December 2005.

January 2006. "UN International Year of Deserts and Desertification Opens," United Nations News Centre, 1 January 2006; "Oil and Gas Zones Cover One-Quarter of the Peruvian Amazon," *Environment News Service*, 4 January 2006; The White House, "Fact Sheet: The Asia-Pacific Partnership on Clean Development and Climate," press release (Washington, DC: 12 January 2006); "Donor Nations Pledge $1.85 Billion to Combat Bird Flu," *Environment News Service*, 18 January 2006; U.S. National Aeronautics and Space Administration (NASA), "2005 Warmest Year in Over a Century," news feature (Washington, DC: 24 January 2006).

February 2006. European Commission, "Maximum Levels Set for Dioxins and PCBs in Feed and Food," press release (Brussels: 3 February 2006); Edward Alden and Jeremy Grant, "WTO Rules Against Europe in GM Food Case," *Financial Times*, 8 February 2006; "Brazil Expands Amazon National Park, Creates Forest Reserves," *Environment News Service*, 15 February 2006; Sebastien Berger, "1,500 Feared Dead in Mudslide Village," *Daily Telegraph* (London), 18 February 2006; Karen R. Lips et al., "Emerging Infectious Disease and the Loss of Biodiversity in a Neotropical Amphibian Community," *Proceedings of the National Academy of Sciences*, 28 February 2006; "Deadly Fungus Wipes Out Central American

Amphibians," *Environment News Service*, 7 February 2006.

March 2006. "Bush and India Reach Pact That Allows Nuclear Sales," *New York Times*, 3 March 2006; NASA, "NASA Survey Confirms Climate Warming Impact on Polar Ice Sheets," press release (Washington, DC: 8 March 2006); Andy White et al., *China and the Global Market for Forest Products* (Washington, DC: Forest Trends, March 2006); International Maritime Organization, "International Rules on Dumping of Wastes At Sea to Be Strengthened with Entry into Force of 1996 Protocol," press release (London: 22 February 2006); "Stronger Rules Take Effect to Govern Dumping of Wastes at Sea," *Environment News Service*, 10 March 2006.

April 2006. "Tokyo Embraces Renewable Energy," *Environment News Service*, 6 April 2006; UNEP, "Restoration of Wetlands Key to Reducing Future Threats of Avian Flu," press release (Nairobi: 11 April 2006); "Record Danube Flooding," *Toronto Sun*, 17 April 2006; "Balkans Battle Flooding," *Calgary Sun*, 23 April 2006; World Bank, "World Bank: Full Debt Cancellation Approved for Some of the World's Poorest Countries," press release (Washington, DC: 21 April 2006); UNEP, "UN Secretary-General Launches 'Principles for Responsible Investment,'" press release (Nairobi: 27 April 2006).

May 2006. IUCN, "Release of the 2006 IUCN Red List of Threatened Species Reveals Ongoing Decline of the Status of Plants and Animals," press release (Gland, Switzerland: 2 May 2006); "Brazil Officially Starts First Uranium Enrichment Facility," *Environment News Service*, 8 May 2006; Lydia Polgreen, "Violent Rebel Rift Adds Layer to Darfur's Misery," *New York Times*, 19 May 2006; John Hagan and Alberto Palloni, "Death in Darfur," *Science*, 15 September 2006, pp. 1,578–79; "Finally Feeling the Heat," *New York Times*, 24 May 2006.

June 2006. "Three Gorges Cofferdam Demolished," NewsGD.com, 7 June 2006, at www.newsgd.com/news/china1/200606070009.htm; "Japan Fails to Reverse Ban on Whaling," *Financial Times*, 20 June 2006; World Urban Forum 3,

"World Urban Forum III – Opens Monday!" press release (Vancouver: 19 June 2006); "Cameroon, France Sign Central Africa's First Debt-for-Nature Swap," *Environment News Service*, 23 June 2006.

July 2006. "More African Ivory Seized in Kaohsiung Harbor," *Central News Agency*, 6 July 2006; "Clean-up Crews Recover Some of Massive Lebanon Oil Spill," *Agence France Presse*, 19 August 2006; Group of Eight, "Global Energy Security" (St. Petersburg, Russia: 16 July 2006); "Governor, Blair Reach Environmental Accord," *Los Angeles Times*, 1 August 2006; "Japan Lifts U.S. Beef Import Ban Imposed Against Mad Cow Disease," *Environment News Service*, 28 July 2006; Brandon F. Keele et al., "Chimpanzee Reservoirs of Pandemic and Nonpandemic HIV-1," *Science*, 28 July 2006.

August 2006. Robyn Dixon, "Ivorians Incensed over Toxins," *Los Angeles Times*, 16 September 2006; "Ivory Coast Toxic Waste Death Toll Rises to 10," *Reuters*, 16 October 2006; Stockholm International Water Institute, "Experts from 140 Countries to Address Water, Environment, Livelihoods and Poverty Reduction in Stockholm," press release (Stockholm: 18 August 2006); Felicity Barringer, "Leaders Accept California Bill to Cut Emissions," *New York Times*, 31 August 2006.

September 2006. BirdLife International, "Bugun Liocichla: A Sensational Discovery in North-east India," press release (Cambridge, UK: 12 September 2006); WWF International, "Bluefin Tuna Overfished in the Mediterranean," press release (Gland, Switzerland; 12 September 2006); U.S. Food and Drug Administration (FDA), "FDA Warning on Serious Foodborne E. coli O157:H7 Outbreak," press release (Washington, DC: 14 September 2006); FDA, "FDA Announces Findings From Investigation of Foodborne E. coli O157:H7 Outbreak in Spinach," press release (Washington, DC: 29 September 2006); B. D. Santer et al., "Forced and Unforced Ocean Temperature Changes in Atlantic and Pacific Tropical Cyclogenesis Regions," *Proceedings of the National Academy of Sciences*, 19 September 2006; UNICEF, "Children Pay the Price for Lack of Safe Water and Sanitation," press release (New

York: 28 September 2006); Anna Dolgov, "Two of Five People Around the World Without Proper Sanitation," *Associated Press*, 29 September 2006.

Chapter I. An Urbanizing World

1. "World's first sustainable city" and projections of growth from Arup, "Dongtan Eco-City, Shanghai, China," 23 August 2005, at www.arup.com/eastasia/project.cfm?pageid=7047; solid waste recycling from Jean-Pierre Langellier and Brice Pedroletti, "China to Build First Eco-city," *Guardian Unlimited*, 5 May 2006; other data from Frank Kane, "Shanghai Plans Eco-metropolis on its Mudflats," *Guardian Unlimited*, 8 January 2006.

2. The fraction of electricity generated by fossil fuels and nuclear fission is estimated from the load served by Consolidated Edison in New York City and reported by the Environmental Disclosure Label Program of the New York State Public Service Commission, at www3.dps.state.ny.us/E/EnergyLabel.nsf/Web+Enviromental+Labels/A3655A006989E8EB8525719C0056A1A5/$File/CONED.PDF?OpenElement, viewed September 2006; waste to landfills and recycling from Kate Ascher, *The Works: Anatomy of a City* (New York: Penguin Press, 2005), pp. 190–94.

3. Ebenezer Howard, *Garden Cities of To-Morrow* (Cambridge, MA: The MIT Press, 1965, reprint); Rudolf Hartog, "Growth without Limits: Some Case Studies of 20th-Century Urbanization," *International Planning Studies*, February 1999, pp. 95–130.

4. Displacement of farmers from "Development Jockeys With Ecology on Shanghai Island," Reuters, 18 April 2006; wetlands disruption from Meg Carter, "Life, But Not as We Know It," *The Independent*, 4 May 2006.

5. United Nations Population Division, *World Urbanization Prospects 2005* (New York: 2006), also available online at esa.un.org/unup.

6. Paul Bairoch, *Cities and Economic Development: From the Dawn of History to the Present*, translated by Christopher Braider (Chicago: Uni-

versity of Chicago Press, 1988), chapters 12–15. Box 1–1 from the following: United Nations Population Division, *World Population Prospects: The 2004 Revision* (New York: 2004); Roy Porter, *The Greatest Benefit to Mankind. A Medical History of Humanity* (New York: W. W. Norton & Company, 1997), chapters 13 and 14; health conditions today from Alan D. Lopez et al., *Global Burden of Disease and Risk Factors* (New York: Oxford University Press for World Bank, 2006); urban penalty from National Research Council, *Cities Transformed: Demographic Change and Its Implications in the Developing World* (Washington, DC: National Academies Press, 2003), pp. 259–60, 271–72, 284, 259–60; per capita income from Angus Maddison, *The World Economy: A Millennial Perspective* (Paris: Organisation for Economic Co-operation and Development, 2003); technological changes from John Eberhard, "A New Generation of Urban Systems Innovations," *Cities: The International Journal of Urban Policy and Planning*, February 1990; George Modelski and Gardner Perry III, "'Democratization in Long Perspective' Revisited," *Technological Forecasting and Social Change*, vol. 69 (2002), pp. 359–76.

7. National Research Council, op. cit. note 6, pp. 89–92.

8. United Nations Population Division, *World Urbanization Prospects: The 2003 Revision* (New York: 2003), Table 14, p. 7; slum dwellers from UN-HABITAT, *State of the World's Cities 2006/7* (London: Earthscan, 2006), p. 16.

9. World Commission on Environment and Development, *Our Common Future* (New York: Oxford University Press, 1987), p. 43.

10. J. T. Houghton et al., eds., *Climate Change 1995: The Science of Climate Change*, Contribution of Working Group I to the Second Assessment of the Intergovernmental Panel on Climate Change (IPCC) (Cambridge, UK: Cambridge University Press, 1996); James J. McCarthy et al., eds., *Climate Change 2001: Impacts, Adaptation, and Vulnerability* (Cambridge, UK: Cambridge University Press, 2001); Habiba Gitay et al., eds., *Climate Change and Biodiversity*, IPCC Technical Paper V (Geneva: 2002); Millennium Ecosystem Assessment, "Summary for Decision-Makers," *Ecosystems and Human Well-Being: Synthesis* (Washington: Island Press, 2005), p. 1.

11. UN-HABITAT, op. cit. note 8, pp. 46–47; National Research Council, op. cit. note 6, pp. 164–80.

12. National Research Council, op. cit. note 6, pp.132–35, with quote from p. 135; implications of changing definition from David Satterthwaite, *Outside the Large Cities: The Demographic Importance of Small Urban Centres and Large Villages in Africa, Asia, and Latin America*, Human Settlements Discussion Paper Series, Urban Change-3 (London: International Institute for Environment and Development (IIED), 2006).

13. Table 1–1 and data in text from United Nations Population Division, op. cit. note 5.

14. Ibid., viewed August 2006. A discussion with David Satterthwaite was helpful in formulating these paragraphs.

15. United Nations Population Division, op. cit. note 5, viewed July 2006; Africa's urban population from Diana Mitlin and David Satterthwaite, eds., *Empowering Squatter Citizen: Local Government, Civil Society, and Urban Poverty Reduction* (Sterling, VA: Earthscan, 2004), p. 6; rate of urbanization from National Research Council, op. cit. note 6, pp. 92–93.

16. National Research Council, op. cit. note 6, p. 107.

17. Share living in settlements below 500,000 from United Nations Population Division, op. cit. note 8, p. 5; Figures 1–1 and 1–2 from United Nations Population Division, op. cit. note 5. Janice Perlman provided the idea for Figure 1–1 on megacities.

18. National Research Council, op. cit. note 6, pp. 95–99; Latin America's urban population share from United Nations Population Division, op. cit. note 5, viewed August 2006.

19. National Research Council, op. cit. note 6,

pp. 99–102.

20. UN-HABITAT, op. cit. note 8, p. 9. A contrasting though not inconsistent account comes from Christine Kessides, who points out that economic growth in sub-Saharan Africa derives largely from industry, including construction and mining, and from service sectors, which are mainly urban. These activities accounted for almost 80 percent of growth in gross domestic product in the region in 1990–2003. Christine Kessides, *The Urban Transition in Sub-Saharan Africa: Implications for Economic Growth and Poverty Reduction*, Africa Region Working Paper Series No. 97 (Washington, DC: World Bank, 2005).

21. National Research Council, op. cit. note 6, pp. 102–06; polluted cities in China from World Bank, cited in "A Great Wall of Waste—China's Environment," *The Economist*, 21 August 2004; India's urban poverty from UN-HABITAT, op. cit. note 8, p. 11.

22. National Research Council, op. cit. note 6, chapter 10.

23. Cecilia Tacoli, "Editor's Introduction," in Cecilia Tacoli, ed., *The Earthscan Reader in Rural-Urban Linkage* (London: Earthscan, 2006), pp. 3–14.

24. National Research Council, op. cit. note 6, pp. 143–46; doubts about rapid growth without economic growth from David Satterthwaite, *The Scale of Urban Change Worldwide 1950–2000 and Its Underpinning*, Human Settlements Discussion Paper Series, Urban Change-1 (London: IIED, 2005).

25. Accra Planning and Development Programme, *Strategic Plan for the Greater Accra Metropolitan Area. Volume I. Context Report* (Accra: Department of Town and Country Planning, Ministry of Local Government, Government of Ghana, draft final report, 1992), pp. 77–81.

26. Health penalty for urban poor from National Research Council, op. cit. note 6, pp. 284–89; Jacob Songsore et al., with the assistance of CERS-GIS, *State of Environmental Health. Report of the*

Greater Accra Metropolitan Area 2001 (Accra: Ghana Universities Press, 2006); Institute for Regional Studies of the Californias, *A Binational Vision for the Tijuana River Watershed* (San Diego, CA: San Diego State University, 2005).

27. Infrastructure spending from UN-HABI-TAT, *The State of the World's Cities, 2004/2005* (London: Earthscan, 2005), p. 28; cost of meeting Millennium Development Goal from UN-HABITAT, op. cit. note 8, p. 162; urban poverty not transitory from ibid., p. 49.

28. Global Urban Observatory from UN-HABI-TAT, op. cit. note 8; National Research Council, op. cit. note 6, chapter 5; Demographic and Health Surveys, at www.measuredhs.com/aboutdhs/whoweare.cfm, viewed August 2006; Shlomo Angel, Stephen C. Sheppard, and Daniel L. Civco, *The Dynamics of Global Urban Expansion* (Washington, DC: World Bank, 2005).

29. Center for Global Development, *When Will We Ever Learn? Improving Lives through Impact Evaluation*, Final Report of the Evaluation Gap Working Group (Washington, DC: 2006); C. S. Holling, ed., *Adaptive Environmental Assessment and Management* (New York: John Wiley & Sons, 1978); Richard Margoluis and Nick Salafsky, *Measures of Success. Designing, Managing, and Monitoring Conservation and Development Projects* (Washington, DC: Island Press, 1998); Kai N. Lee, "Appraising Adaptive Management," *Conservation Ecology*, vol. 3, no. 2 (1999), article 3.

30. Table 1–2 from the following: First five rows of data and row seven on per capita health expenditures are taken from U.N. Development Programme, *Human Development Report 2005* (New York: Oxford University Press, 2005). Probability of dying before age five from Lopez et al., op. cit. note 6, annex 2A, pp. 36–42. Energy use from World Bank, *World Development Indicators* (2006), at devdata.worldbank.org/dataonline. Municipal populations, estimated for 2005, are taken from United Nations Population Division, op. cit. note 8, Table A.12, pp. 266, 268, 270. The population figures are current, but the information on water, sanitation, and energy was collected at earlier dates; Accra, in particular, was considerably smaller in

1991–92, and it is possible that the levels of access to improved water and sanitation have declined since then. The estimates of populations without "improved" sanitation and water supply need to be read with caution, because the definitions used to define "improved" vary across the cities, reflecting their different origins. These differences not only are unavoidable consequences of the way data are collected but also reflect differing histories and cultural standards, as explained in UN-HABITAT, *Water and Sanitation in the World's Cities: Local Action for Global Goals* (London: Earthscan, 2003), pp. 2–5, as well as in the water and sanitation chapter of this book. The figures for Accra are taken from Gordon McGranahan et al., *The Citizens at Risk: From Urban Sanitation to Sustainable Cities* (Sterling, VA: Earthscan, for Stockholm Environment Institute, 2001), chapter 4 (with sanitation considered unimproved if a household shares a toilet with more than 10 other households and water supply considered unimproved if a household lacks access to piped water inside the house). The figures for Tijuana are drawn from *Potable Water and Wastewater Master Plan for Tijuana and Playas de Rosarito* (Comisión Estatal de Servicios Públicos de Tijuana, 2003), p. 2-37, quoting census data for 2000 (with unimproved sanitation defined as residences without sewage services and unimproved water supply meaning a residence without piped water). Figures from Singapore are taken from World Bank, op. cit. this note. Other material in text from Carolyn Stephens et al., "Urban Equity and Urban Health: Using Existing Data to Understand Inequalities in Health and Environment in Accra, Ghana and São Paulo, Brazil," *Environment and Urbanization*, April 1997, pp. 181–202, and from Songsore et al., op. cit. note 26.

31. Gordon McGranahan and Frank Murray, eds., *Air Pollution and Health in Rapidly Developing Countries* (London: Earthscan, 2003).

32. Thomas L. Friedman, *The World Is Flat* (New York: Farrar, Straus and Giroux, 2005).

33. Figure 1–3 from Millennium Ecosystem Assessment, *Ecosystems and Human Well-Being. Vol. 1: Current State and Trends* (Washington, DC: Island Press, 2005), p. 807; Xuemei Bai and

Hidefumi Imura, "A Comparative Study of Urban Environment in East Asia: Stage Model of Urban Environmental Evolution," *International Review for Environmental Strategies*, summer 2000, pp. 135–58; McGranahan et al., op. cit. note 30. The environmental Kuznets curve was initially proposed in World Bank, *World Development Report 1992* (New York: Oxford University Press, 1992), p. 11; see also Kirk R. Smith and Majid Ezzati, "How Environmental Health Risks Change with Development: The Epidemiologic and Environmental Risk Transitions Revisited," *Annual Review of Environment and Resources*, November 2005, pp. 291–333.

34. O. Alberto Pombo, "Water Use and Sanitation Practices in Peri-Urban Areas of Tijuana: A Demand-side Perspective," in Lawrence A. Herzog, ed., *Shared Space: Rethinking the U.S.-Mexico Border Environment* (La Jolla, CA: Center for U.S.-Mexican Studies, University of California, San Diego, 2000); Serge Dedina, "The Political Ecology of Transboundary Development: Land Use, Flood Control, and Politics in the Tijuana River Valley," *Journal of Borderlands Studies*, vol. 10, no. 1 (1995), pp. 89–110; Accra from Songsore et al., op. cit. note 26; Singapore, where the implementation of public housing on a very large scale since 1960 has dramatically altered the environmental risks facing households, from Tan Sook Yee, *Private Ownership of Public Housing in Singapore* (Singapore: Times Academic Press, 1998), but compare the wider range of views in Giok Ling Ooi and Kenson Kwok, eds., *The City & The State: Singapore's Built Environment Revisited* (Singapore: Oxford University Press for the Institute of Policy Studies, 1997); community-level organizations from Mitlin and Satterthwaite, op. cit. note 15.

35. Sumerians from Sandra Postel, *Pillar of Sand* (New York: W. W. Norton & Company, 1999) pp. 14–21; Mayans from Jared Diamond, *Collapse: How Societies Choose to Fail or Succeed* (New York: Viking, 2005), pp. 156–77.

36. Definition from Global Footprint Network, at www.footprintnetwork.org/gfn_sub.php?content=footprint_overview, viewed August 2006; relative sizes of per capita footprints come from national footprint estimates prepared by the European Environment Agency, at org.eea.europa.eu/news/Ann1132753060, viewed August 2006. The definition may make it seem that a country's footprint is related to the area within and adjacent to the national border, but all countries depend on distant sources for resources such as oil or manufactured products and sometimes water, so the footprint is actually an abstract measure based on globally averaged estimates of ecological productivity.

37. Elinor Ostrom, *Governing the Commons. The Evolution of Institutions for Collective Action* (Cambridge, U.K.: Cambridge University Press, 1990), especially chapter 3; J. Stephen Lansing, *Priests and Programmers: Technologies of Power in the Engineered Landscape of Bali* (Princeton, NJ: Princeton University Press, 1991); Daniel Pauly et al., "The Future for Fisheries," *Science*, 21 November 2003, pp. 1359–61.

38. Joel A. Tarr, *The Search for the Ultimate Sink: Urban Pollution in Historical Perspective* (Akron, OH: University of Akron Press, 1996), pp. 179–217.

39. UN-HABITAT, op. cit. note 30, pp. 8–12; John Thompson et al., "Waiting at the Tap: Changes in Urban Water Use in East Africa over Three Decades," *Environment and Urbanization*, October 2000, pp. 37–52.

40. Table 1–3 from the following: Edem Dzidzienyo and Kai N. Lee, unpublished field notes, Accra, Ghana, January 2006, and Thompson et al., op. cit. note 39 (prices of water in Accra in Ghanaian cedis (GHC) were converted to dollars at the 2006 market rate of 8,900 GHC to the dollar; these prices were projected back to 1997 dollars using the implicit price deflator in Executive Office of the President, *Economic Report of the President*, 2006, Table B-3; the value for the fourth quarter of 2005 was combined with the 1997 value for an overall deflator of 113.369/95.414 = 1.188); Ethiopia from UN-HABITAT, op. cit. note 30, pp. 67–68, and from UN-HABITAT, op. cit. note 8, p. 75; scientist from Dr. Frederick Amu-Mensah, Accra, Ghana, discussion with author, 9 February 2006.

41. Association between human development and urbanization from UN-HABITAT, op. cit. note 8, p. 46; scale of urban population growth and need for urban habitat from United Nations Population Division, op. cit. note 5.

42. Nineteenth-century engineers from Hans van Engen, Dietrich Kampe, and Sybrand Tjallingii, eds., *Hydropolis: The Role of Water in Urban Planning* (Leiden, Netherlands: Backhuys Publishers, 1995); Christer Nilsson et al., "Fragmentation and Flow Regulation of the World's Large River Systems," *Science*, 15 April 2005, pp. 405–08.

43. Chester L. Arnold, Jr., "Impervious Surface Coverage: The Emergence of a Key Environmental Indicator," *Journal of the American Planning Association*, spring 1996, p. 243ff; Thomas Dunne and Luna B. Leopold, *Water in Environmental Planning* (New York: W. H. Freeman and Company, 1998, 15th ed.); New York from Toni Nelson, "Closing the Nutrient Loop," *World Watch*, November/December 1996; Vivian Toy, "Planning to Close Its Landfill, New York Will Export Trash," *New York Times*, 30 November 1996; nutrient cycle from Gary Gardner, *Recycling Organic Waste: From Urban Pollutant to Farm Resource*, Worldwatch Paper 135 (Washington, DC: Worldwatch Institute, August 1997).

44. Millennium Project, *Halving Hunger: It Can Be Done*, Summary of Report of the Task Force on Hunger (New York: Earth Institute, Columbia University, 2005); people without electricity from V. Modi et al., *Energy and the Millennium Development Goals* (New York: U.N. Development Programme, UN Millennium Project, and World Bank, 2006).

45. UN-HABITAT, op. cit. note 30, chapter 5.

46. Arif Hasan, *Working with Government: The Story of the Orangi Pilot Project's Collaboration with State Agencies for Replicating its Low Cost Sanitation Programme* (Karachi, Pakistan: City Press, 1997).

47. UN-HABITAT, op. cit. note 8, pp. 160–67.

48. Herbert Girardet, *Cities People Planet* (Chichester, U.K.: John Wiley & Sons, 2004), pp. 123–25; Herbert Girardet, *The Gaia Atlas of Cities* (London: Gaia Books, 1992), pp. 22–23; Box 1–2 from City of Stockholm, "Hammarby Sjöstad: The Best Environmental Solutions in Stockholm," undated, from Timothy Beatley, *Green Urbanism: Learning from European Cities* (Washington, DC: Island Press, 2000), and from site visits and interviews with city officials by Timothy Beatley, 2000–06.

49. Ken Yeang, *Bioclimatic Skyscrapers* (London: Ellipsis London Press, 2000); Condé Nast from "Urban Sustainability at the Building and Site Scale," in Stephen M. Wheeler and Timothy Beatley, eds., *The Sustainable Urban Development Reader* (London: Routledge, 2004), p. 300.

50. Steven Peck and Chris Callaghan, "Gathering Steam: Eco-Industrial Parks Exchange Waste for Efficiency and Profit," *Alternatives Journal*, spring 1997.

51. Mark A. Benedict and Edward T. McMahon, *Green Infrastructure: Linking Landscapes and Communities* (Washington, DC: Island Press, 2006); Laura Kitson, "Green Spaces Widen Growth Area Purpose," *Planning*, 9 June 2006.

52. International Telecommunications Union, "Mobile Cellular, Subscribers per 100 People" and "Main Telephone Lines, Subscribers per 100 People," at www.itu.int/ITU-D/ict/statistics, viewed 21 July 2006; William McDonough, "China as a Green Lab," *Harvard Business Review*, February 2006; Box 1–3 from Jeremy Harris, discussion with Molly O'Meara Sheehan and author, 19 June 2006, with additional information provided by Karl Hausker, adjunct fellow, Center for Strategic and International Studies, Washington, DC, e-mail to Molly Sheehan, 19 September 2006, and by Peter Chaffey, manager, Business Melbourne, City of Melbourne, e-mail to Molly Sheehan, 20 September 2006.

53. Hon. Kodjo Gbli-Boyetey, discussion with author, Edem Dzidzienyo, and Dana Lee, 24 January 2006.

54. Tim Campbell and Travis Katz, "The Politics of Participation in Tijuana, Mexico: Inventing a New Style of Governance," in Tim Campbell and Harald Fuhr, *Leadership and Innovation in Subnational Government: Case Studies from Latin America* (Washington, DC: World Bank Institute, 2004), section 5-1, pp. 69–97.

55. W. G. Huff, *The Economic Growth of Singapore: Trade and Development in the Twentieth Century* (Cambridge, U.K.: Cambridge University Press, 1994); Tan, op. cit. note 34; energy from World Bank, op. cit. note 30 (which shows energy use at 5,359 kilograms of oil equivalent per capita for Singapore in 2003 and 7,843 kilograms for the United States).

Timbuktu: Greening the Hinterlands

1. Estimate of 100,000 described in Richard W. Franke and Barbara H. Chasin, *Seeds of Famine: Ecological Destruction and the Development Dilemma in the West African Sahel* (Lanham, MD: Rowman and Littlefield, 1980), p. 11; population figures from Jan Lahmeyer, Population Statistics, at www.library.uu.nl/wesp/populstat/populframe.html, viewed July 2006.

2. National income and poverty data from Government of Mali, *Poverty Reduction Strategy Paper* (Bamako, Mali: 29 May 2002), pp. 7, 13.

3. Aly Bacha Konaté and Mamadou Diakité, *Etude de la filière de l'Eucalyptus dans la vallée du Yamé* (Mopti, Mali: GDRN5 and USAID/FRAME, 2006), p. 28; quote from Aly Bacha Konaté and Aly Bocoum, *Lutte contre la desertification, reduction de la pauvreté: Etude de cas du Mali* (Mopti, Mali: GDRN5 and USAID/FRAME, 2005), p. 36.

4. Konaté and Bocoum, op. cit. note 3, p. 36.

5. Yield from Konaté and Bocoum, op. cit. note 3, p. 36; net value from Konaté and Diakité, op. cit. note 3, p. 28.

6. Regional director, Forest Service, discussion with authors, March 2005.

7. Yield from Konaté and Bocoum, op. cit. note 3, p. 33; quote from ibid., p. 32.

8. Konaté and Bocoum, op. cit. note 3, p. 31.

9. Quote from Konaté and Bocoum, op. cit. note 3, p. 32.

Loja: Ecological and Healthy City

1. Marlon Cueva, discussion with author, 23 April 2004; Fernando Montesinos, discussion with author, 15 April 2004.

2. Dr. José Bolívar Castillo, discussion with author, 3 May 2004.

3. Dr. Ermel Salinas, discussion with author, 29 April 2004.

4. Jorge Muños Alvarado, discussion with author, 21 April 2004; Dr. Humberto Tapia, discussion with author, 28 April 2004.

5. Wilson Jaramillo, discussion with author, 16 April 2006.

6. Fernando Montesinos, discussion with author, 19 April 2004.

7. Ibid.

8. Fabián Álvarez, municipal engineer, discussion with author, 19 April 2004; Fernando Montesinos, discussion with author, 26 April 2004.

9. Lolita Samaniego Idrovo, discussion with author, 6 May 2004; Fernando Montesinos, discussion with author, 6 May 2004.

10. Cueva, op. cit. note 1.

11. Montesinos, op. cit. note 6.

12. Tierramérica, *Ecobreves*, 2001, at www.tierramerica.net/2001/1216/iecobreves.shtml; LivCom, *LivCom Awards–Previous Winners 2001*, at www.livcomawards.com/previous-winners/2001.htm; "Turismo Nacional se Reunirá en Loja," *El Mercurio*, 16 September 2003.

Chapter 2. Providing Clean Water and Sanitation

1. U.N. Human Settlements Programme, *Water and Sanitation in the World's Cities: Local Action for Global Goals* (London: Earthscan, 2003).

2. Simon Szreter, *Health and Wealth: Studies in History and Policy* (Rochester, NY: University of Rochester Press, 2005); U.N. Human Settlements Programme, op. cit. note 1.

3. U.N. Human Settlements Programme, op. cit. note 1.

4. David Satterthwaite, *The Under-estimation of Urban Poverty in Low and Middle-Income Nations*, Working Paper 14 on Poverty Reduction in Urban Areas (London: International Institute on Environment and Development (IIED), 2004); see also African Population and Health Research Center, *Population and Health Dynamics in Nairobi's Informal Settlements* (Nairobi: 2002); Table 2–1 from U.N. Human Settlements Programme, op. cit. note 1.

5. See, for instance, U.N. Development Programme (UNDP), *Human Development Report 2006* (New York: Oxford University Press, 2006), which focuses on water and sanitation; see also UNICEF and World Health Organization (WHO), *Meeting the MDG Drinking Water and Sanitation Target: A Mid-Term Assessment of Progress*, WHO/UNICEF Joint Monitoring Programme for Water Supply and Sanitation (New York: 2004).

6. Rualdo Menegat, "Participatory Democracy and Sustainable Development: Integrated Urban Environmental Management in Porto Alegre, Brazil," *Environment and Urbanization*, October 2002, pp. 181–206; U.N. Human Settlements Programme, op. cit. note 1; Lo Heller, "Access to Water Supply and Sanitation in Brazil: Historical and Current Reflections; Future Perspectives," background paper for UNDP, op. cit. note 5; U.N. Human Settlements Programme, *Meeting Development Goals in Small Urban Centres: Water and Sanitation in the World's Cities 2006* (London: Earthscan, 2006).

7. Jorge E. Hardoy, Diana Mitlin, and David Satterthwaite, *Environmental Problems in an Urbanizing World: Finding Solutions for Cities in Africa, Asia and Latin America* (London: Earthscan, 2001); U.N. Human Settlements Programme, op. cit. note 1.

8. Figure of 900 million from U.N. Human Settlements Programme, *The Challenge of Slums: Global Report on Human Settlements 2003* (London: Earthscan, 2003).

9. UNICEF and WHO, op. cit. note 5, p. 23.

10. U.N. Human Settlements Programme, op. cit. 1; U.N. Human Settlements Programme, op. cit. note 6.

11. WHO and UNICEF, *Global Water Supply and Sanitation Assessment, 2000 Report* (Geneva: WHO, UNICEF, and Water Supply and Sanitation Collaborative Council, 2002); high infant and child mortality rates from Satterthwaite, op. cit. note 4.

12. See, for instance, Arif Hasan, *Understanding Karachi: Planning and Reform for the Future* (Karachi, Pakistan: City Press, 1999); Arif Hasan, "Orangi Pilot Project: The Expansion of Work beyond Orangi and the Mapping of Informal Settlements and Infrastructure," *Environment and Urbanization*, October 2006, pp. 451–80.

13. Celine D'Cruz and David Satterthwaite, *Building Homes, Changing Official Approaches: The Work of Urban Poor Federations and Their Contributions to Meeting the Millennium Development Goals in Urban Areas*, Working Paper 16 on Poverty Reduction in Urban Areas (London: IIED, 2005); Mohini Malhotra, "Financing Her Home, One Wall at a Time," *Environment and Urbanization*, October 2003, pp. 217–28; Lula da Silva et al., "The Programme for Land Tenure Legalization on Public Land," *Environment and Urbanization*, October 2003, pp. 191–200; Jessica Budds, with Paulo Teixeira and SEHAB, "Ensuring the Right to the City: Pro-poor Housing, Urban Development and Land Tenure Legalization in São Paulo, Brazil," *Environment and Urbanization*, April 2005, pp. 89–114.

14. Table 2–2 adapted from U.N. Human Settlements Programme, op. cit. note 1, and from U.N. Human Settlements Programme, op. cit. note 6.

15. J. C. Melo, *The Experience of Condominial Water and Sewerage Systems in Brazil: Case Studies from Brasilia, Salvador and Parauapebas* (Lima, Peru: Water and Sanitation Program Latin America, 2005).

16. This draws on case studies prepared by Vincentian Missionaries Social Development Foundation Incorporated in Manila and provided to UN-HABITAT by the Asian Coalition for Housing Rights for the preparation of U.N. Human Settlements Programme, op. cit. note 6.

17. Hardoy, Mitlin, and Satterthwaite, op. cit. note 7.

18. Table 2–3 based on a table in UNDP, op. cit. note 5.

19. Edi Medilanski et al., "Wastewater Management in Kunming, China: A Stakeholder Perspective on Measures at the Source," *Environment and Urbanization*, October 2006, pp 353–68.

20. This draws on a case study by Arif Hasan of the Orangi Pilot Project–Research & Training Institute (OPP–RTI) prepared for a Research Project of the Max Lock Centre, Westminster University, London, UK, April 2003; on Arif Hasan, *Working with Government: The Story of the Orangi Pilot Project's Collaboration with State Agencies for Replicating Its Low Cost Sanitation Programme* (Karachi, Pakistan: City Press, 1997); on Perween Rahman, *Katchi Abadis of Karachi: A Survey of 334 Katchi Abadis* (Karachi, Pakistan: OPP-RTI, 2004); on Hasan, "Orangi Pilot Project," op. cit. note 12; and on OPP-RTI staff, Karachi, discussions with David Satterthwaite, July 2006.

21. Hasan, "Orangi Pilot Project," op. cit. note 12.

22. Mimi Jenkins and Steven Sugden, "Rethinking Sanitation—Lessons and Innovation for Sustainability and Success in the New Millennium,"

background paper for UNDP (London: London School of Hygiene and Tropical Medicine, 2006).

23. Box 2–1 from Sundar Burra, Sheela Patel, and Tom Kerr, "Community-designed, Built and Managed Toilet Blocks in Indian Cities," *Environment and Urbanization*, October 2003, pp. 11–32.

24. International agencies in the 1990s from Matthias Finger and Jeremy Allouche, *Water Privatisation: Trans-National Corporations and the Re-Regulation of the Water Industry* (London: Spon Press, 2002).

25. For proponents, see, for example, Walter Stottman, "The Role of the Private Sector in the Provision of Water and Wastewater Services in Urban Areas," in Juha Uitto and Asit Biswas, eds., *Water for Urban Areas* (Tokyo: United Nations University Press, 2000).

26. Clare Joy and Peter Hardstaff, *Dirty Aid, Dirty Water: The UK Government's Push to Privatise Water and Sanitation in Poor Countries* (London: World Development Movement, 2005).

27. For large reviews, see, for example, George R. G. Clarke, Katrina Kosec, and Scott Wallsten, *Has Private Participation in Water and Sanitation Improved Coverage? Empirical Evidence from Latin America*, Policy Research Working Paper 3445 (Washington, DC: World Bank, 2004), and Colin Kirkpatrick, David Parker, and Yin-Fang Zhang, "State versus Private Sector Provision of Water Services in Africa: An Empirical Analysis," presented at Pro-Poor Regulation and Competition: Issues, Policies and Practices, Cape Town, South Africa, 7–9 September 2004; J. Budds and G. McGranahan, "Are the Debates on Water Privatization Missing the Point? Experiences from Africa, Asia and Latin America," *Environment and Urbanization*, October 2003, pp. 87–113.

28. A. J. Loftus and D. A. McDonald, "Of Liquid Dreams: A Political Ecology of Water Privatization in Buenos Aires," *Environment and Urbanization*, October 2001, pp. 179–99; Ricardo Schusterman et al., *Public Private Partnerships and the Poor: Experiences with Water Provision in*

Four Low-income Barrios in Buenos Aires (Loughborough, U.K.: Water, Engineering and Development Centre, Loughborough University, 2002); A. Harsono, *Water and Politics in the Fall of Suharto* (Washington, DC: International Consortium of Investigative Journalists, 2003).

29. David Hall and Emanuele Lobina, *Pipe Dreams: The Failure of the Private Sector to Invest in Water Services in Developing Countries* (London: Public Services International Research Unit, 2006).

30. K. Bakker, "Archipelagos and Networks: Urbanization and Watrer Privatization in the South," *Geographical Journal*, vol. 169 (2003), pp. 328–41.

31. Gordon McGranahan and David Lloyd Owen, *Local Water and Sanitation Companies and the Urban Poor*, Water Discussion Paper 3 (London: IIED, 2006).

32. Bernard Collignon and Marc Vezina, *Independent Water and Sanitation Providers in African Cities*, Water and Sanitation Program (Washington, DC: World Bank, 2000); Herve Conan and Maria Paniagua, *The Role of Small Scale Private Water Providers in Serving the Poor* (Manila: Asian Development Bank, 2003); Tova Maria Solo, *Independent Water Entrepreneurs in Latin America: The Other Private Sector in Water Services* (Washington, DC: World Bank, 2003).

33. Silver Mugisha and Sanford V. Berg, "Turning Around Struggling State-owned Enterprises in Developing Countries: The Case of NWSC-Uganda," prepared for seminar on Reforming Public Utilities to Meet the Water and Sanitation Millennium Development Goals (London: World Development Movement and WaterAid, 2006).

34. Gordon McGranahan, *Demand-Side Water Strategies and the Urban Poor*, Poverty, Inequality and Environment Series No. 4 (London: IIED, 2002).

35. Gordon McGranahan et al., *The Citizens at Risk: From Urban Sanitation to Sustainable Cities* (London: Earthscan, 2001); José Esteban Castro, *Water, Power and Citizenship: Social Struggles in the Basin of Mexico* (New York: Palgrave Macmillan, 2006).

36. For examples of cities with difficulties in water supply, see Etienne von Bertrab, "Guadalajara's Water Crisis and the Fate of Lake Chapala: A Reflection of Poor Water Management in Mexico," *Environment and Urbanization*, October 2003, pp. 127–40; J. Wolf et al., "Urban and Peri-urban Agricultural Production in Beijing Municipality and Its Impact on Water Quality," *Environment and Urbanization*, October 2003, pp. 141–56; aggregate water withdrawls from Peter H. Gleick, "Water Use," *Annual Review of Environment and Resources*, Vol. 28 (2003), pp. 275–315.

37. For examples of the possible contribution to water supplies of rainwater harvesting, see, for instance, Anil Agarwal and Sunita Narain, *Dying Wisdom: Rise, Fall and Potential of India's Traditional Water-harvesting Systems* (New Delhi: Centre for Science and Environment, 1997).

38. UNDP, op. cit. note 5; U.N. Human Settlements Programme, op. cit. note 1.

39. Jenkins and Sugden, op. cit. note 22.

40. Antonio Miranda, *Developing Public-Public Partnerships: Why and How Not-for-profit Partnerships Can Improve Water and Sanitation Services Worldwide*, prepared for seminar on Reforming Public Utilities to Meet the Water and Sanitation Millennium Development Goals (London: World Development Movement and WaterAid, 2006).

41. Boost to improvements from U.N. Human Settlements Programme, op. cit. note 1; Jose Esteban Castro and Leo Heller, "The Historical Development of Water and Sanitation in Brazil and Argentina" in Petri Juuti, Tapio Katko, and Heikki Vuorinen, eds., *Environmental History of Water: Global View of Community Water Supply and Sanitation* (Tampere, Finland: Department of History, University of Tampere, forthcoming).

42. D'Cruz and Satterthwaite, op. cit. note 13.

43. Satterthwaite, op. cit. note 4.

44. David Satterthwaite, "Reducing Urban Poverty: Constraints on the Effectiveness of Aid Agencies and Development Banks and Some Suggestions for Change," *Environment and Urbanization*, April 2001, pp. 137–57; U.N. Human Settlements Programme, op. cit. note 6.

45. Julie Crespin, "Aiding Local Action: The Constraints Faced by Donor Agencies in Supporting Effective, Pro-poor Initiatives on the Ground," *Environment and Urbanization*, October 2006, pp. 433–50.

46. U.N. Human Settlements Programme, op. cit. note 6.

47. Meera Bapat and Indu Agarwal, "Our Needs, Our Priorities: Women and Men from the 'Slums' in Mumbai and Pune Talk about Their Needs for Water and Sanitation," *Environment and Urbanization*, October 2003, pp. 71–86.

48. Ibid.

Lagos: Collapsing Infrastructure

1. Matthew Gandy, "Planning, Anti-Planning and the Infrastructure Crisis Facing Metropolitan Lagos," *Urban Studies*, vol. 43, no. 2 (2006), p. 372; Charisma Acey, "Towards Sustainability in an African Mega-City: A Spatial Analysis of Potable Water Service Areas in Lagos, Nigeria," Final Project Paper, Department of Urban Planning, University of California at Los Angeles, spring 2005, p. 6.

2. Dele Olowu, *Lagos State: Governance, Society and Economy* (Lagos, Nigeria: Malthouse, 1990); Ayodeji Olukoju, *Infrastructure Development and Urban Facilities in Lagos, 1861–2000* (Ibadan, Nigeria: French Institute for Research in Africa, 2003); Rem Koolhaas, "Fragments of a Lecture on Lagos," in O. Enwezor et al. (eds.), *Under Siege: Four African Cities: Freetown, Johannesburg, Kinshasa, Lagos* (Ostfildern-Ruit, Germany: 2003), pp. 181, 183.

3. Olukoju, op. cit. note 2; Matthew Gandy, "Learning from Lagos," *New Left Review*, May-June 2005, pp. 37–52; Doyin Abiola, "Lagos

Megacity: On the Way to Recovery?" *Sunday PUNCH* (Lagos), 30 July 2006, p. 2; Tunde Alao, "NIESV Seeks Roles in Lagos Mega-city Implementation," *The Guardian* (Lagos), 31 July 2006, p. 33.

4. Bola Olaosebikan, *Lagos State Water Corporation: Dawn of a New Era in Water Supply* (Lagos: Lagos State Water Corporation, 1999); Acey, op. cit. note 1; Gandy, op. cit. note 1.

5. Olukoju, op. cit. note 2; Samuel Shofuyi, "Study X-rays Poor Quality of 'Pure Water,'" *The PUNCH* (Lagos), 4 February 2003, p. 46; J. W. K. Duncan and A. O. Olawale, "Properties of Water from 9 Existing Wells in Shomolu, a Suburb of Lagos," *The Nigerian Engineer*, vol. 6, no. 2 (1970), pp. 17–19.

6. Acey, op. cit. note 1; Coker cited in Chinedu Uwaegbulam, "Four Firms Jostle for Lagos Water Corporation IPP Project," *The Guardian* (Lagos), 31 July 2006.

7. Olukoju, op. cit. note 2; Gandy, op. cit. note 1; Alao, op. cit. note 3.

8. Abiola, op. cit. note 3; Madu Onuorah, "FEC Okays Police Reforms, N26b Lagos Facelift Loan," *The Guardian* (Lagos), 10 August 2006.

Chapter 3. Farming the Cities

1. Resource Centers on Urban Agriculture and Food Security (RUAF), "Cities," at www.ruaf.org/node/486, viewed August 2006; Ghana from World Bank, *World Development Indicators Database*, 1 July 2005, at siteresources.worldbank.org/DATASTATISTICS/Resources/GNIPC.pdf; Beijing from Embassy of the People's Republic of China, United States of America and China and Beijing Statistic Services, at www.china-embassy.org/chn/gyzg/t234138.htm, 2 February 2006, translated by Zijun Li; Vancouver Board of Trade, Vancouver, BC, Canada, e-mail to Danielle Nierenberg, September 2006.

2. RUAF, op. cit. note 1.

3. Ibid.

4.   Ibid.

5.   Ibid.

6.   Michael Levenston, executive director, City Farmer, Vancouver, BC, Canada, discussion with authors, April 2006; City Farmer, "44% of Vancouver Households Grow Food Says City Farmer" (Vancouver, BC, Canada: September 2002); International Development Research Centre (IDRC), *Shaping Livable Cities: Stories of Progress Around the World* (Ottawa, ON, Canada: 2006).

7.   Jac Smit, Urban Agriculture Network, Bethesda, MD, discussion with Brian Halweil, February 2004, cited in Brian Halweil, *Eat Here: Reclaiming Homegrown Pleasures in a Global Supermarket* (New York: W. W. Norton & Company, 2003).

8.   Ibid.

9.   U.N. Development Programme, *Urban Agriculture: Food, Jobs, and Sustainable Cities* (New York: 1996), p. 26.

10.   U.N. Food and Agriculture Organization (FAO), "Food Insecurity in an Urban Future," FAO Newsroom (Rome: 2004); FAO, *State of World Food Insecurity 2005* (Rome: 2005); Fred Pearce, "Cultivating the Urban Scene," in Paul Harrison and Fred Pearce, *Atlas of Population & Environment* (Washington, DC: American Association for the Advancement of Science and University of California Press, 2000).

11.   Kameshwari Pothukuchi and Jerome L. Kaufman, "Placing the Food System on the Urban Agenda: The Role of Municipal Institutions in Food Systems Planning," *Agriculture and Human Values*, vol. 16 (1999), pp. 213–24; FAO, "Feeding Asian Cities," Proceedings of the Regional Seminar, Food Supply and Distribution to Cities Programme, Bangkok, Thailand, 27–30 November 2000.

12.   "Technical Overview: The Challenge of Feeding Asian Cities," in FAO, op. cit. note 11.

13.   Olivio Argenti, *Feeding the Cities: Food Supply and Distribution*, 2020 Focus 3, Brief 5 (Washington, DC: International Food Policy Research Institute (IFPRI), 2000).

14.   Maurizio Aragrande and Olivio Argenti, *Studying Food Supply and Distribution Systems to Developing Countries and Countries in Transition: Methodological and Operational Guide* (Rome: FAO, 2001).

15.   RUAF, *Cities Farming for the Future* (Philippines: RUAF, International Institute of Rural Reconstruction, and ETC Urban Agriculture, 2006), p. 3.

16.   Diana Lee-Smith and Gordon Prain, *Understanding the Links Between Agriculture and Health*, Focus 13 (Washington, DC: IFPRI, 2006).

17.   RUAF, op. cit. note 15, p. 402.

18.   Pothukuchi and Kaufman, op. cit. note 11, p. 214; FAO, op. cit. note 11, p. 11; FAO, *The State of Food and Agriculture 1998* (Rome: 1998).

19.   FAO, op. cit. note 11, p. 11; Capital Area Food Bank, Washington, DC, at www.capital areafoodbank.org/programsresources/fmp.cfm.

20.   Nelso Companioni et al., "The Growth of Urban Agriculture," in Fernando Funes et al., eds., *Sustainable Agriculture and Resistance: Transforming Food Production in Cuba* (Oakland, CA: Food First Books, 2002), pp. 221–22.

21.   Ibid., pp. 223, 228–29.

22.   RUAF, op. cit. note 15, p. 177.

23.   The Food Project, Boston, at www.the foodproject.org; Cairo from Jac Smit, Urban Agriculture Network, Washington, DC, discussion with authors, 20 July 2006.

24.   Anne C. Bellow, "Health Benefits of Urban Agriculture, An Overview," *Community Food Security News*, winter 2006, p. 6.

25.   Wayne Roberts, Toronto Food Policy Council, Toronto, ON, discussion with Brian Halweil,

June 2002, cited in Halweil, op. cit. note 7.

26. Stacia and Kristof Nordin, "Improving Permaculture Through Nutrition in Malawi," ProNutrition, at www.pronutrition.org/archive/200606/msg00013.php, June 2006; Anne Bellows, Food Policy Institute, Rutgers University, interview with Dana Artz, Worldwatch Institute, June 2006.

27. Wayne Roberts, Toronto Food Policy Council, discussion with Brian Halweil, 6 June 2002; Pothukuchi and Kaufman, op. cit. note 11; unpublished studies from Robert Sommer, discussion with Brian Halweil, 23 February 2002.

28. Donna Armstrong, "A Survey of Community Gardens in Upstate New York: Implications for Health Promotion and Community Development," *Health and Place*, vol. 6, no. 4 (2000), pp. 319–27; Anne C. Bellows, Katherine Brown, and Jac Smit, "Health Benefits of Urban Agriculture," Community Food Security Coalition, Venice, CA, undated.

29. Bellows, Brown, and Smit, op. cit. note 28.

30. International Water Management Institute (IWMI), *Confronting the Realities of Wastewater Use in Agriculture*, Water Policy Briefing, Issue 9 (Colombo, Sri Lanka: 2003). Box 3–1 from the following: "Waste Not Want Not," *New Agriculturalist On-Line*, Issue 28, 2002; P. Drechsel et al., *Informal Irrigation in Urban West Africa: An Overview*, Research Report Series (Colombo, Sri Lanka: IWMI, in press); IWMI–Global Water Partnership, *Recycling Realities: Managing Health Risks to Make Wastewater an Asset*, Water Policy Briefing, Issue 17 (Colombo, Sri Lanka: 2006); E. Obuobie et al., *Irrigated Urban Vegetable Production in Ghana: Characteristics, Benefits and Risks* (Accra, Ghana: IWMI, 2006).

31. Table 3–1 from the following: Cambodia from "Take It Personally," *The Earth Report Series 7*, Program 4, at www.handsontv.info/series7/programme_4.html, viewed 10 August 2006; Cuba from Companioni et al., op. cit. note 20, pp. 227–29; Sierra Leone from Thomas Winnebah and Raymond Alfredson, *Food Security Situation in Sierra Leone Since 1961*, Food Security Monograph No. 2 (United Nations World Food Programme Sierra Leone, Technical Support Unit, rev. March 2006); Replant New Orleans, at www.replantneworleans.org, viewed 10 August 2006; Los Angeles from Kate H. Brown and Andrew L. Jameton, "Public Health Implications of Urban Agriculture," *Journal of Public Health Policy*, vol. 21, no. 1 (2000), pp. 20–39; St. Petersburg from Erio Ziglio et al. , eds., *Health Systems Confront Poverty* (Geneva: World Health Organization, 2003), p. 137; "Analysis of Tropical Storm Stan in El Salvador," Centro de Intercambio y Solidaridad, 16 November 2005, Relief Web, at www.reliefweb.int/rw/RWB.NSF/db900SID/RMOI-6JL56Z?OpenDocument, viewed August 2006.

32. Pearce, op. cit. note 10; Smit, op. cit. note 23.

33. RUAF, op. cit. note 15, p. 386.

34. Ibid.

35. Devi quoted in IDRC, op. cit. note 6.

36. RUAF, op. cit. note 15, Chapter 13, "Urban Aquatic Production."

37. Ibid., p. 396.

38. Challenge of dealing with garbage in developing-world cities from Gisèle Yasmeen, *Urban Agriculture in India: A Survey of Expertise, Capacities and Recent Experience*, Cities Feed People Report 32 (Ottawa, ON, Canada: IDRC, 2001), p. 9. Box 3–2 from the following: Christine Furedy, "Urban Waste and Rural Farmers: Enabling Low-Cost Organic Waste Reuse in Developing Countries," presentation given at "R'2002: Recovery, Recycling, Reintegration," the 6th World Congress on Integrated Resource Management, Geneva, 12–15 February 2002; Eduardo Spiaggi, "Urban Agriculture and Local Sustainable Development in Rosario, Argentina: Integration of Economic, Social, Technical and Environmental Variables," in Luc J. A. Mougeot, ed., *AGROPOLIS: The Social, Political, and Environmental Dimensions of Urban Agriculture* (London: Earthscan/IDRC, 2005); "Supermarket Composting in California," *Bio-*

Cycle, July 1997, pp. 70–71.

39. FAO, *FAOSTAT Statistical Database*, at apps.fao.org, updated 20 December 2005; Danielle Nierenberg, "Meat Consumption and Output Up," in Worldwatch Institute, *Vital Signs 2006–2007* (New York: W. W. Norton & Company, 2006), p. 24; RUAF, op. cit. note 15, p. 352; Box 3–3 from Spiaggi, op. cit. note 38.

40. Michael Greger, *Bird Flu: A Virus of Our Own Hatching* (draft), (Washington, DC: Humane Society of the United States, 2006), p. 106.

41. World Bank, *Managing the Livestock Revolution: Policy and Technology to Address the Negative Impacts of a Fast-Growing Sector* (Washington, DC: 2005), p. 6.

42. Ibid.

43. "Cleaning Up Its Act: Recycling Livestock Waste," *New Agriculturalist On-Line*, at www.new-agri.co.uk/06-2/focuson/focuson3.html, 1 March 2006.

44. Ibid.

45. FAO, *Pollution from Industrialized Livestock Production*, Policy Brief 2 (Rome: Livestock Information, Sector Analysis, and Policy Branch, Animal Production and Health Division, undated).

46. Ebenezer Howard, *Garden Cities of To-Morrow* (Cambridge, MA: The MIT Press, 1965), pp. 33–35.

47. A. A. Sorenson et al., *Farming on the Edge* (DeKalb, IL: American Farmland Trust, 1997).

48. Pothukuchi and Kaufman, op. cit. note 11; FAO, op. cit. note 11.

49. American Farmland Trust, "Fact Sheet: Cost of Community Services Studies," Farmland Information Sheet (Washington, DC: June 1986); FAO, op. cit. note 11, pp. 38–39.

50. Seidler, cited in FAO, op. cit. note 11, pp. 45–46.

51. Edward Seidler, senior officer, Marketing Group, FAO, e-mail to Brian Halweil, 11 July 2002.

52. IRDC, op. cit. note 6.

53. Ibid.

54. "Introduction," in RUAF, op. cit. note 15.

55. IRDC, op. cit. note 6.

56. The Food Trust, *Food Geography: How Food Access Affects Diet and Health* (Philadelphia: 2004).

57. Leslie Hoffman, Earth Pledge, New York, discussion with Brian Halweil, 21 April 2004.

58. Lee Rood, "Praise Grows for Lush Roof in Chicago," *Des Moines Register*, 29 July 2002; Chang-Ran Kim, "Tokyo Turns to Rooftop Gardens to Beat the Heat," *Reuters*, 8 August 2002; Marty Logan and Mark Foss, *Urban Agriculture Reaches New Heights Through Rooftop Gardening*, IDRC Reports (Ottawa, ON, Canada: 2004); Frederic Perron, "Jardins Suspendus," *La Presse* (Montreal), 8 March 2004.

59. Pothukuchi and Kaufman, op. cit. note 11; Neil Hamilton, "Putting a Face on Our Food: How State and Local Food Policies Can Promote the New Agriculture," *Drake Journal of Agricultural Law*, November 2002.

60. Mark Winne, Hartford Food System, discussion with Brian Halweil, 4 April 2002; Hartford Food System, at www.hartfordfood.org, viewed 1 September 2002.

61. Winne, op. cit. note 60.

62. Smit, op. cit. note 23.

63. Matovu quote from "More Investment in Agriculture Will Reduce Migration, Improve Urban Life: UN Agency," *UN Daily News Digest*, 5 June 2006.

64. Ibid.

65. FAO study from ibid.

66. RUAF, op. cit. note 15, p. 17.

Freetown: Urban Farms After a War

1. For early agricultural history, see Donald Davies, *Historical Development of Agricultural Land Use in Greater Freetown* (M.Sc. in rural development dissertation submitted in 2004), Department of Geography and Rural Development, Faculty of Environmental Sciences, Njala University College, University of Sierra Leone, Freetown; Christopher Fyfe, "The Foundation of Freetown," in Christopher Fyfe and Eldred Jones, eds., *Freetown: a Symposium* (Freetown: Sierra Leone University Press, 1968), pp. 1–8; and E. G. Ingham, *Sierra Leone after a Hundred Years* (London: Seeley and Co. Ltd, 1894). For connection to the slave trade, see Joe A. D. Alie, *A New History of Sierra Leone* (Malaysia: Macmillan Publishers, 1990), and R. J. Olu-Wright, "The Physical Growth of Freetown," in Fyfe and Jones, op. cit. this note, pp. 24–37.

2. Growth in population and land area from GOPA-Consultants, "Solid Waste Management Study, June," Freetown Infrastructure Rehabilitation Project, Sierra Leone, 1995, and from Statistics Sierra Leone, *Final Results 2004 Population and Housing Census Report* (Freetown: 2006).

3. Government of Sierra Leone, *Poverty Reduction Strategy Paper: A National Programme for Food Security, Job Creation, and Good Governance (2005–2007)* (Freetown: 2005).

4. For health status, see Mohamed Mankay Sesay, *Occupational Health and Safety Hazards and Production in Small Scale Agriculture: A Case Study of Inland Valley Farmers in the Eastern Part of Greater Freetown, Western Area, Sierra Leone* (B.Sc. in education project submitted in 1998), Department of Geography and Rural Development, Faculty of Environmental Sciences, Njala University College, University of Sierra Leone, Freetown; for technical studies, see Josie Abraham Scott-Manga, *Agricultural Practices at Dumpsites: Case Study of Bormeh and Grandville Brook in Freetown* (B.A. in education project submitted in

2000), Department of Geography and Rural Development, Faculty of Environmental Sciences, Njala University College, University of Sierra Leone, Freetown; Abrassac Abu Kamara, *Water Mangement Studies of Okra Under Drip Irrigation System* (M.Sc. in soil and water engineering dissertation submitted in 2006), Department of Agricultural Engineering, School of Technology, Njala University, Freetown.

5. Effects of rebel war on food production from Thomas R. A. Winnebah, *Food Security Situation in Sierra Leone Since 1961*, Food Security Monograph No. 2, World Food Programme Sierra Leone, Technical Support Unit, Freetown, revised March 2003.

6. GOPA-Consultants, op. cit. note 2; Statistics Sierra Leone, op. cit. note 2.

7. Report on Training Workshop on Multistakeholder Processes for Action Planning and Policy Formulation in Urban Agriculture, Freetown, Sierra Leone, 12–17 June 2006.

Chapter 4. Greening Urban Transportation

1. The Perth story, as described in the opening paragraphs, is told in P. Newman, "Railways and Reurbanisation in Perth," in J. Williams and R. Stimson, eds., *Case Studies in Planning Success* (New York: Elsevier, 2001).

2. Michael Renner, "Vehicle Production Continues to Expand," in Worldwatch Institute, *Vital Signs* (New York: W. W. Norton & Company, 2006), pp. 64–65.

3. The data are from a comparative study of 100 global cities for the International Union of Public Transport (UITP) and conducted by the Institute for Sustainability and Technology Policy (ISTP) at Murdoch University, involving 27 parameters using highly controlled processes to ensure comparability of data. The study took five years and builds on previous data collection since 1980. See J. Kenworthy and F. Laube, *The Millennium Cities Database for Sustainable Transport* (Brussels: UITP/ISTP, 2001), and J. Kenworthy et al., *An International Sourcebook of Automobile Depen-*

dence in Cities, 1960–1990 (Boulder, CO: University Press of Colorado, 1999). The 2005 data collection will commence shortly. Complete data on the cities can be viewed at www.sustainabil ity.murdoch.edu.au (16 cities were incomplete, so the data are mainly for 84 cities).

4. Figure 4–1 from Kenworthy and Laube, op. cit. note 3. These data are for city regions in 1995 and include all the gasoline and diesel for private passenger travel. Note that 1 liter of gasoline equals about one quarter of a U.S. gallon.

5. The lack of strong correlation between city wealth and car use is shown in Kenworthy and Laube, op. cit. note 3.

66. Figure 4–2 from Kenworthy and Laube, op. cit. note 3.

7. Figure 4–3 from ibid. Box 4–1 from Lester R. Brown, Plan B 2.0 (New York: W. W. Norton & Company, 2006), p. 10, and from J. Kenworthy and C. Townsend, "An International Comparative Perspective on Motorisation in Urban China: Problems and Prospects," IATSS Research, vol. 26, no. 2 (2002), pp. 99–109.

8. Figures 4–4 and 4–5 from Kenworthy and Laube, op. cit. note 3; for studies on the significance of factors other than urban form, see O. Mindali, A. Raveh, and I. Saloman, "Urban Density and Energy Consumption: A New Look at Old Statistics," Transportation Research Record, Part A 38 (2004), pp. 143–62, and R. E. Brindle, "Lies, Damned Lies and 'Automobile Dependence'— Some Hyperbolic Reflections," Australian Transport Research Forum 94 (1994), pp. 117–31; Figure 4–6 from P. Newman and J. Kenworthy, "Urban Design to Reduce Automobile Dependence," Opolis, vol. 2, no. 1 (2006), pp. 35–52.

9. For auto use 10 times as high in some cities, see Figures 4–1 and 4–5.

10. See, for example, D. Simon, Transport and Development in the Third World (New York: Routledge, 1996); Jorge E. Hardoy, Diana Mitlin, and David Satterthwaite, Environmental Problems in an Urbanizing World: Finding Solutions for Cities in Africa, Asia and Latin America (London: Earthscan, 2001); P. A. Barter, J. R. Kenworthy, and F. Laube, "Lessons from Asia on Sustainable Urban Transport," in N. Low and B. Gleeson, eds., Making Urban Transport Sustainable (Basingstoke, U.K.: Palgrave-Macmillan, 2003), pp. 252–70.

11. C. Marchetti, "Anthropological Invariants in Travel Behavior," Technical Forecasting & Social Change, September 1994, pp. 75–88; P. Newman and J. Kenworthy, Sustainability and Cities: Overcoming Automobile Dependence (Washington, DC: Island Press, 1999); Standing Advisory Committee for Trunk Road Assessment, Trunk Roads and the Generation of Traffic (London: Department for Transport, 1994).

12. Britton from www.newmobility.org; for more on resilience in cities, see Jane Jacobs, Cities and the Wealth of Nations (Harmondsworth, U.K.: Penguin, 1984), and L. Sandercock, Mongrel Cities: Cosmopolis2 (London: Continuum, 2003).

13. K. S. Deffeyes, Beyond Oil: The View from Hubbert's Peak (New York: Hill and Wang, 2005); Table 4–2 from Newman and Kenworthy, op. cit. note 11, pp. 74–77, discussed further in P. Newman and J. Kenworthy, "Transportation Energy in Global Cities: Sustainable Transportation Comes in from the Cold?" Natural Resources Forum, vol. 25 (2001), pp. 91–107.

14. Kenworthy and Laube, op. cit. note 3.

15. Jeremy Leggett, Half Gone: Oil, Gas, Hot Air and the Global Energy Crisis (London: Portobello Books, 2006); trends in driving from Kenworthy and Laube, op. cit. note 3.

16. U.S. data from U.S. Environmental Protection Agency, Light Duty Automotive Technology and Fuel Economy Trends, 1975–2006 (Ann Arbor, MI: 2006); Australian data in P. Laird et al., Back on Track: Rethinking Australian and New Zealand Transport Policy (Sydney, Australia: University of Sydney, 2001); technological advances from Amory B. Lovins and David R. Cramer, "Hypercars, Hydrogen and the Automotive Transition," International Journal of Vehicle Design, vol. 35, no. 1/2 (2004), pp. 50–85.

17. Three-wheeler taxis from Kenworthy and Laube, op. cit. note 3; Mexico City data from International Mayors Forum, *Sustainable Urban Energy Development*, Kunming, China, 10–11 November 2004; surpassing recommended levels from U.N. Population Fund, *The State of the World Population* (New York: 2001), and from Clean Air Initiative for Asian Cities, at www.cleanairnet.org/caiasia.

18. Clean Air Initiative for Asian Cities, op. cit. note 17; A. Rosencranz and M. Jackson, *The Delhi Pollution Case*, 2002, at indlaw.com; S. Jain, "Smog City to Clean City: How Did Delhi Do It?" *Mumbai Newsline*, 26 May 2004.

19. Figure 4–7 from Kenworthy and Laube, op. cit. note 3; this argument is expanded in P. Newman and J. Kenworthy, "The Transport Energy Trade-off: Fuel Efficient Traffic vs Fuel Efficient Cities," *Transportation Research Record*, vol. 22A, no. 3 (1988), pp. 163–74.

20. Surface Transportation Policy Project (STPP), *An Analysis of the Relationship Between Highway Expansion and Congestion in Metropolitan Areas: Lessons from the 15 Year Texas Transportation Institute Study* (Washington, DC: 1998).

21. Table 4–3 from Kenworthy and Laube, op. cit. note 3.

22. Congestion charging covered in detail in European Transport Conference, Strasbourg, France, 18–20 September 2006: see, for example, S. Kearns, "Congestion Charging Trials in London," J. Eliasson and M. Beser, "The Stockholm Congestion Charging System," and J. Baker and S. Kohli, "Challenges in the Development and Appraisal of Road User Charging Schemes."

23. See Newman and Kenworthy, op. cit. note 11, pp. 191–231.

24. R. Gordon, "Boulevard of Dreams," 8 September 2005, at www.sfgate.com.

25. See www.metro.seoul.kr/kor2000/chungae home/en/seoul/2sub; Institute for Transport and Development, "Seoul to Raze Elevated High-

way, Giving Way to Revitalized Center," *Sustainable Transport e-update*, May 2003; L. Gemsoe, "Turning the Downside Up: Creating Value for People," Profitable Places Conference, Sheffield Hallam University, Sheffield, U.K., 19–20 September 2006.

26. David Burwell, "Way to Go! Three Simple Rules to Make Transportation a Positive Force in the Public Realm," *Making Places Bulletin* (Project for Public Spaces), June 2005; Aarhus from Gemsoe, op. cit. note 25; Melissa Mean and Charlie Tims, *People Make Places: Growing the Public Life of Cities* (London: Demos, 2005).

27. Andy Wiley-Schwartz, "A Revolutionary Change in Transportation Planning: The Slow Road Movement," *New York Times*, 10 July 2006; Jan Gehl et al., *New City Life* (Copenhagen: Danish Architectural Press, 2006).

28. Kenworthy and Laube, op. cit. note 3.

29. Data analyzed in Newman and Kenworthy, op. cit. note 8; see also N. Mercat, "Evaluating Exposure to the Risk of Accident in the Grenoble Conurbation," European Transport Conference, Strasbourg, France, 18–20 September 2006, particularly suggesting the cycling figure.

30. Greater Vancouver Regional District, *Livable Region Strategic Plan* (Vancouver, BC: 1996); New South Wales Government, *City of Cities: A Plan for Sydney's Future* (Sydney: Department of Planning, 2005); Denver Region Council of Governments, *Metro Vision 2030 Plan* (Denver, CO: 2005).

31. Kenworthy and Laube, op. cit. note 3; U.S. data from Janette Sadik-Khan, Center for Transit-Oriented Development, ReconnectingAmerica.org, e-mail to and discussion with authors.

32. C. Hass Klau, *Bus or Light Rail: Making the Right Choice*, 2nd ed. (Brighton, U.K.: Environmental and Transport Planning, 2004); Brian Goodknight and Peter Buryk, "Along the Tracks: A Tale of Transit and Development," *The Next American City*, spring 2006; R. Cervero et al., *Transit-Oriented Development in America: Expe-*

riences, Challenges and Prospects (Washington DC: Transportation Research Board, National Research Council, 2004); J. Renne and J. S. Wells, "Transit-Oriented Development: Developing a Strategy to Measure Success," *TRB Research Results Digest 294* (Washington DC: Transportation Research Board, 2005).

33. P. Newman, "Transport Greenhouse Gas and Australian Suburbs: What Planners Can Do," *Australian Planner*, vol. 43, no. 2 (2006), pp. 6–7; P. Calthorpe, *The Next American Metropolis: Ecology, Community and the American Dream* (Cambridge, MA: Harvard University Press, 1993).

34. For information on Denver, see www.green printdenver.org/landuse/index.php and Denver Regional Council of Governments, *2030 Metro Vision Regional Transportation Plan* (Denver: 2004).

35. City of Vancouver, *Downtown Transportation Plan—Progress Report, City of Vancouver* (Vancouver, BC: 2006).

36. City of Vancouver, *Transportation Plan* (Vancouver, BC: 1997); Newman and Kenworthy, op. cit. note 11, pp. 174–77, 217–23.

37. J. Michaelson, "Lessons from Paris," *Making Places Bulletin* (Project for Public Spaces), June 2005.

38. Box 4–2 from the following: data on Curitiba's BRT from Urbanizacao de Curitiba, SA, and Bogotá TransMilenio from TransMilenio SA, both discussions with Walter Hook of Institute for Transportation and Development Policy (ITDP); history of the two systems from Arturo Ardila-Gómez, *Transit Planning in Curitiba and Bogotá: Roles in Interaction, Risk, and Change*, PhD dissertation, Massachusetts Institute of Technology, 2004; bus operators blocking change and Quito from Cesar Arias, former Director of City Planning, City of Quito, discussions with Walter Hook, ITDP; Bogotá quality of life and transit speeds and loads from Lloyd Wright and Lewis Fulton, "Climate Change Mitigation and Transport in Developing Nations," *Transport Reviews*, November 2005, pp. 691–717;

Jakarta traffic data from PT TransJakarta; critique of TransJakarta from *Making TransJakarta a World Class BRT System* (New York: ITDP, 2005); new systems in other cities from field interviews, ITDP staff.

39. V. Vuchic, *Transportation for Livable Cities* (New Brunswick, NJ: Center for Urban Policy Research, Rutgers University, 1999); V. Vuchic, *Urban Transit: Planning, Operations and Economics* (Indianapolis, IN: Wiley Press, 2005).

40. International Mayors Forum, op. cit. note 17; L. Fulton and L. Schipper, *Bus Systems for the Future: Achieving Sustainable Transport Worldwide* (Paris: International Energy Agency and Organisation for Economic Co-operation and Development, 2002).

41. Barter, Kenworthy, and Laube, op. cit. note 10.

42. American Public Transportation Association and Millar from "Light Rail and Buses Beckon. But Will Americans Really Abandon Their Cars?" *The Economist*, 31 August 2006.

43. Kobenhavns Kommune, *Copenhagen City Green Accounts and Environmental Report: Indicators for Traffic* (Copenhagen: 2002); Roads and Parks Department, *Bicycle Accounts 2004* (Copenhagen: City of Copenhagen, 2005); J. Watts, "China Backs Bikes to Kick Car Habit," *Guardian* (London), 15 June 2006; tensions between drivers and pedestrians from John Whitelegg and Nick Williams, "Non-motorised Transport and Sustainable Development: Evidence from Calcutta," *Local Environment*, February 2000, pp. 7–18.

44. Charles Surjadi and Haryatiningsih Darrundono, *Review of Kampung Improvement Program Evaluation in Jakarta*, Final Report for Water and Sanitation Program by the Regional Water and Sanitation Group for East Asia and the Pacific (Jakarta: U.N. Development Programme/World Bank, 1998).

45. Glendening and Whitman quoted in "Stalling Sprawl: CT Must Take Leadership Role in Shaping Development," *Hartford Courant*, 17 July

2006. Box 4–3 from the following: population data from Instituto Brasilera de Geografía e Estatística, at ww.ibge.gov.br, viewed 7 August 2006; vehicle contribution to smog from "Brazil: Environmental Issues," Energy Information Administration, U.S. Department of Energy, August 2003, at www.eia.doe.gov/cabs/brazenv.html; Nelson Gouveia and Tony Fletcher, "Respiratory Diseases in Children and Outdoor Air Pollution in São Paulo, Brazil: A Time Series Analysis," *Occupational and Environmental Medicine*, July 2000, pp. 477–83; helicopter fleet from Adhemar Altieri, "Letter: Sao Paulo's Balancing Act," 18 August 2004, at news.bbc.co.uk/2/hi/programmes/3570402 .stm, viewed 7 August 2006; Laura Ceneviva, discussion with Jonas Hagen, 6 August 2006.

46. Kenworthy and Laube, op. cit. note 3; L. Schipper, C. Marie-Lilliu, and R. Gorham, *Flexing the Link Between Transport and Greenhouse Gas Emissions: A Path for the World Bank* (Paris: International Energy Agency, 2000); World Bank, *Sustainable Transport: Priorities for Policy Reform* (Washington, DC: 1996).

47. For the history of this legislation, see D. Camph, *Transportation, the ISTEA and American Cities* (Washington, DC: STPP, 1996); for reauthorization, see STPP, *Transfer Bulletin*, at www.transact.org.

48. Newman and Kenworthy, op. cit. note 11, pp. 191–97; A. Eichi et al., *A History of Japanese Railways, 1872–1999* (Tokyo: East Japanese Railway Culture Foundation, 2000).

49. Data on city wealth and transport from Newman and Kenworthy, op. cit. note 11; data on transportation costs from House of Representatives, *Sustainable Cities* (Canberra: Parliament of Australia, 2005), and from G. Glazebrook, "Taking the Con Out of Convenience: The True Cost of Transport Modes in Sydney," *Journal of Urban Policy and Planning*, forthcoming; Vuchic, *Transportation for Livable Cities*, op. cit. note 39; STPP and Center for Neighborhood Technology, *Driven to Spend: Pumping Dollars Out of Our Households and Communities* (Washington, DC: 2005).

50. Vuchic, *Urban Transit*, op. cit. note 39; data on parking calculated by J. Kenworthy, *Transport Energy in Australian Cities*, Honours Thesis (Perth, Australia: Murdoch University, 1979).

51. Calculations are by the authors and are partly published in Peter Vintila, John Phillimore, and Peter Newman, eds., *Markets, Morals and Manifestos* (Perth, Australia: Institute for Science and Technology Policy, Murdoch University, 1992); see also H. Frumkin, L. Frank, and R. Jackson, *Urban Sprawl and Public Health: Designing, Planning and Building for Healthy Communities* (Washington, DC: Island Press, 2004).

52. Center for Transit-Oriented Development, for the Federal Transit Administration, *Hidden in Plain Sight: Capturing the Demand for Housing Near Transit* (Oakland, CA: 2004).

53. Free Congress Foundation, *Conservatives for Mass Transit* (Washington, DC: 2003).

54. House of Representatives, op. cit. note 49.

55. Newman and Kenworthy, op. cit. note 11, pp. 191–204.

56. Perth data from Department of Transport, Perth, Australia, discussion with authors; Porto Alegre from D. Recondo, "Local Participatory Democracy in Latin America," New Frontiers of Social Policy, Arusha Conference, 12–15 December 2005; Milwaukee from "Public Opinion and Transportation Priorities in Southeastern Wisconsin," *Regional Report* (Public Policy Forum), June 2006; Oregon from InterACT, *Findings of the Transportation Priorities Project* (Vancouver, WA: 2003).

57. Center for Transportation Excellence, *Transportation Finance at the Ballot Box* (Washington, DC: 2006).

58. Beltline loop details and Calthorpe quote in Shaila Dewan, "The Greening of Downtown Atlanta," *New York Times*, 6 September 2006.

59. Data from Center for Transit-Oriented Development, at www.reconnectingamerica.org/

html/TOD; J. Hartz-Karp and P. Newman, "The Participative Route to Sustainability," in S. Paulin, ed., *Communities Doing It for Themselves: Creating Space for Sustainability* (Perth: University of Western Australia Press, 2006); A. Curry et al., *Intelligent Infrastructure Futures, The Scenarios–Towards 2055* (London: Foresight Directorate, Office of Science and Technology, U.K. Government, 2006).

Los Angeles: End of Sprawl

1. In the whole Los Angeles metropolitan region, 16 million people reside in five counties containing 177 cities covering about 36,000 square kilometers (along with 54,000 square kilometers of government-owned land in the region). Southern California Studies Center and the Brookings Institution Center on Urban and Metropolitan Policy, *Sprawl Hits the Wall* (Los Angeles: University of Southern California, 2001), p. 6; for 2005 census estimate, see quickfacts.census.gov/qfd/states/06000.html; population density calculations taken from *Demographia: World Urban Areas*, at www.demographia.com/db-worldua.pdf, viewed 13 February 2006, and from the U.S. Census Bureau 2000 census, in Blaine Harden, "Out West, a Paradox: Densely Packed Sprawl," *Washington Post*, 11 August 2005.

2. For an extensive description of suburban postwar expansion, zoning, and racial restrictions in Los Angeles, see Dana Cuff, *The Provisional City* (Cambridge, MA: The MIT Press, 2000).

3. U.S. Bureau of the Census, *Census of Housing, Historical Census of Housing Tables* (Suitland, MD: revised 2 December 2004).

4. Southern California Studies Center and Brookings Institution, op. cit. note 1.

5. Population density calculations taken from *Demographia*, op. cit. note 1.

6. Allen J. Scott and E. Richard Brown, *South-Central Los Angeles: Anatomy of an Urban Crisis*, Lewis Center for Regional Policy Studies, Working Paper No. 6 (Los Angeles: University of California at Los Angeles, 1993).

7. John Landis, H. Hood, and C. Amado, "The Future of Infill Housing in California," *Frameworks* (University of California, Berkeley), spring 2006, pp. 14–21.

Melbourne: Reducing a City's Carbon Emissions

1. Queen Victoria Market from City of Melbourne Web site, at www.melbourne.vic.gov.au.

2. Geoff Lawler, director, Sustainability and Innovation, "Towards a Thriving and Sustainable City," presentation to the Seattle Trade Alliance, May 2006.

3. City of Melbourne, *Zero Net Emissions by 2020–A Roadmap for a Climate Neutral City* (Melbourne: 2003); five milestones from "How It Works," Cities for Climate Protection, at iclei.org/index.php?id=810.

4. Quote from Ben Heywood, *Melbourne Age*, 14 August 2006.

5. Ibid.

6. City of Melbourne Web site, op. cit. note 1.

7. Greenfleet from www.greenfleet.org.au; Mayor's offset from Sara Gipton, acting CEO, Greenfleet, discussion with author.

8. City of Melbourne Web site, op. cit. note 1.

9. T. Roper, "Greening Major Public Events," *Global Urban Development*, March 2006.

10. So quote from *WME Weekly, Environment Business Media*, 3 August 2006.

11. Broad support indicated by original decision in July 1998 after a joint presentation to Council by Ian Carruthers (Australian Greenhouse Office) representing the Liberal Minister of the Environment and the author; National Australia Bank, *Corporate Social Responsibility Report 2005* (Melbourne: 2005), p. 53.

12. City of Melbourne, *City Plan 2010* (Mel-

bourne: 2005), pp. 60, 65.

Chapter 5. Energizing Cities

1. Real limits from John Byrne et al., "An Equity- and Sustainability-Based Policy Response to Global Climate Change," *Energy Policy*, March 1998, pp. 335–43.

2. Bridging to the Future, "How Have Energy Systems Shaped Cities Through History? Human Food Cities; Wood and Hay Cities; Coal Cities," at www.bridgingtothefuture.org, viewed 8 August 2006.

3. Tetsunari Iida, Institute for Sustainable Energy Policies, Tokyo, e-mail to Janet Sawin, 29 August 2006; New York most efficient from GreenHomeNYC, "For Tenants," 2003, at www.greenhomenyc.org/page/tenants.

4. India from "Underpowering," *The Economist*, 22 September 2005; China from Arno Rosemarin, United Nations Development Programme (UNDP), *China Human Development Report 2002: Making Green Development a Choice* (New York: Oxford University Press, 2002), p. 57.

5. Box 5–1 from the following: Emissions and 40 percent from U.N. Environment Programme (UNEP), International Environmental Technology Centre, *Energy and Cities: Sustainable Building and Construction* (Osaka, Japan: 2003); embodied energy of concrete from Alex Wilson, "Cement and Concrete: Environmental Considerations," *Environmental Building News*, March 1993; embodied energy of steel from Center for Building Performance Research, University of Wellington, "Table of Embodied Energy Coefficients," 7 July 2004, at www.vuw.ac.nz/cbpr/resources/index.aspx; San Francisco home based on 66.6 million Btus per household in Pacific region for space and water heating, cooling, refrigerators, lighting, and appliances, from U.S. Department of Energy (DOE), Energy Information Administration (EIA), "Total Energy Consumption in U.S. Households by West Census Region, 2001," updated 18 November 2004, at www.eia.doe.gov/emeu/recs/recs2001/ce_pdf/enduse/ce1-12c-westregion2001.pdf; embodied energy of

cement from Wilson, op. cit. this note; emissions from cement industry from Nadav Malin, "The Fly Ash Revolution: Making Better Concrete with Less Cement," *Environmental Building News*, June 1999; emissions in Japan from Carbon Dioxide Information Analysis Center, cited in UNDP, "Human Development Reports 2005—Indicators: Carbon Dioxide Emissions, Share of World Total," at hdr.undp.org/statistics/data/indicators.cfm?x=212&y=1&z=1; 15 percent from U.S. Environmental Protection Agency, "Cement and Concrete," updated 15 August 2006, at www.epa.gov/cpg/products/cement.htm; potential savings from Malin, op. cit. this note; Germany from World Resources Institute (WRI), *Climate Analysis Indicators Tool* (Washington, DC: 2003); lighter, stronger bricks from "Superior Building Products," University of New South Wales, New South Innovations (Sydney, Australia: June 2006); completely replacing cement from Doug Cross, Jerry Stephens, and Jason Vollmer, *Structural Applications of 100 Percent Fly Ash Concrete* (Billings, MT: Montana State University, 2005); benefits of local materials from Ibrahim Togola, Mali Folkecentre, "Sustainable Building of Local Materials in Sahel Countries of West Africa" (draft paper), sent to Janet Sawin, 4 August 2006; transportation share of embodied energy from Wilson, op. cit. this note; Big Dig waste from Raphael Lewis, "End Nears for Elevated Artery," *Boston Globe*, 14 April 2002; "Man Builds Home From Big Dig Scrap Materials," *Associated Press*, 30 July 2006.

6. Effect on 50 million people and blackout in Italy from Alan Katz, "Maintaining Facility Power in the Age of the Blackout," *Electrical Construction and Maintenance*, 1 June 2004; cause and costs of August blackout from Electricity Consumers Resource Council, *The Economic Impacts of the August 2003 Blackouts* (Washington, DC: 2004).

7. Line losses from World Bank, *World Development Report 1997* (New York: Oxford University Press, 1997), from M. S. Bhalla, "Transmission and Distribution Losses (Power)," in *Proceedings of the National Conference on Regulation in Infrastructure Services: Progress and Way Forward* (New Delhi: The Energy and Resources Institute, 2000), and from Seth Dunn, *Micropower: The Next Electrical Era*, Worldwatch Paper 151 (Washington, DC: Worldwatch Institute, 2000), p. 46; New Delhi from John Lancaster, "Sniffing Out the Freeloaders Who Stress the Grid," *Washington Post*, 12 June 2006.

8. One fifth from World Bank, *Energy Poverty Issues and G8 Actions*, Discussion Paper (Washington, DC: 2 February 2006), p. 1; number could be higher from "Power to the Poor," *The Economist*, 8 February 2001; Africans from Bereket Kebede and Ikhupuleng Dube, "Chapter 1: Introduction," in Bereket Kebede and Ikhupuleng Dube, eds., *Energy Services for the Urban Poor in Africa: Issues and Policy Implications* (London: Zed Books in Association with the African Energy Policy Research Network, 2004), p. 1.

9. Millions of deaths from "Power to the Poor," op. cit. note 8; India, Sri Lanka, and Thailand from Emily Matthews et al., *The Pilot Analysis of Global Ecosystems: Forest Ecosystems* (Washington, DC: WRI, 2000); Khartoum from Business in Africa, "Energy in Africa: Is There Energy for All?" 4 November 2005, p. 1.

10. Coal use from "Chapter 5: World Coal Markets," in EIA, *International Energy Outlook 2006* (Washington, DC: DOE, 2006); coal projections from International Energy Agency (IEA), *Key World Energy Statistics 2006* (Paris: 2006), p. 46; impact in China from Bill McKibben, "The Great Leap: Scenes from China's Industrial Revolution," *Harper's Magazine*, December 2005.

11. Cities account for 75 percent of world's fossil fuel consumption (and hence about 75 percent of energy-related emissions); see World Council for Renewable Energy (WCRE), "Renewable Energy and the City," discussion paper for World Renewable Energy Policy and Strategy Forum, Berlin, Germany, 13–15 June 2005.

12. Buildings more than 40 percent from UNEP, International Environmental Technology Centre, *Energy and Cities: Sustainable Building and Construction* (Osaka, Japan: 2003), p. 1; U.S. buildings from Greg Franta, "High-Performance Buildings Through Integrated Design," *RMI Solutions*, summer 2006, p. 6, and from Greg Franta, Rocky Mountain Institute, e-mail to Stephanie Kung, Worldwatch Institute, 20 September 2006.

13. Shanghai from David Barboza, "China Builds Its Dreams, and Some Fear a Bubble," *New York Times*, 18 October 2005; China urban infrastructure from McKibben, op. cit. note 10.

14. Lighting from IEA, "Light's Labour's Lost—Policies for Energy-Efficient Lighting," press release (Paris: 29 June 2006); Eric Corey Freed, "Ask the Green Architect: Mirrors for Lighting; Radiant Heating for Floors; Efficient Exit Signs," GreenBiz.com, undated; Jonathan Rider, *Light Shelves*, Advanced Buildings, Technologies and Practices (Ottawa, ON: Natural Resources Canada and Public Works and Government Services Canada); glass from Rick Cook, Partner, Cook + Fox Architects, interview on "The Green Apple," *Design E²*, U.S. Public Broadcasting System series, summer 2006.

15. Incandescent bulbs and compact fluorescents from DOE, Office of Energy Efficiency and Renewable Energy (EERE), "Technology Fact Sheet: Improved Lighting," GHG Management Workshop, 25–26 February 2003, pp. 1–2; DOE, EERE, "Energy Efficient Lighting and Light Emitting Diodes," fact sheet (Richland, WA: May 2006); DOE, EERE, "LED Traffic Lights Save Energy in Idaho," *Conservation Update*, May-June 2004.

16. Ceilings in South Africa from Randall Spalding-Fecher et al., "The Economics of Energy Efficiency for the Poor—A South Africa Case Study," *Energy*, December 2002, pp. 1,099–117.

17. Combined heat and power efficiency from DOE, EERE, "Distributed Energy Resources: Combined Heat & Power Program for Buildings, Industry and District Energy," at www.eere.energy.gov/de/pdfs/chp_buildings_industry_district.pdf, viewed 15 July 2006; Verdesian, 80 percent or higher, and typical fossil fuel plant from Robin Pogrebin, "Putting Environmentalism on the Urban Map," *New York Times*, 17 May 2006; typical plant efficiency also from DOE, Office of Fossil Energy, "DOE Launches Project to Improve

Materials for Supercritical Coal Plants," press release (Pittsburgh, PA: 16 October 2001).

18. Radiant heating from DOE, EERE, Office of Energy Efficiency and Renewable Energy, "Radiant Heating," updated 12 September 2005; Romans from Freed, op. cit. note 14; Hewlett building from The William and Flora Hewlett Foundation, "The Hewlett Foundation Building: Energy Efficiency," (Menlo Park, CA: updated 23 August 2005), and from Stephanie Kung, Worldwatch Institute, personal observations.

19. Cooling in China from National Renewable Energy Laboratory, *Renewable Energy in China: Development of the Geothermal Heat Pump Market in China* (Golden, CO: 2006); Tokyo and Houston from David J. Sailor and Chittaranjan Vasireddy, "Correcting Aggregate Energy Consumption Data to Account for Variability in Local Weather," *Environmental Modelling & Software*, May 2006, p. 733.

20. Susan Roaf and Mary Hancock, "Future-Proofing Buildings Against Climate Change Using Traditional Building Technologies in the Mediterranean Region," *EuroSun 98*, II.1.13, pp. 1–7.

21. M. Santamouris, "Special Issue of the Solar Energy Program Devoted to Natural Ventilation in Urban Areas" (editorial), *Solar Energy*, April 2006, pp. 369–70; EPA, "Heat Island Effect— What Can Be Done—Trees & Vegetation," fact sheet (Washington, DC: 9 June 2006).

22. Cool roof savings from Hashem Akbari, "Estimating Energy Saving Potentials of Heat Island Mitigation Measures," Heat Island Group, Lawrence Berkeley National Laboratory, powerpoint presentation, updated 16 June 1999.

23. EPA, op. cit. note 21; Madrid study and heat island reductions from Susana Saiz et al., "Comparative Life Cycle Assessment of Standard and Green Roofs," *Environmental Science & Technology*, 1 July 2006, pp. 4312–16; heat island and smog reductions from Akbari, op. cit. note 22.

24. Less than half from DOE, EERE, "Technology Fact Sheet: Resources for Whole Building Design," GHG Management Workshop, 25–26 February 2003, p. 11; savings up to 80 percent from James Read, associate director, Arup Communications, on "Deeper Shades of Green," *Design E²*, op. cit. note 14; savings in energy and construction costs from Franta, "High-Performance Buildings," op. cit. note 12, p. 7; Accord 21 Building from Robert Watson, senior scientist, Natural Resources Defense Council, on "China: From Red to Green?" *Design E²*, op. cit. note 14.

25. Healthier, more comfortable occupants and greater worker productivity from Gregory H. Kats, "Green Building Costs and Financial Benefits," Massachusetts Technology Collaborative, 2003, p. 6, from Judith Heerwagen, "Sustainable Design Can Be an Asset to the Bottom Line," *Environmental Design + Construction*, 15 July 2002, and from DOE, op. cit. note 24, p. 11; reduced turnover from "Study: Environmentally Friendly Buildings Also Most Market Friendly," Greenbiz.com, 31 October 2005; Heschong Mahone Group, "Daylighting in Schools: An Investigation Into the Relationship Between Daylighting and Human Performance," prepared for the California Board for Energy Efficiency (Fair Oaks, CA: 20 August 1999); Warren E. Hathaway et al., *A Study into the Effects of Light on Children of Elementary School Age—A Case Study of Daylight Robbery* (Edmonton, AB: Policy and Planning Branch, Planning and Information Services Division, Alberta Education, 1992); bank building from Nicholas Lenssen and David Roodman, *A Building Revolution*, Worldwatch Paper 124 (Washington, DC: Worldwatch Institute, 1994), p. 45; more than 10 times the benefits from Kats, op. cit. this note, p. 8; and from U.S. Green Building Council, "Green Buildings by the Numbers," 2006, at www.usgbc.org/DisplayPage.aspx?CMSPageID=1442; costs falling from Kats, op. cit. this note, p. 3.

26. One year payback from Franta, "High-Performance Buildings," op. cit. note 12, p. 7; retrofit projects from Jiang Lin et al., *Developing an Energy Efficiency Service Industry in Shanghai* (Berkeley, CA: Lawrence Berkeley National Laboratory, 2004), pp. 2, 17.

27. Anuj Chopra, "Low-cost Lamps Brighten

the Future of Rural India," *Christian Science Monitor*, 3 January 2006.

28. John Perlin, "Solar Evolution: The History of Solar Energy," California Solar Center, 2005, at www.californiasolarcenter.org/history_pas sive.html; Babylon from The Garland Company, "History of Green Roofs," at www.garlandco .com/green-roof-history.html, viewed 9 August 2006.

29. Edison and evolution of industry from Dunn, op. cit. note 7, pp. 6, 11, 13–14.

30. For microturbines, see, for example, Wilson TurboPower, Inc., "The Wilson Microturbine," at www.wilsonturbopower.com; fuel cells already producing power from Joel N. Swisher, *Cleaner Energy, Greener Profits: Fuel Cells as Cost-Effective Distributed Energy Resources* (Snowmass, CO: Rocky Mountain Institute, 2002), p. 12; Susan Nasr, "More Powerful Fuel Cells Get Closer to Market," *Technology Review*, 13 June 2006.

31. Energy & Enviro Finland, "Utilizing Biogas as a Fuel: Wärtsilä Fuel Cell Unit to Power the City of Vaasa," 15 June 2006.

32. Renewables from REN21 Renewable Energy Policy Network, *Executive Summary: Renewables Global Status Report 2006 Update* (Washington, DC: Worldwatch Institute, 2006); wind, solar, and biofuels from Worldwatch Institute, *Vital Signs 2006–2007* (New York: W. W. Norton & Company, 2006), pp. 36–41.

33. Cheaper for building facades from Steven Strong, "Solar Electric Buildings: PV as a Distributed Resource," *Renewable Energy World*, July-August 2002, p. 171; Europe from "BIPV Technology," Wisconsin Public Service, University of Wisconsin, at www.buildingsolar.com/tech nology.asp; use elsewhere from Natural Resources Canada, *Technologies and Applications—Photovoltaic: Integrating Photovoltaic Arrays in Buildings* (Ottawa, ON: updated 26 July 2006); potential in Finland, Australia, and United States (adjusted for the fact that it was based on 1998 electricity consumption data and conservative estimates for available rooftops and façades and for

solar resources) from IEA, *Summary: Potential for Building Integrated Photovoltaics* (Paris: 2002), p. 8.

34. UNDP, Equator Initiative, "Solar City—Germany," August 2000, at www.tve.org/ho/ doc.cfm?aid=657, viewed 23 September 2006; David Faiman, "Solar Energy in Israel," Ben-Gurion University of the Negev, Sde Boker, Israel, 26 November 2002; fuel savings from Environmental and Energy Study Institute, "Renewable Energy Fact Sheet: Solar Water Heating—Using the Sun's Energy to Heat Water" (Washington, DC: May 2006); China solar energy from REN21, *Renewables Global Status Report 2006 Update* (Paris and Washington, DC: REN21 Secretariat and Worldwatch Institute, 2006); China's driver from Zijun Li, "Solar Energy Booming in China," *China Watch*, 23 September 2005.

35. P. J. Hughes and J. A. Shonder, *The Evaluation of a 4000 Home Geothermal Heat Pump Retrofit at Fort Polk, Louisiana: Final Report* (Oak Ridge, TN: Oak Ridge National Laboratory, March 1998), p. 2; Beijing Linked Hybrid Project from Li Hu, Partner, Steven Holl Architects, on "China: From Red to Green?" *Design E²*, op. cit. note 14.

36. Pompeii and list of countries from Geothermal Education Office, "Geothermal Energy," slideshow, funded by DOE, undated; Paris from European Renewable Energy Council, "Joint Declaration for a European Directive to Promote Renewable Heating and Cooling," Brussels, undated, p. 8.

37. New York waste and shipped to Ohio from Timothy Gardner, "Hot Trash-to-Fuel Technology Gathering Steam," *Reuters*, 27 February 2004; disposal costs as of 2002 from Steven Cohen, "Putting Garbage to Good Use," *New York Times*, 15 August 2002; increasing waste in industrial countries from Euiyoung Yoon and Sunghan Jo, "Municipal Solid Waste Management in Tokyo and Seoul," Proceedings of Workshop of IGES/APN Mega-City Project, Kitakyushu, Japan, 23–25 January 2002, p. 1; developing countries, garbage burned or left to rot, and impacts from "Hazardous Waste: Special Reference to Municipal Solid Waste Management," in The Energy and Resources Institute, *India: State of the Environment 2001* (Delhi: 2001), pp. 133–41; 90 percent not collected from UNEP, "At a Glance: Waste," *Our Planet*, 1999.

38. Methane potency from EPA, "Global Warming—Emissions," at yosemite.epa.gov/OAR/globalwarming.nsf/content/Emissions.html; U.S. cities from Daniela Chen, "Converting Trash Gas into Energy Gold," CNN.com, 17 July 2006; São Paulo and Riga from Carl R. Bartone, Horacio Terraza, and Francisco Grajales-Cravioto, "Opportunities for LFGTE Projects in LAC Utilizing International Carbon Financing," World Bank, presentation at LMOP 8th Annual Conference, Baltimore, MD, 10–11 January 2005; Cheryl Smith, "Monterrey Plans to Turn Rotting Garbage into Electricity," *Christian Science Monitor*, 21 March 2002.

39. Incomes spent on cooking fuel from Christopher Flavin and Molly Hull Aeck, *Energy for Development: The Potential Role of Renewable Energy in Meeting the Millennium Development Goals* (Washington, DC: Worldwatch Institute, 2005), p. 17; Innocent Rutamu, "Low Cost Biodigesters for Zero Grazing Smallholder Diary Farmers in Tanzania," *Livestock Research for Rural Development*, July 1999.

40. European cities and San Francisco from "San Francisco to Test Turning Dog Waste into Power," *Reuters*, 23 February 2006; Alister Doyle, "Oslo's Sewage Heats Its Homes," *Reuters*, 10 April 2006; Helsingborg from Michael D. Lemonick, "Cleaner Air Over Scandinavia," *Time*, 3 April 2006, p. 47; hospital and industrial wastes to electricity from Timothy Gardner, "Hot Trash-to-Fuel Technology Gathering Steam," *Reuters*, 27 February 2006; "$84 Million for the First Tires-to-Ethanol Facility," RenewableEnergyAccess.com, 23 March 2006.

41. "Tokyo Embraces Renewable Energy," *Environment News Service*, 6 April 2006; "Boston's First Wind Turbine Serves as Example," RenewableEnergyAccess.com, 18 May 2005; Middelgrunden Wind Turbine Cooperative, "The Middelgrunden Offshore Wind Farm—A Popular Initiative," undated, at www.middelgrunden.dk/MG_UK/project_info/mg_pjece.htm.

42. New York and San Francisco from Jeff Johnson, "Power from Moving Water," *Chemical and Engineering News*, 4 October 2004, and from Adam Aston, "Here Comes Lunar Power," *Business Week*, 6 March 2006; Paris from Doyle, op. cit. note 40; "Deep Lake Water Cooling: Chilled Water for Cooling Toronto's Buildings," at www.enwave.com/enwave/view.asp?/dlwc/energy, viewed 6 August 2006.

43. Environment Department, City of Malmö, "100 Percent Locally Renewable Energy in the Western Harbour of Malmö in Sweden," *ICLEI in Europe: Cities in Action—Good Practice Examples*, at www.iclei-europe.org; Dongtan from Fred Pearce, "Eco-cities Special: A Shanghai Surprise," *New Scientist*, 21 June 2006; expected population from Jean-Pierre Langellier and Brice Pedroletti, "China to Build First Eco-city," *The Guardian Weekly*, 2006.

44. Janet L. Sawin, "National Policy Instruments: Policy Lessons for the Advancement & Diffusion of Renewable Energy Technologies Around the World," background paper prepared for Secretariat of the International Conference of Renewable Energies, Bonn, Germany, January 2004, p. 24.

45. Blackout from Richard Perez et al., "Solution to the Summer Blackouts? How Dispersed Solar Power-Generating Systems Can Help Prevent the Next Major Outage," *Solar Today*, July/August 2005, and from Richard Perez, Atmospheric Sciences Research Center, State University of New York at Albany, e-mail to Janet Sawin, 3 October 2006.

46. UNDP survey from Molly O'Meara, *Reinventing Cities for People and the Planet*, Worldwatch Paper 147 (Washington, DC: 1999), p. 57; Germany from Preben Maegaard, "Wind, Not Nuclear!–Why Does the UK Not Take This Opportunity?" WCRE, July 2006; China's solar heating industry from REN21 Renewable Energy Policy Network, *Renewables 2005 Global Status Report* (Washington, DC: Worldwatch Institute,

2005), pp. 24–25; India's biogas industry from Institute of Science in Society, "Biogas Bonanza for Third World Development," press release (London: 20 June 2005).

47. Box 5–2 from the following: event statistics and Olympics overview from The Hon. Tom Roper, "The Environmental Challenge of Major Events," presented at the Eighth World Congress Metropolis, Berlin, Germany, 2005, pp. 1–2, 6, and from Tom Roper, "Producing Environmentally Sustainable Olympic Games and Greening Major Public Events," *Global Urban Development Magazine*, March 2006; additional Beijing data from Environment News Service, "Beijing Enlists U.S. Help to Green the 2008 Olympic Games," 18 April 2005, p. 1; World Cup data from Fédération Internationale de Football Association (FIFA), "Green Goal™: Environmental Protection Targets," Zurich, 2005; renewable sources from FIFA, "Sunny Days Ahead in Kaiserslautern," Zurich, 26 May 2006.

48. Forum Barcelona, "Imma Mayol: The Closure of Nuclear Power Stations in Catalonia is One of the Prime Objectives to Be Carried Out in the Next 15 Years, in Terms of Defining a New Energy Model," 2004; Ajuntament de Barcelona, *Plan for Energy Improvement in Barcelona* (Barcelona: 2002); elections, sunlight and energy equivalent, demonstration, and timeline from Pamela Stirzaker, "Spain's Chain Reaction: Municipal Obligations Spur on Solar Thermal Growth," *Renewable Energy World*, September-October 2004, pp. 2–3, and from Josep Puig i Boix, "The Barcelona Solar Ordinance: A Case Study About How the Impossible Became Reality," presented at the International Sustainable Energy Organization Special Session, World Summit for Sustainable Development, Johannesburg, South Africa, 28 August 2002.

49. Toni Pujol, Barcelona Energy Agency, "The Barcelona Solar Thermal Ordinance: Evaluation and Results," presented at the 9th Annual Conference of Energie-Cités, Martigny, 22–23 April 2004; expanded requirements from REN21, op. cit. note 34, p. 10; more than 70 cities and national government from European Solar Thermal Industry Federation, "Spain Approves National Solar Thermal Obligation," at www.estif.org, viewed 7 July 2006, and from REN21, op. cit. note 34.

50. Sacramento Municipal Utility District (SMUD), "EBSS Solicitations–Solicitation Detail," at www.bids.smud.org/sDsp/sDsp004.asp?solicitation_id=2195, viewed 19 September 2006; SMUD, "Solar for Your Home: PV Pioneers," at www.smud.org/green/solar/index.html, viewed 21 September 2006; from SMUD, "Solar Power for Your Business," at www.smud.org/green/solar/compv.html, viewed 21 September 2006.

51. Heat wave and ComEd settlement from Ken Regelson, *Sustainable Cities: Best Practices for Renewable Energy & Energy Efficiency* (Denver, CO: Sierra Club–Rocky Mountain Chapter, 2005), pp. 10–15; 20 percent agreement from City of Chicago, Office of the Mayor and Department of Environment, *Energy Plan* (Chicago, IL: 2001); target year moved to 2010, for budgetary and contractual reasons, from Mike Johnson, project coordinator, City of Chicago Department of Energy, Environment and Air Quality, discussion with Kristen Hughes, 4 October 2006, and from SustainLane, "#4 Chicago: The Wind at its Back," at www.sustainlane.com/article/846, viewed 7 October 2006.

52. "Most environmentally friendly" from City of Chicago, "A Message from the Mayor," at www.cityofchicago.org/Transportation/bikemap, viewed 26 September 2006; Regelson, op. cit. note 51.

53. Kevin McCarthy, "Chicago Approves Big Grants for Green Roof Retrofits," Construction.com, 19 July 2006; trees from J. Slama, "Chicago Will Be America's Greenest City," *Conscious Choice*, April 2002.

54. David Engle, "With the Power at Hand: Examining the Merits of Distributed Energy," *Planning Magazine* (American Planning Association), July 2006; community aggregation impacts from Donald Aitken, Donald Aitken Associates, e-mail to Kristen Hughes, 4 September 2006.

55. Jong-dall Kim, Dong-hi Han, and Jung-gyu Na, "The Solar City Daegu 2050 Project: Visions for a Sustainable City," *Bulletin of Science, Technology & Society*, April 2006, pp. 99–100; financial crisis from EIA, "South Korea: Environmental Issues," at www.eia.doe.gov/emeu/cabs/skoren .html, viewed 9 August 2006.

56. Renewables goal from REN21, op. cit. note 46, p. 28; 2050 planning from Eric Martinot, "Solar City Case Study: Daegu, Korea," Renewable Energy Information on Markets, Policy, Investment, and Future Pathways, 2004, p. 1; Kim, Han, and Na, op. cit. note 55, pp. 98–99.

57. Population from Dejan Sudjic, "Making Cities Work: Mexico City," *BBC News*, 21 June 2006; haze from Michelle Hibler, "Taking Control of Air Pollution in Mexico City," *IDRC Reports*, 12 August 2003; "most dangerous" title from "New Center a Breath of Fresh Air for Mexico City," *Environment News Service*, 3 June 2002; remains among most polluted from EIA, "Mexico: Environmental Issues," at www.eia.doe.gov/ emeu/cabs/mexenv.html, viewed 8 August 2006.

58. Proaire campaign and installation figures from The Climate Group, *Less Is More: 14 Pioneers in Reducing Greenhouse Gas Emissions* (Woking, Surrey, U.K.: 2004), p. 29; supporting organizations from The Climate Group, "Mexico City–Municipal Government," at www.thecli mategroup.org/index.php?pid=427, viewed 5 August 2006.

59. The Climate Group, *Cape Town–Municipal Government* (Woking, Surrey, U.K.: undated), pp. 1–3; Gold Standard from Renewable Energy and Energy Efficiency Partnership (REEEP), "CDM Housing Project to Become Replicable Energy Savings Model for South Africa," at www.reeep.org/index.cfm?article id=1198&ros=1, viewed 6 September 2006.

60. Table 5–1 from the following: "China's Capital Launches Plan to Save Sparse Energy for Sustainable Development," *People's Daily Online*, 7 June 2005; Berlin, Copenhagen, Melbourne, and Tokyo from The Climate Group, *Low Carbon Leader: Cities Oct. 2005* (Woking, Surrey, U.K.: 2005); Freiburg, Oxford, and Portland from Eric Martinot, "Index of Solar Cities," at www.martin

ot.info/solarcities, viewed 19 July 2006; Leicester from ICLEI, "Profiting from Energy Efficiency: 7.0 Best Municipal Practices for Energy Efficiency," at www.iclei.org/index.php?id= 1677&0, viewed 11 July 2006; City of Portland, Resolution Adopted 27 April 2005, at www.port landonline.com/osd/index.cfm?a=112681&c=41 701, viewed 7 August 2006; Toronto Environmental Alliance, *Getting Green Power On-line in Toronto* (Toronto, ON: 2005); Tokyo Metropolitan Government Bureau of Environment, "Tokyo Renewable Energy Strategy," 3 April 2006, at www.isep.or.jp/e/Eng_project/TokyoREstrat egy060526.pdf, viewed 20 July 2006.

61. U.S. Mayors' agreement and ICLEI from Wilson Rickerson and Kristen Hughes, "The Policy Framework for Greenhouse Gas Reductions in New York City," presented at the 2006 International Solar Cities Congress, Oxford, U.K., 4 April 2006; ICLEI, "About CCP," at www.iclei.org/ index/php?id=811, viewed 19 September 2006; Office of the Mayor, *U.S. Mayors Climate Protection Agreement* (Seattle, WA: 2005); Australia ICLEI from Tom Roper, Project Leader, Global Sustainable Energy Islands Initiative, discussion with Kristen Hughes, 17 August 2006, and from Cities for Climate Protection® Australia, "About CCP Australia," at www.iclei.org/index.php?id =about, viewed 19 September 2006.

62. Target for pathfinders from International Solar Cities Initiative, "International Solar Cities Congress 2006," at www.solarcities.org.uk, viewed 19 July 2006; 3.3 tons from Byrne et al., op. cit. note 1; Argentina and China from United Nations Statistics Division, "Millennium Development Goals Indicators: Data Availability by Country," at millenniumindicators.un.org/unsd/mdg/default .aspx, viewed 19 September 2006.

63. Investment priorities from WCRE and Asia Pacific & Renewable Energy Foundation Limited, *Asia Pacific Renewable Energy and Sustainable Development Agenda 2004* (Bonn, Germany: 2004), pp. 4–6; initiatives in global programs from Practical Action, *Power to the People: Sustainable Energy Solutions for the World's Poor* (Rugby, U.K.: 2002), p. 8.

64. Subsidies from WCRE, *Action Plan for the Global Proliferation of Renewable Energy* (Bonn, Germany: 2002), p. 8, and from WCRE, "First World Renewable Energy Policy and Strategy Forum Successfully Carried Out," press release (Bonn, Germany: 18 June 2002); The White House, *Department of Energy–Overview: The Budget for Fiscal Year 2005* (Washington, DC: 2005); India and China from Jon Gertner, "Atomic Balm?" *New York Times*, 16 July 2005; leadership from WCRE and Asia Pacific & Renewable Energy Foundation Limited, op. cit. note 63; Janet Sawin, "Charting a New Energy Future," in Worldwatch Institute, *State of the World 2003* (New York: W. W. Norton & Company, 2003), p. 105.

65. Lori Bird et al., "Policies and Market Factors Driving Wind Power Development in the United States," *Energy Policy*, July 2005, p. 1,405; electricity privatization from Kirsty Hamilton, "Finance and Investment: A Challenge of Scale," *Renewable Energy World*, September-October 2002.

66. Negawatts from Amory Lovins, "The Negawatt Revolution—Solving the CO_2 Problem," presented at Green Energy Conference, Montreal, 14–17 September 1989.

67. Anchor tenants from David Roeder, "Eco-Friendly Builders Starting to Grow," *Chicago Sun Times*, 20 February 2006; energy costs small share of expenses from Board of Governors of the Federal Reserve System, *Monetary Policy Report to the Congress* (Washington, DC: 19 July 2006), p. 20; savings not reflected in conventional accounting from Massachusetts Public Interest Research Group, "Testimony on Senate Bill 360: 'An Act To Promote An Energy Efficient Massachusetts,' Cost-Effective Solutions to Protect Consumers, Reduce Pollution, and Boost Our Economy," submitted 27 March 2001.

68. Juliet Eilperin, "22 Cities Join Clinton Anti-Warming Effort," *Washington Post*, 2 August 2006.

69. DOE, "Energy Savers: Homeowners," at www.energysavers.gov/homeowners.html, viewed 6 September 2006, and "Building Energy Efficient New Homes," at www.eere.energy.gov/buildings/info/homes/newconstruction.html, viewed 6 Sep-

tember 2006.

70. Local conditions and expertise from UNEP, op. cit. note 12, pp. 1, 3, 8–9; institutional capacity from Tom Roper, "5 Star Housing–Victoria, Australia: Performance Based Building Regulation Delivers Major Sustainability Outcomes," presented to Greenbuild 2005, Atlanta, GA, 9–11 November 2005, pp. 1, 5–7.

71. Chicago from Tom Roper, Project Leader, Global Sustainable Energy Islands Initiative, e-mail to Kristen Hughes, 21 August 2006, and from Chicago Department of Construction and Permits, *Green Permit Program* (Chicago: date unknown), p. 3; "New Urbanism: Creating Livable Sustainable Communities," at www.newurbanism.org, viewed 5 September 2006.

72. Table 5–2 from the following: Cooperatives from Paul Gipe, *Community Wind: The Third Way* (Toronto: Ontario Sustainable Energy Association, 2004); fuelwood pricing and replanting from Practical Action, *Energy: Working with Communities to Provide Appropriate Solutions* (Rugby, U.K.: 2006), p. 5; secondary power and lifeline tariffs from Kebede and Dube, op. cit. note 8, pp. 1, 4, 6–8; energy service companies from Lin et al., op. cit. note 26, p. 2; microfinance and loans from Abhishek Lal and Betty Meyer, "An Overview of Microfinance and Environmental Management," Global Development Research Center, at www.gdrc.org/icm/environ/abhishek.html, viewed 19 April 2006, p. 9; bundling from Hamilton, op. cit. note 65, pp. 5–6; solar water heaters from REEEP, "Innovative Financing for Solar Water Heating Increases Affordability," press release (Vienna, Austria: April 2006); importance of champions and utility involvement from Roper, op. cit. note 71; demonstration projects and awareness from Practical Action, op. cit. note 63, pp. 3–4; state and city de facto policy from Aitken, op. cit. note 54.

73. Partnerships and commitments from Practical Action, op. cit. note 63, p. 1; importance of private-sector allies from Innovest Strategic Value Advisors, *Climate Change & The Financial Services Industry: Module 1–Threats and Opportunities*, prepared for the UNEP Finance Initiatives Climate

Change Working Group (Toronto, ON: 2002).

Rizhao: Solar-Powered City

1. Population from Rizhao City Government, *2005 Rizhao Economic and Social Development Statistic Bulletin*, at www.rizhao.gov.cn/rztj/show .asp?ListName=YDTJ&ID=15, viewed 10 June 2006.

2. Usage data from Rizhao City Construction Committee, internal statistics.

3. Li Zhaoqian, Presentation at World Urban Forum III, Vancouver, 19–23 June 2006; Wang Shuguang, e-mail to author, 15 October 2006.

4. Li Zhaoqian, discussion with author, 17 June 2006.

5. Costs from Li, op. cit. note 3; average salaries from Rizhao City Government, op. cit. note 1.

6. Installations first for government buildings and homes of city leaders from Li, op. cit. note 3; free installation for some employees from Li, op. cit. note 4; Wang Shuguang, e-mails and discussions with author, August and October 2006.

7. Li, op. cit. note 4.

8. State Environmental Protection Agency, *2005 Annual Report of Urban Environmental Management and Comprehensive Pollution Control* (Beijing: 2006).

9. Wang, op. cit. note 6.

10. Tourist numbers from Rizhao City Government, op. cit. note 1; Rizhao City Government, *2004 Rizhao Economic and Social Development Statistic Bulletin*, at www.rizhao.gov.cn/rztj/ show.asp?ListName=YDTJ&ID=5., viewed 10 June 2006.

11. "More than 300 Peking University Professors Bought Houses in Rizhao," *Beijing Youth Daily* (*Beijing Qingnian Bao*), 11 August 2006; Qufu from Wang Shuguang, discussion with author, 11 October 2006.

Malmö: Building A Green Future

1. Height of Turning Torso from Sam Lubbell, "Just Opened...." *Architectural Record*, January 2006, p. 36; population from City of Malmö Web site, at www.malmo.se; Santiago Calatrava from Paul Goldberger, "The Sculptor," *The New Yorker*, 31 October 2005, pp. 88–90.

2. Earliest settlements from City of Malmö, at www.malmo.se/turist/inenglish/malmocityofdi versityandpossibilities.4.33aee30d103b8f1591680 0021971.html; foreign-born population from City of Malmö, op. cit. note 1; Kockum's history from Christer Persson, "Sweden Learns the Harsh Lessons of Regeneration," Special Supplement, (London) *Guardian*, 19 January 2005; related industries and economic downturn from Kevin Done with Hilary Barnes, "Survey: Malmo and Southern Sweden," *Financial Times*, 25 May 1983; number leaving from Malmö, at Wikipedia.

3. Mats Olsson, "Bo01 as a Strategic Project," in FORMAS (Swedish Research Council for Environment, Agricultural Sciences and Spatial Planning), *Sustainable City of Tomorrow, Bo01–Experiences of a Swedish Housing Exposition* (Stockholm: 2005).

4. Daniel Nilsson, "The LIP Programme—A Prerequisite for the Environmental Initiatives," in FORMAS, op. cit. note 3.

5. Li Löverhed, "100 Per Cent Local Renewable Energy," in FORMAS, op. cit. note 3.

6. Information from Tor Fossum, Environmental Strategy Unit, Environmental Department, City of Malmö, discussion with author, 8 October 2006.

7. Carbon dioxide reduction target from "58 Environmental Objectives for the City of Malmö," at www.malmö.se; further information on SECURE available at www.secureproject.org.

8. Current population for the Västra Hamnen area from City of Malmö Web site, at www.malmo.se/faktaommalmopolitik/statistik/ eomradesfaktaformalmo/omradesfakta2005; pro-

jected population from "Västra Hamnen: The Bo01-Area, A City for People and the Environment," information folder, City of Malmö, at www.ekostaden.com/pdf/vhfolder_malmostad_0308_eng.pdf.

9. Housing construction and education from "Malmö 2006, Facts and Figures," leaflet, City of Malmö, Serviceförvaltningen, 2006; joblessness from *København Malmö 2006* (Malmö and Copenhagen: City Office of Malmö and Statistical Office of Copenhagen, 2006).

10. Population for Rosengård area from City of Malmö Web site, op. cit. note 8.

11. Jeanette Andersson, discussion with author, 27 September 2006.

Chapter 6. Reducing Natural Disaster Risk in Cities

1. John Noble Wilford, "Scientists Unearth Urban Center More Ancient Than Plato," *New York Times*, 2 December 2003; Helike Foundation, "Appendix B: Helike and Atlantis," at www.helike.org/atlantis.shtml.

2. Helike Foundation, "The Lost Cities of Ancient Helike," at www.helike.org/index.shtml; John Noble Wilford, "Ruins May Be Ancient City Swallowed by Sea," *New York Times*, 17 October 2000; "Helike—The Real Atlantis," *BBC*, 10 January 2002; Steven Soter, research scientist, American Museum of Natural History, e-mail to author, 5 October 2006.

3. Worldwatch calculations based on data from "EM-DAT: The OFDA/CRED International Disaster Database," Université Catholique de Louvain, Brussels, Belgium, at www.em-dat.net, viewed 16 September 2006. This chapter includes the following hazards under the umbrella term "natural disaster": drought, earthquake, extreme temperature, flood, slides, volcano, wildfire, and wind storm. Note: due to a change in disaster recording methods in 2003, recent figures on the number of disasters may appear artificially inflated compared with historical figures; see Center for Research on the Epidemiology of Disasters

(CRED), "EM-DAT Data Entry Procedures," at www.em-dat.net/guidelin.htm. Definition of natural disaster from CRED, "EM-DAT Criteria and Definition," at www.em-dat.net/criteria.htm, viewed 16 September 2006.

4. Data in Figure 6–1 are a Worldwatch calculation based on "EM-DAT: The OFDA/CRED International Disaster Database," op. cit. note 3, viewed 4 October 2006; city and national populations from United Nations Population Division, "File 14: The 30 Largest Urban Agglomerations Ranked by Population Size, 1950–2015," and "File 1: Total Population at Mid-Year by Major Area, Region and Country, 1950–2030 (thousands)," *World Urbanization Prospects 2005* (New York: 2006); definition of "total affected" from CRED, "EM-DAT Criteria and Definition," op. cit. note 3, viewed 4 October 2006.

5. Worldwatch calculations based on data from "EM-DAT: The OFDA/CRED International Disaster Database," op. cit. note 3, viewed 21 July 2006, and from United Nations, "File 1: Total Population at Mid-Year," op. cit. note 4.

6. United Nations, "File 2: Urban Population at Mid-Year by Major Area, Region and Country, 1950–2030 (thousands)," *World Urbanization Prospects 2005* (New York: 2005); urban population at risk of earthquakes from GeoHazards International/UN Center for Regional Development, *Global Earthquake Safety Initiative Pilot Project: Final Report* (Palo Alto, CA: 2001), p. 1.

7. Physical surroundings as cause of death from Jaime Valdés, "Disaster Risk Reduction: A Call to Action," *@local.glob 3* (Delnet Journal), 2006. Box 6–1 from the following: Charlotte Benson and John Twigg, "Measuring Mitigation: Methodologies for Assessing Natural Hazard Risks and the Net Benefits of Mitigation—A Scoping Study" (Geneva: International Federation of Red Cross and Red Crescent Societies/ ProVention Consortium, December 2004); U.N. International Strategy for Risk Reduction, "Terminology: Basic Terms of Disaster Risk Reduction," at www.unisdr.org/eng/library/lib-terminology-eng-p.htm, viewed 13 September 2006. Box 6–2 from the following: Mark Pelling, *The Vulnerability of Cities:*

Natural Disasters and Social Resilience (London: Earthscan, 2003), p. 57; Inter-agency Secretariat for the International Strategy for Disaster Reduction (UN/ISDR), *Women, Disaster Reduction, and Sustainable Development* (Geneva: undated), p. 4; Ian Davis and Yasamin Izadkhah, "Building Resilient Urban Communities," *Open House International*, March 2006, pp. 11–21; James Morrissey and Anna Taylor, "Fire Risk in Informal Settlements: A South African Case Study," *Open House International*, March 2006, pp. 98–104.

8. Independent Evaluation Group, *Hazards of Nature, Risks to Development: An IEG Evaluation of World Bank Assistance for Natural Disasters* (Washington DC: World Bank, 2006), p. xx.

9. Worldwatch "slum" growth calculation based on UN-HABITAT, *State of the World's Cities 2006/7* (London: Earthscan, 2006), p. 16; Mike Davis, "Slum Ecology," *Orion Magazine*, March/April 2006; traditional networks from Munich Re, *Megacities—Megarisks: Trends and Challenges for Insurance and Risk Management* (Munich: 2004), p. 23; Mark Pelling, *The Vulnerability of Cities: Natural Disasters and Social Resilience* (London: Earthscan, 2003), p. 7.

10. Christine Wamsler, "Managing Urban Risk: Perceptions of Housing and Planning as a Tool for Reducing Disaster Risk," *Global Built Environment Review*, vol. 4, no. 2 (2004), p. 15; building standards in Mexico City from United Nations University, "Mexico City, Mexico: We're Still Here," video, at www.unu.edu/env/urban/social-vulnerability. Table 6–2 from the following: populations from United Nations, "File 14: The 30 Largest Urban Agglomerations," op. cit. note 4; disaster risk from "Megacities and Natural Hazards," insert in Munich Re, op. cit. note 9.

11. Morrissey and Taylor, op. cit. note 7; Davis, op. cit. note 9.

12. Henny Vidiarina, "The Challenges and Lessons of Working with Communities in Urban Areas: The ACF Experience in Kampung Melayu," in *Partnerships for Disaster Reduction–Southeast Asia 3 Newsletter*, April 2006, pp. 3–4.

13. "New York: A Documentary Film, Episode 1: The Country and the City," Public Broadcasting System, 1999.

14. Terry McPherson and Wendell Stapler, "Tropical Cyclone 05B," in *1999 Annual Tropical Cyclone Report* (Pearl Harbor, HI: U.S. Naval Pacific Meteorology and Oceanography Center/Joint Typhoon Warning Center, 1999). Table 6–2 from the following: Unless otherwise noted, deaths and economic damages from Munich Re, op. cit. note 9, p. 21; Hurricane Katrina deaths from Gary Younge, "Gone with the Wind," *The Guardian* (London), 26 July 2006; Hurricane Katrina economic losses from Munich Re, "Two Natural Events Play a Prominent Role in the 2005 Catastrophe Figures," press release (Munich: 29 December 2005); Mumbai deaths from Somini Sengupta, "Torrential Rain Reveals Booming Mumbai's Frailties," *New York Times*, 3 August 2005; Mumbai economic losses from Radhika Menon, "A 'Disastrous' Year for Insurers," *Hindu Business Line*, 1 January 2006; Bam deaths from IRIN News, "Iran: Tehran Lowers Bam Earthquake Toll," press release (New York: U.N. Office for the Coordination of Humanitarian Affairs, 30 March 2004); Bam economic losses from International Federation of Red Cross and Red Crescent Societies, *World Disasters Report 2005* (Geneva: 2005); Bhuj deaths and economic losses from Anil Kkumar Sinha, *The Gujarat Earthquake 2001* (Kobe, Japan: Asian Disaster Reduction Center, undated); Tangshan earthquake deaths from "EM-DAT: The OFDA/CRED International Disaster Database," op. cit. note 3, viewed 8 September 2006; Dhaka flood data from *1988 Global Register of Extreme Flood Events* (Dartmouth, MA: Dartmouth Flood Observatory, 1988); deflator from Robert Sahr, "Consumer Price Index Conversion Factors 1800 to estimated 2016 to Convert to Dollars of 2005," Oregon State University, revised 11 April 2006, at oregonstate.edu/cla/polisci/faculty/sahr/sahr.htm.

15. Tsunami distance from Chuck Herring, "Images Help Rebuilding," *Planning*, August/September 2005; "Aid Arrives for Volcano Victims," *BBC News*, 22 January 2002; deaths from "Health Update: WHO Activities—Nyiragongo Eruption," press release (Geneva: World Health

Organization, 31 January 2002); Nelson F. Fernandes et al., "Topographic Controls of Landslides in Rio de Janeiro: Field Evidence and Modeling," *Catena*, 20 January 2004, pp. 163–81. Box 6–3 from the following: human toll of tsunami from U.N. Office of the Special Envoy for Tsunami Recovery, "The Human Toll," at www.tsunami specialenvoy.org/country/humantoll.asp; population and deaths in Banda Aceh from Official Website of the Banda Aceh Government, at www.bandaaceh.go.id/tsunamifx.htm, viewed 31 August 2006; on-the-ground observations from Michael Renner's visit to Aceh, 15–23 December 2006; houses rebuilt from World Bank, *Aceh Public Expenditure Analysis: Spending for Reconstruction and Poverty Reduction* (Washington, DC: September 2006).

16. Temperature 10 degrees Celsius higher from Munich Re, op. cit. note 9, p. 25; Cynthia Rosenzweig et al., "Mitigating New York City's Heat Island with Urban Forestry, Living Roofs, and Light Surfaces," Presentation at 86th American Meteorological Society Annual Meeting, 31 January 2006, Atlanta, GA.

17. Munich Re, op. cit. note 9, p. 25.

18. Galle land use from Ranjith Premalal De Silva, "Reciprocation by Nature," *Tsunami in Sri Lanka: Genesis, Impact and Response* (Colombo, Sri Lanka: Geo Informatics Society of Sri Lanka, 2006); Jane Preuss, "Why 'Tsunami' Means 'Wake Up Call': What Planners Can Learn from Sri Lanka," *Planning*, August/September 2005.

19. Kobe economic damages from Munich Re, op. cit. note 9, p. 21; Hurricane Katrina economic losses from Munich Re, op. cit. note 14; deflator from Sahr, op. cit. note 14.

20. Munich Re, op. cit. note 9, p. 23.

21. Dan Roberts, "Buffett Raises Climate Cover Insurance," *Financial Times*, 6 March 2006.

22. "AIG Adopts First Policy on Global Climate Change," *Reuters*, 17 May 2006; Lorna Victoria, "Networking for CBDRM among Practitioners in the Philippines: An NGO Perspective," in *Part-*

nerships for Disaster Reduction–Southeast Asia 3 Newsletter, April 2006, pp. 5–6; "Corporate Citizenship," Makati Business Club, at www.mbc.com.ph/corporate_citizenship/default.htm.

23. Losses from power disruption from National Association of Home Builders, "Natural Disaster Survival Helped by Renewable Energy," *Nation's Building News*, 10 April 2006; business motivations from Howard Kunreuther, "Interdependent Disaster Risks: The Need for Public Private Partnerships," in World Bank, *Building Safer Cities* (Washington, DC: 2003), p. 86.

24. Cities account for 75 percent of world's fossil fuel consumption (and hence about 75 percent of energy-related emissions); see World Council on Renewable Energy, "Renewable Energy and the City," discussion paper for World Renewable Energy Policy and Strategy Forum, Berlin, Germany, 13–15 June 2005; duration of heat waves from Juliet Eilperin, "More Frequent Heat Waves Linked to Global Warming," *Washington Post*, 4 August 2006; "Weather Related Natural Disasters in 2003 Cost the World Billions," press release (Nairobi: U.N. Environment Programme, 10 December 2003).

25. Coastal wetlands from John Young, "Black Water Rising," *World Watch*, September/October 2006, p. 26; Boston from Jane Holtz Kay, "Shore Losers: US Leaders, Residents Turn Backs on Impending Coastal Chaos," *Grist*, 15 June 2005; U.S. Environmental Protection Agency, *Climate Change and Massachusetts* (Washington, DC: September 1997); New York City from Climate Change Information Resources—New York Metropolitan Region, "What Are the Projected Costs of Climate Change in the Region's Coastal Communities and Coastal Environments?" Issue Brief, at ccir.ciesin.columbia.edu/nyc/ccir-ny_q2e.html, viewed 27 July 2006.

26. Gareth Davies, "Dutch Answer to Flooding: Build Houses that Swim," *Der Spiegel*, 26 September 2005; Hilary Clarke, "Tide of Opinion Turns Against Venice Dam," *Daily Telegraph* (London), 29 January 2006.

27. Young, op. cit. note 25, p. 30.

28. Chen Ganzhang, "Roof-Garden: Getting Popular in China," *China Economic Net*, 18 July 2006.

29. Chicago tree plantings from Keith Schneider, "To Revitalize a City, Try Spreading Some Mulch," *New York Times*, 17 May 2006; Casey Trees Endowment, "The Case for Trees," at www.casey trees.org/resources/casefortrees.html, viewed 6 August 2006.

30. Juliet Eilperin, "22 Cities Join Clinton Anti-Warming Effort," *Washington Post*, 2 August 2006; "Clinton Climate Initiative," at www.clinton foundation.org/cf-pgm-cci-home.htm, viewed 5 October 2006.

31. "US Mayors Climate Protection Agreement," at www.seattle.gov/mayor/climate, viewed 5 October 2006.

32. Sengupta, op. cit. note 14; "As Toll Rises to 749 in India Monsoon, Mumbai Goes Back to Work," *New York Times*, 30 July 2005.

33. Sengupta, op. cit. note 14; 92 percent from International Federation of Red Cross and Red Crescent Societies, *World Disasters Report 2004* (Geneva: 2004), p. 150.

34. Sri Lanka from Preuss, op. cit. note 18; individual feeling of control over risk from Kunreuther, op. cit. note 23, p. 86.

35. London from Munich Re, op. cit. note 9, p. 20; El Salvador from Kurt Rhyner, "Disaster Prevention: Are We Really Trying," *Basin News* (Building Advisory Service and Information Network), June 2002, pp. 2–5.

36. Quote from Sengupta, op. cit. note 14; cities merging from Munich Re, op. cit. note 9, p. 23.

37. CITYNET Web site at www.citynet-ap .org/en; member numbers from www.citynet-ap .org/en/Members/member.html, viewed 4 August 2006; Banda Aceh and community centers in Sri Lanka from Bernadia Irawati Tjandradewi, CITYNET, discussion with author, 8 August 2006.

38. Hurricane Mitch deaths from "Mitch: The Deadliest Atlantic Hurricane since 1780," National Climatic Data Center, at www.ncdc.noaa.gov/ oa/reports/mitch/mitch.html, updated 29 August 2006; Jorge Gavidia and Annalisa Crivellari, "Legislation as a Vulnerability Factor," *Open House International*, vol. 31, no. 1, 2006, p. 84.

39. Benson and Twigg, op. cit. note 7; Dominica and Jamaica from Jan Vermeiren and Steven Stichter, *Costs and Benefits of Hazard Mitigation for Building and Infrastructure Development: A Case Study in Small Island Developing States* (Washington, DC: Organization of American States and U.S. Agency for International Development, 1998); Philippines and U.S. Geological Survey from Reinhard Mechler, "Natural Disaster Risk and Cost-Benefit Analysis," in World Bank, op. cit. note 23, p. 52. Box 6–4 from the following: Colombia from "Double Jeopardy: The Impact of Conflict and Natural Disaster on Cities," in UN-HABITAT, op. cit. note 9, p. 157; China flood control from Mechler, op. cit. this note, p. 52; China population from United Nations, "File 1: Total Population at Mid-Year," op. cit. note 4; Seattle earthquake from Robert Freitag, "The Impact of Project Impact on the Seattle Nisqually Earthquake," *Natural Hazards Observer*, May 2001; "Seattle Project Impact," at www.seat tle.gov/projectimpact/; Pacific Northwest Seismograph Network, at www.ess.washington.edu/ SEIS/EQ_Special/WEBDIR_01022818543p/wel come.html.

40. Davis, op. cit. note 9.

41. Box 6–5 based on field work in February and March 2006 by Christine Wamsler; see Christine Wamsler, "Understanding Disasters from a Local Perspective," *TRIALOG* (Journal for Planning and Building in the Third World), special issue on "Building on Disasters," December 2006, and Christine Wamsler, "Bridging the Gaps: Stakeholder-based Strategies for Risk Reduction and Financing for the Urban Poor," forthcoming.

42. "Slum Communities and Municipality Cooperate in Santo Domingo," and "Tuti Islanders Fight Floods Together," in International Federation of Red Cross and Red Crescent Societies, op. cit. note 33, pp. 24–25.

43. Guilliame Chantry and John Norton, "Local Confidence and Partnership to Strengthen Capacity for Community Vulnerability Reduction: Development Workshop France in Central Vietnam," in *Partnerships for Disaster Reduction–Southeast Asia 3 Newsletter*, April 2006, pp. 7–8.

44. Development Marketplace Team, *Scrap Tires Save Homes in Turkey* (Washington, DC: World Bank, October 2005); safer shelters from Independent Evaluation Group, op. cit. note 8, p. xxiii.

45. Reinhard Mechler and Joanne Linnerooth-Bayer with David Peppiatt, *Disaster Insurance for the Poor?* (Geneva: ProVention Consortium/International Institute for Applied Systems Analysis, 2006), p. 6; Independent Evaluation Group, op. cit. note 8, p. xx.

46. Muhammad Yunus, "Rebuilding Through Social Entrepreneurship," *Changemakers News Feature*, October 2005/February 2006.

47. Mechler and Linnerooth-Bayer with Peppiatt, op. cit. note 45, p. 18.

48. Zijun Li, "Shanghai Completes Massive Underground Bunker to Protect Citizens from Disasters," *China Watch*, 8 August 2006, at www.worldwatch.org/node/4424; death toll from "EM-DAT: The OFDA/CRED International Disaster Database," op. cit. note 3, viewed 8 Sept 2006. The death toll of 242,000 is considered an official figure, with other estimates going as high as 655,000.

49. For information on the International Charter, "Space and Major Disasters," see www.disastercharter.org; charter activations from International Charter on Space and Major Disasters Webmaster, discussion with author, 7 August 2006; "Flooding in Central Europe" and "Floods in Senegal," at www.disasterscharter.org/disasters.

50. "Expert Predicted Volcano Eruption," *BBC News*, 23 January 2002.

51. Li, op. cit. note 48.

52. Official pledges from Adam Blenford, "Concern over Urban Evacuation Plans," *BBC News*, 27 September 2005; Matthew L. Wald, "Deadly Bus Fire is Focus of Safety Hearings," *New York Times*, 5 August 2006.

53. "Run, Tell Your Neighbor! Hurricane Warning in the Caribbean," in International Federation of Red Cross and Red Crescent Societies, op. cit. note 14.

54. Ted Koppel, "Preparing for a Disaster is Neither Rocket Science nor Brain Surgery," *Morning Edition*, National Public Radio, 3 August 2006.

55. "To the People," *Oakland Tribune*, 18 April 1906; Egeland quoted in "Disaster Reduction: Changes Since the Kobe Conference and the Tsunami," in "Disaster Reduction and the Human Cost of Disaster," IRIN Web Special, 2005.

56. Dilley quoted in Quirin Schiermeier, "The Chaos to Come," *Nature*, 15 December 2005, p. 906.

57. "Hyogo Framework for Action 2005–2015: Building the Resilience of Nations and Communities to Disasters," International Strategy for Disaster Reduction, 2005, at www.unisdr.org/eng/hfa/hfa.htm; number of governments and visitors from "Statistics of the Public Forum," and "Brief History of the WCDR Process," at www.unisdr.org/wcdr/, viewed 2 October 2006.

58. "Hyogo Framework for Action 2005–2015," op. cit. note 57.

59. Christine Wamsler, *Operational Framework for Integrating Risk Reduction: For Aid Organizations Working in Human Settlement Development*, Working Paper No. 14 (Lund, Sweden: Benfield Hazard Research Centre/Lund University, 2006).

Jakarta: River Management

1. Number of zones from Accion Contra la Faime, *Report of Area Selection for Flood Disaster*

Preparedness Programme Development in DKI Jakarta Province (Jakarta: 2006), p. 11; Damar Harsanto, "Jakarta Braces for Annual Floods," *Jakarta Post*, 28 October 2005; official city population census data for 2000 from Badan Pusat Statisik, at www.bps.go.id, although the population of metropolitan Jakarta is significantly larger, particularly when contiguous urban areas within greater Jabotabek are included; quote from Bambang Nurbianto, "Residents Stick by Riverbank Despite Floods," *Jakarta Post*, 26 September 2005.

2.   Harsanto, op. cit. note 1; Adianto P. Simamora, "Floodwaters Swell as Residents Take to Rooftops," *Jakarta Post*, 14 January 2006.

3.   Financial Times Information–Global News Wire, "Household, Chemical Waste Pollutes Rivers," *Jakarta Post*, 20 August 2001; quote from Financial Times Information–Asia Intelligence Wire, "Integrated Waste Management," *Jakarta Post*, 12 June 2004.

4.   River length from Adianto P. Simamora, "Cleaned-Up Ciliwung Set to Go With the Flow," *Jakarta Post*, 13 April 2006; Damar Harsanto, "City Seeks Closer Ties With Neighboring," *Jakarta Post*, 5 September 2005.

5.   Canal details from Departemen Pekerjaan Umum (Department of Public Works), "Proyek Pembangunan Banjir Kanal Timur Mulai Dikerjakan," PU-Net, 10 December 2003, at www.pu.go.id/index.asp?link=Humas/news2003/ppw160606gt.htm; Moch. N. Kurniawan, "Government to Rehabilitate 17 Catchment Areas Across Indonesia," *Jakarta Post*, 7 March 2003; improved flood warning system from Edward Turvill, Accion Contra la Faime, Jakarta, discussion with author, 4 August 2006; community groups from Alwi Shahab, "Mengamankan Bantaran Ciliwung," *Republika Online*, 15 April 2006, at www.republika.co.id/detail.asp?id=256616, from Bambang Parlupi, "Ciliwung Ecosystem Restored by Local People," *Jakarta Post*, 25 April 2006, from Financial Times Information–Asia Intelligence Wire, "Jakarta's Rivers Scoured Every Three Months," *Antara*, 19 March 2006, and from Adianto P. Simamora and Anissa S. Febrina,

"Riverbank Community Takes Pollution of Waterways to Heart," *Jakarta Post*, 12 April 2006.

6.   Financial Times Information–Asia Intelligence Wire, "Dutch Government Donates Rp.3 Billion to Control Jakarta Floods," *Antara*, 20 February 2003; Carin Bobeldijk, Karin Bosman, and Robert-Jan de Jonge, eds., *Made in Holland* (The Hague, Netherlands: Land and Water International, Netherlands Agency for International Business and Cooperation, and NEDECO, 2004); Friso Roest, NEDECO, Jakarta, letters to author, 28 June–11 July 2006.

7.   Richel Dursin, "Environment: For Indonesian Villagers, Floods are a Part of Life," *Inter Press Service*, 8 July 2005; Kampung Improvement Program, ArchNet Digital Library at archnet.org/library/one-site.tcl?site_id=63; Roest, op. cit. note 6.

8.   Damar Harsanto, "Low-Cost Apartments to be Built, But Only for a Chosen Few," *Jakarta Post*, 6 January 2005; residents challenging from Hexa Rahmawati and Suryanto, Sanggar Ciliwung, Jakarta, discussions with author, 7–25 July 2006; Financial Times Information–Asia Intelligence Wire, "Bridges a Hive of Activity for the Poor," *Jakarta Post*, 15 April 2006.

9.   Azas Tigor Nainggolan, "Jakarta in 2001: No Room for the Poor," *Jakarta Post*, 11 December 2001; Evi Mariani, "City Gears Up to Make More Jakartans Homeless," *Jakarta Post*, 28 November 2003; Muninggar Sri Saraswati and Annastashya Emmanuelle, "Riverbank Squatters Demand Fair Compensation," *Jakarta Post*, 5 November 2001; Rahmawati and Suryanto, op. cit. note 8; Evi Mariani, "Fishermen Challenge Eviction Notice," *Jakarta Post*, 13 October 2003.

10.   Questioning of government's sincerity from Selamet Daroyni, WALHI (Indonesia Forum for Environment), Jakarta, discussion with author, 25 July 2006; Rahmawati and Suryanto, op. cit. note 8; Kurniawan, op. cit. note 5; Devi Asmarani, "Jakarta Floods Back With a Vengeance," *Straits Times*, 15 February 2002.

Mumbai: Policing by the People

1. Census of India, 2001, at www.censusindia .net/results/slum/slum2.html.

2. A. N. Roy, Commissioner of Police, Mumbai, India, discussion with author, 12 August 2006.

3. A. Jockin, president, National Slum Dwellers' Federation, Mumbai, India, discussion with author, August 2006.

4. Ibid.

5. Ibid.

6. Roy, op. cit. note 2.

7. Malti Ambre, Mankhurd, India, discussion with author, 15 August 2006.

8. Ibid.

9. Roy, op. cit. note 2.

10. Ibid.; Jockin, op. cit. note 3.

11. Roy, op. cit. note 2.

Chapter 7. Charting a New Course for Urban Public Health

1. C. Ofili, *Children's Experiences of Water and Health in Low Income Urban Settlements—A Case Study in Asaba, Nigeria*, University of London PhD Thesis, London School of Hygiene & Tropical Medicine, 2006.

2. Diseases and deficiency in early cities from Mark Nathan Cohen, *Health and the Rise of Civilization* (New Haven, CT: Yale University Press, 1989), pp. 122–23; Richard R. Paine and Glenn R. Storey, "Epidemics, Age at Death, and Mortality in Ancient Rome," in Glenn R. Storey, ed., *Urbanism in the Preindustrial World: Cross-Cultural Approaches* (Tuscaloosa, AL: University of Alabama Press, 2006), p. 82; life expectancy in London from Roger Finlay, *Population and Metropolis: The Demography of London* (Cambridge, U.K.: Cambridge University Press, 1981), p.108.

3. Mortality for urban poor from U.S. Agency for International Development (USAID), "Urban Health and Poverty," at www.makingcities work.org, viewed 22 August 2006; Box 7–1 from U.N. Population Division, *World Urbanization Prospects: The 2003 Revision* (New York: 2003), p. 5, and from C. McCord and H. P. Freeman, "Excess Mortality in Harlem," *New England Journal of Medicine*, 18 January 1990, pp. 173–77.

4. C. Stephens, "Inequalities in Environment, Health and Power—Reflections on Theory and Practice," in C. Pugh, ed., *Urban Sustainable Development in Developing Countries* (London: Earthscan, 2000), pp. 91–115; Anthony McMichael, "The Urban Environment and Health in a World of Increasing Globalization: Issues for Developing Countries," *Bulletin of the World Health Organization*, vol. 78, no. 9 (2000), pp. 1117–26. Box 7–2 from the following: Pedro Pirez, "Buenos Aires: Fragmentation and Privatization of the Metropolitan City," *Environment & Urbanization*, April 2002, pp. 145–58; Centre for Urban Studies and U.S. National Institute of Population Research and Training, *Slums in Bangladesh Cities: Mapping and Census 2005* (Chapel Hill, NC: USAID, 2005); people living between toxic factory and poisoned lake from J. Seabrook, *In the Cities of the South* (New York: Verso, 1996).

5. N. Ohenjo et al., "Health of Indigenous People in Africa," *The Lancet*, 10 June 2006, pp. 1937–46; J. Suzman, *An Introduction to the Regional Assessment of the San in Southern Africa* (Windhoek, Namibia: Legal Assistance Centre, 2001).

6. N. Thapar and I. Sanderson, "Diarrhoea in Children: An Interface between Developing and Developed Countries," *The Lancet*, 21 February 2004, pp. 641–53.

7. J. Clauson-Kaas et al., "Urban Health: Human Settlement Indicators of Crowding," *Third World Planning Review*, vol. 18, no. 3 (1996), pp. 349–63; E. Drucker et al., "Childhood Tuberculosis in the Bronx, New York," *The Lancet*, 11 June 1994, pp. 1482–85.

8. S. Ghosh and D. Shah, "Nutritional Problems in Urban Slum Children," *Indian Pediatrics*, July 2004, pp. 682–96; J. C. Fotso, "Child Health Inequities in Developing Countries: Differences across Urban and Rural Areas," *International Journal for Equity in Health*, 11 July 2006.

9. World Health Organization (WHO), *Diet, Nutrition and the Prevention of Chronic Diseases*, Report of a Joint WHO/Food and Agriculture Organization Expert Consultation (Geneva: 2002).

10. WHO, *Fuel for Life: Household Energy and Health* (Geneva: 2006).

11. Table 7–1 from J. Hardoy, D. Mitlin, and D. Satterthwaite, *Environmental Problems in an Urbanizing World: Finding Solutions for Cities in Africa, Asia and Latin America* (London: Earthscan, 2001), pp. 89–107; 800,000 deaths from J. Kenworthy and F. Laube, "Urban Transport Patterns in a Global Sample of Cities and Their Linkages to Transport Infrastructures, Land Use, Economics and Environment," *World Transport Policy & Practice*, vol. 8, no. 3 (2002), pp. 5–20.

12. E. Broughton, "The Bhopal Disaster and Its Aftermath: A Review," *Environmental Health*, 10 May 2005, p. 6.

13. WHO, *World Report on Road Traffic Injury Prevention* (Geneva: 2004), p. 3; Hardoy, Mitlin, and Satterthwaite, op. cit. note 11, p. 112; H. Nizamo et al., "Mortality Due to Injuries in Maputo City, Mozambique," *International Journal of Injury Control and Safety Promotion*, March 2006, pp. 1–6; young people's cause of death from C. Stephens, "The Urban Environment and Health–A Review of the Evidence." in S. Atkinson, J. Songsore, and E. Werna, eds., *Urban Health in Developing Countries* (Oxford, Avebury Press, 1996), pp. 115–35.

14. Buddha Basnyat and Lalini C. Rajapaksa, "Cardiovascular and Infectious Diseases in South Asia: The Double Whammy," *BMJ* (British Medical Journal), 3 April 2004, p. 781; Figure 7–1 from C. Stephens, "What Has Health Got to Do With It? Using Health to Guide Urban Priority-setting Processes Towards Equity," in J. Davila et

al., eds., *Environmental Management in Metropolitan Areas* (London: University College London Press, 1999), pp. 88–95.

15. Letícia Legay Vermelho and Maria Helena P. de Mello Jorge, "Mortalidade de Jovens: Análise do Período de 1930 a 1991 (a Transição Epidemiológica para a Violência)," *Rev. Saúde Pública*, August 1996; C. Stephens, M. Akerman, and P. Borlina-Maia, "Health and Environment in São Paulo, Brazil: Methods of Data Linkage and Questions for Policy," *World Health Statistics Quarterly*, vol. 48 (1995), pp. 95–107.

16. Edinilsa Ramos de Souza and Maria Luiza Carvalho de Lima, "The Panorama of Urban Violence in Brazil and Its Capitals," *Ciênc. Saúde Coletiva*, April/June 2006; L. A. Teplin et al., "Early Violent Death among Delinquent Youth: A Prospective Longitudinal Study," *Pediatrics*, June 2005, pp. 1586–93.

17. De Souza and Carvalho de Lima, op. cit. note 16.

18. E. Krug et al., eds., *World Report on Violence and Health* (Geneva: WHO, 2002); National Research Council, *Cities Transformed: Demographic Change and Its Implications in the Developing World* (Washington, DC: National Academies Press, 2003), p. 267.

19. McMichael, op. cit. note 4.

20. Intergovernmental Panel on Climate Change, *Climate Change 2001: The Scientific Basis. Contribution of Working Group I to the Third Assessment Report of the Intergovernmental Panel on Climate Change* (New York: Cambridge University Press, 2001); Janet Larsen, "Setting the Record Straight: More than 52,000 Europeans Died from Heat in Summer 2003," *Earth Policy Indicator* (Washington, DC: Earth Policy Institute, July 2006); "Heat Island Effect: Basic Information," U.S. Environmental Protection Agency, at www.epa.gov.

21. Quote from Siddharth Agarwal, Arti Bhanot, and Geetanjali Goindi, "Understanding and Addressing Childhood Immunization Coverage in Urban Slums," *Indian Pediatrics*, July 2005, pp.

653–63; Jean-Frédéric Levesque et al., "Outpatient Care Utilization in Urban Kerala, India," *Health Policy and Planning*, July 2006, pp. 289–301.

22. Robert Woods and John Woodward, eds., *Urban Disease and Mortality in Nineteenth-Century England* (London: Batsford Academic and Educational, 1984).

23. M. Daunton, ed., *The Cambridge Urban History of Britain, Vol. 3* (Leicester, U.K.: Centre for Urban History, 2001), p. 632.

24. Lisa C. Smith, Marie T. Ruel, and Aida Ndiaye, "Why Is Child Malnutrition Lower in Urban Than in Rural Areas?" *IFPRI Discussion Paper Briefs* (Washington, DC: International Food Policy Research Institute, 2002); USAID, "Urban Health and Poverty," at www.makingcities work.org/, viewed 22 August 2006.

25. National Research Council, op. cit. note 18, p. 263.

26. A. T. Geronimus, J. Bound, and T. A. Waidmann, "Poverty, Time, and Place: Variation in Excess Mortality across Selected US Populations, 1980–1990," *Journal of Epidemiological Community Health*, June 1999, pp. 325–34; T. A. Houweling et al., "Rising Under-5 Mortality in Africa: Who Bears the Brunt?" *Tropical Medicine and International Health*, August 2006, pp. 1218–27.

27. Kolkata from Stephens, op. cit. note 14.

28. Ivo Imparato, *Slum Upgrading and Participation: Lessons from Latin America* (Washington, DC: World Bank, 2003), p. 10.

29. Brazil from Marion Gret, "The Emergence of a New Urban Structure Thanks to the Public Policies of Civil Participation," 2006, *Development Gateway*, at topics.developmentgateway.org/poverty/rc/ItemDetail.do~1059017, viewed 12 August 2006; Julio Diaz Palacios and Liliana Miranda Sara, "Concertación (Reaching Agreement) and Planning for Sustainable Development in Ilo, Peru," in S. Bass et al., eds., *Reducing Poverty and Sustaining the Environment: The Pol-*

itics of Local Engagement (London: Earthscan, 2005), pp. 255–79.

30. Grupo Cultural AfroReggae, *Favela Rising*, 2006, at favelarising.com, viewed 14 August 2006.

31. Alex Roberto et al., *Curitiba, Capital Social—Social?* (Curitiba, Brazil: 2006).

32. Stephens, op. cit. note 14.

33. American Forests, "Setting Urban Tree Canopy Goals," at www.americanforests.org, viewed 22 August 2006.

34. D. Nowak and D. Crane, "The Urban Forest Effects (UFORE) Model: Quantifying Urban Forest Structure and Functions," in M. Hansen, ed., *Second International Symposium: Integrated Tools for Natural Resources Inventories in the 21st Century* (Washington, DC: Forest Service, U.S. Department of Agriculture, 2000); Atlanta from C. Cardelino and W. Chameides, "Natural Hydrocarbons, Urbanization, and Urban Ozone," *Journal of Geophysical Research*, August 1990, pp.13,971–79; U.S. Environmental Protection Agency (EPA), Air Quality Strategy and Standards Division, *Incorporating Emerging and Voluntary Measures in a State Implementation Plan* (Washington, DC: 2004).

35. F. Kuo and W. Sullivan, "Environment and Crime in the Inner City: Does Vegetation Reduce Crime?" *Environment and Behavior*, May 2001, pp. 343–67.

36. Enrique Peñalosa, "High Achievements," *Our Planet* (U.N. Environment Programme), vol. 16, no. 1 (2005).

37. Charlie Pye-Smith, "Building Green Islands in Bombay," *peopleandplanet.net*, 12 June 2001.

38. Mardie Townsend, "Feel Blue? Touch Green! Participation in Forest/Woodland Management as a Treatment for Depression," *Urban Forestry & Urban Greening*, vol. 5, issue 3 (2006), pp. 111–20.

39. Peñalosa, op. cit. note 36.

40. Figure of three quarters from Peñalosa, op. cit. note 36; Michelle Hibler, "Taking Control of Air Pollution in Mexico City," *Reports* (International Development Research Centre), 12 August 2003; Julia Preston, "The World: Mexico City's Air; A Fatal Case of Fatalism," *New York Times*, 14 February 1999.

41. Sheng-Yong Wang et al., "Trends in Road Traffic Crashes and Associated Injury and Fatality in the People's Republic of China, 1951–1999," *Injury Control and Safety Promotion*, April 2003, pp. 83–87; WHO, "Road Safety is No Accident," press release (Geneva: 7 April 2004); Road Traffic Injuries Research Network, cited in *Detroit Free Press*, 24 September 2002.

42. WHO, *World Health Report 2002* (Geneva: 2002); trips by foot in poorer countries from Ken Gwilliam, *Cities on the Move* (Washington, DC: World Bank, 2002), pp. 25–38; EPA, *Travel and Environmental Implications of School Siting* (Washington, DC: 2003).

43. James Sallis and Karen Glanz, "The Role of Built Environments in Physical Activity, Eating, and Obesity in Childhood," *The Future of Children*, spring 2006, pp. 89–108.

44. John Donne, Divine Meditations XVII (1624), in John Donne, *The Complete English Poems* (Troy, MI: Phoenix Press, 1994).

Nairobi: Life in Kibera

1. Size from Government of Kenya and U.N. Centre for Human Settlements (UNCHS), *Nairobi Situation Analysis, Consultative Report on the Collaborative Nairobi Slum Upgrading Initiative* (Nairobi: 2001), Table 9, p. 36; history from John Mbaria, "Kibera and the Politics of Dispossession," *The East African*, 15–21 July 2002; Nairobi population from United Nations, *World Urbanization Prospects: The 2003 Revision* (New York: 2003).

2. Number of slums from Government of Kenya and UNCHS, op. cit. note 1, p. 1; official figures for the population of Kibera are notoriously hard to find, although the Government of Kenya and UNCHS report estimated it in 1999 to be 377,624—however, a more recent document (Government of Kenya, Kibera-Soweto Slum Upgrading Project, Nairobi, December 2004) states that the population of Kibera is 600,000 and some nongovernmental organizations have estimated the figure to be as high as 900,000; recent survey from Government of Kenya and UN-HABITAT, *Kibera Social and Economic Mapping. Document RI/4733—Executive Summary Report* (Nairobi: 2004–05), pp. 7–9; surveys and socioeconomic mapping by Research International as part of the Kenya Slum Upgrading Programme; illegal structures from W. Olima and S. Karirah-Gitau, *Land Tenure and Tenancy Concerns and Issues in Kibera*, KUESP Preparatory Group Task, Final Report (Washington, DC: World Bank, unpublished); average rents from Government of Kenya and UN-HABITAT, "A Study to Conduct Kibera Social and Economic Mapping: Executive Summary Report RI/4733" (Nairobi: Research International 2004/5). Figures based on exchange rate on 7 September 2006 of $1 equals 72.7 Kenya shillings.

3. Share who are tenants from Central Bureau of Statistics, *Kenya Demographic and Health Survey 2003* (Nairobi: 2004); apartment rents and unauthorized tenements from Marie Huchzermeyer, *Slum Upgrading Initiatives in Kenya within the Basic Services and Wider Housing Market: A Housing Rights Concern*, draft discussion paper No. 1/2006, Kenya Housing Rights Project (Geneva: Centre on Housing Rights and Evictions, unpublished); average earnings from Government of Kenya and UN-HABITAT, *Kibera Social and Economic Mapping*, op. cit. note 2, p. 4.

4. Rasna Warah, "Bill Bryson Tour: I Let Kibera Speak for Itself," *The East African*, 20 December 2002–5 January 2003. Obwaya interviewed by the author in March 2002.

5. Emily Lugano and Geoff Sayer, "Girls' Education in Nairobi's Informal Settlements," *Links* (Oxfam GB), October 2003.

6. The Kenya Slum Upgrading Programme was initiated in 2000 through an agreement between the previous government of Kenya (under President Moi) and UN-HABITAT. It was renewed in January 2003 with the new National Rainbow Coalition government under President Kibaki. A pilot project in the Soweto "village" in Kibera, which houses approximately 60,000 people, was launched in October 2004, but actual construction has not yet begun. Stalemate from Dauti Kahura, "Sh.880 Billion Housing Dream for the Poor Turns Sour," *The Standard* (Nairobi), 14 November 2005.

7.   Huchzermeyer, op. cit. note 3.

8.   UN-HABITAT, *State of the World's Cities Report 2006/7* (London: Earthscan, 2006).

9.   Ibid. The report shows that in many countries, living in a slum is as hazardous to a person's health as living in a deprived rural area.

Petra: Managing Tourism

1.   "The Mysterious Nabateans," The Hashemite Kingdom of Jordan, at www.kinghussein.gov.jo/his_nabateans.html.

2.   Jordan Department of Statistics, "Housing and Population Census 1979" and "Housing and Population Census 1994," at www.dos.gov.jo; estimated population in 2006 from The Hashemite Kingdom of Jordan, Ministry of Tourism and Antiquities, *Petra Priority Action Plan Study, Phase One Report, Outline Development and Growth Scenario Petra Region* (Amman: Dar Al Handasah, 1996).

3.   Hashemite Kingdom of Jordan, op. cit. note 2.

4.   Ibid.; Aysar Akrawi, executive director, Petra National Trust, Amman, discussion with author, 13 September 2006.

5.   Quote from Hashemite Kingdom of Jordan, op. cit. note 2, p. 8-2.

6.   Akrawi, op. cit. note 4.

7.   Ibid.

8.   Ibid.

9.   Ibid.

10.   Ibid.

11.   Challenge for many cities from UN-HABITAT, *Our Future: Sustainable Cities—Turning Ideas into Action*, Background Paper, World Urban Forum III, Vancouver, Canada, 19–23 June 2006, p. 3.

12.   Mohammad Ajlouni, "Dana Nature Reserve: Jordan," in U.N. Development Programme Special Unit for SSC, *Sharing Innovative Experiences—Volume 9: Examples of the Successful Conservation and Sustainable Use of Dryland Biodiversity* (New York: 2004), pp. 23–28.

Chapter 8. Strengthening Local Economies

1.   United Nations Population Division, *World Urbanization Prospects: The 2005 Revision*, online database at esa.un.org/unup, viewed September 2006.

2.   UN-HABITAT, *State of the World's Cities, 2006/7* (London: Earthscan, 2006); International Council for Local Environmental Initiatives (ICLEI), *Accelerating Sustainable Development: Local Action Moves the World* (New York: United Nations Economic and Social Council, 2002).

3.   Mark Magnier, "Huge Environmental Battle Leaves Legacy of Rage," *Los Angeles Times*, reprinted in the *Vancouver Sun*, 6 September 2006; polluted cities from UN-HABITAT, op. cit. note 2.

4.   Magnier, op. cit. note 3.

5.   *China Daily* cited in ibid.

6.   UN HABITAT, op. cit. note 2.

7.   For Millennium Development Goals, see www.un.org/millenniumgoals.

8.   UN Millennium Project, *Investing in Development: A Practical Plan to Achieve the Millennium*

Development Goals (London: Earthscan, 2005); Millennium Villages Project, *Annual Report: Millennium Research Villages–First Year July 2004 to June 2005* (New York: Earth Institute at Columbia University, 2005); Earth Institute Millennium Villages Project, at www.earthinstitute.colum bia.edu/mvp, viewed 30 September 2006.

9. Lee Scott, Wal-Mart CEO, "Wal-Mart: 21st Century Leadership," speech, 24 October 2005; for critics, see Wal-Mart, "Sustainability: Starting the Journey," at walmartstores.com/GlobalWM StoresWeb/navigate.do?catg=345.

10. October 2005 announcement from Jad Mouawad, "The Greener Guys," *New York Times*, 30 May 2006; Scott, op. cit. note 9; Michael Pollan, "Mass Natural," *New York Times Magazine*, 4 June 2006.

11. Sean Markey et al., *Second Growth: Community Economic Development in Rural British Columbia* (Vancouver: University of British Columbia Press, 2005).

12. Stephan J. Goetz and Hema Swaminathan, *Wal-Mart and County-Wide Poverty*, Staff Paper No. 371 (State College, PA: Department of Agricultural Economics and Rural Sociology, Pennsylvania State University, 2004); Michael H. Shuman, *The Small-Mart Revolution: How Local Businesses Are Beating The Global Competition* (San Francisco: Berrett-Koehler Publishers, 2006).

13. Shuman, op. cit. note 12.

14. Civic Economics, *Economic Impact Analysis: A Case Study: Local Merchants vs. Chain Retailers*, prepared for Liveable City (Austin, TX: 2002); Goetz and Swaminathan, op. cit. note 12, p. 12.

15. For more detail, see the Web sites of Local Government Commission, Rocky Mountain Institute, and Smart Growth America.

16. Box 8–1 from John Restakis, *The Emilian Model—Profile of a Co-operative Economy*, British Columbia Co-operative Association (Vancouver, BC: undated), and from John Restakis, *The Lessons of Emilia Romagna* (Vancouver, BC: British Columbia Co-operative Association, 2005).

17. Ian MacPherson, "Into the Twenty-First Century: Co-operatives Yesterday, Today and Tomorrow," in British Columbia Institute for Co-operative Studies, *Sorting Out: A Selection of Papers and Presentations, 1995–2005* (Victoria, BC: 2004); quote from Johnston Birchall, *Rediscovering the Co-operative Advantage: Poverty Reduction through Self-help* (Geneva: Co-operative Branch, International Labour Office, 2003), p. 3; Box 8–2 from International Co-operative Alliance, Statement on the Co-operative Identity, at www.ica.coop/coop/principles.html, viewed 30 September 2006.

18. Julia Smith, *Worker Co-operatives: A Glance Around the World* (Victoria, BC: British Columbia Institute for Co-operative Studies, 2003).

19. Benjamin Dangl, "Worker-Run Cooperatives in Buenos Aires," *Z Magazine*, April 2005; Geoff Olson, "The Take—A Story of Hope," *Common Ground*, November 2004.

20. Smith, op. cit. note 18.

21. Birchall, op. cit. note 17; United Nations, "Cooperatives Are Significant Actors in Development, says Secretary-General," press release (New York: 7 July 2001).

22. For general information on microfinance, see Consultative Group to Assist the Poor, at www.cgap.org.

23. Microcredit Summit Campaign, at www.microcreditsummit.org, viewed 16 September 2006; Sam Daley-Harris, *State of the Microcredit Summit Campaign Report 2005* (Washington, DC: Microcredit Summit Campaign, 2005).

24. Daley-Harris, op. cit. note 23.

25. International Year of Microcredit 2005, *Microfinance and the Millennium Development Goals* (New York: U.N. Capital Development Fund, 2005).

26. Grameen—Banking for the Poor, at www.grameen-info.org/index.html, viewed 30 September 2006.

27. Celia W. Dugger, "Peace Prize to Pioneer of Loans for Those Too Poor to Borrow," *New York Times*, 14 October 2006; Grameen—Banking for the Poor, op. cit. note 26; Alexandra Bernasek, "Banking on Social Change: Grameen Bank Lending to Women," *International Journal of Politics, Culture and Society*, spring 2003, pp. 369–85.

28. International Year of Microcredit 2005, op. cit. note 25; Box 8–3 from Unitus—Innovative Solutions to Global Poverty at www.unitus.com/sections/impact/impact_css_kenya.asp; Daley-Harris, op. cit. note 23.

29. Bernasek, op. cit. note 27; Rosintan D. M. Panjaitan-Drioadisuryo and Kathleen Cloud, "Gender, Self-employment and Microcredit Programs: An Indonesian Case Study," *The Quarterly Review of Economics and Finance*, vol. 39 (1999), pp. 769–79.

30. World Council of Credit Unions, at www.woccu.org, viewed 15 October 2006.

31. Information and Box 8–4 from Vancouver City Savings Credit Union, at www.vancity.com/MyCommunity, viewed 30 September 2006.

32. Shorebank Pacific, at www.eco-bank.com, viewed 30 September 2006.

33. Table 8–1 from Fair Trade Labelling Organizations International, at www.fairtrade.net/30.html, viewed 30 September 2006; William Young and Karla Utting, "Fair Trade, Business and Sustainable Development," *Sustainable Development*, vol. 13 (2005), pp. 139–42.

34. Fair Trade Labelling Organizations International, op. cit. note 33.

35. TransFair Canada, at www.transfair.ca/en/fairtrade, viewed 30 September 2006; Fair Trade Labelling Organizations International, op. cit. note 33.

36. Box 8–5 from Centre for Development in Central America, at www.fairtradezone.jhc-cdca.org/story.htm.

37. Business Alliance for Local Living Economies, at www.livingeconomies.org, viewed 14 October 2006.

38. Local campaigns from ibid.

39. ICLEI, International Development Research Centre (IDRC), and United Nations Environment Programme, *The Local Agenda 21 Planning Guide* (Toronto and Ottawa, ON: ICLEI and IDRC, 1996); Mark Roseland, *Toward Sustainable Communities: Resources for Citizens and Their Governments* (Gabriola Island, BC: New Society Publishers, 2005).

40. Shuman, op. cit. note 12.

41. Zane Parker, "Unravelling the Code: Aligning Taxes and Community Goals," *Focus on Municipal Assessment and Taxation*, June 2005, pp. 46–47.

42. Mark Roseland, ed., *Tax Reform as If Sustainability Mattered: Demonstrating Ecological Tax-Shifting in Vancouver's Sustainability Precinct* (Vancouver, BC: Simon Fraser University, 2005).

43. Judy Wicks, *Local Living Economies: The New Movement for Responsible Business* (San Francisco: Business Alliance for Local Living Economies, 2006).

44. Social Venture Network, Standards of Corporate Social Responsibility, 1999, at www.svn.org/initiatives/standards.html, viewed 30 September 2006.

45. E. F. Schumacher, *Small is Beautiful* (New York: Harper & Row, 1973).

46. Markey et al., op. cit. note 11, p. 2.

47. Roseland, op. cit. note 39.

48. All from ibid.

49. Lucy Stevens, Stuart Coupe, and Diana Mitlin, eds., *Confronting the Crisis in Urban Poverty: Making Integrated Approaches Work* (Warwickshire, UK: Intermediate Technology Publications, 2006); Robert Chambers and Gordon Conway, "Sustainable Rural Livelihoods: Practical Concepts for the 21st Century," *IDS Discussion Paper No. 296* (Brighton, U.K.: Institute of Development Studies, December 1991).

50. Stevens, Coupe, and Mitlin, op. cit. note 49.

51. International Labour Office, *Global Employment Trends Brief*, February 2005.

52. Roseland, op. cit. note 39.

Brno: Brownfield Redevelopment

1. Karel Kuča, *BRNO—Vývoj Města, Předměstí a Připojených Vesnic* (Prague-Brno: BASET, 2000); *Die Hunderjahrige Geschichte der Erste Brunner Maschinen-Fabriks-Gesellschaft in Brunn von 1821 bis 1921* (Leipzig, Germany: Von Eckert & Pflug); Alstom Power s.r.o., *100 Let Parních Turbín v Brně* (2002).

2. Kovoprojekta Brno a.s., *Územní Generel Výroby Města Brna—Obecná Analýza* (2001–03).

3. Karel Stránský, "Obnova a rozvoj města na příkladu projektu Jižní centrum," in Brno—Město Uprostřed Evropy, proceedings of international conference, 2–4 December 1993, pp. 155–60; www.pps.org/info/projects/international_projects/czech_placemaking; "Krok za Krokem," video about activity of PPS in Czech Republic, Místa v srdci Foundation, Prague, 1996.

4. B.I.R.T. Consulting, s.r.o., *Variantní Ekonomická Studie Využití Areálu Vaňkovka*, September 1996; Ilos Crhonek, PhD, "Areál Strojírenského Závodu Vaňkovka na Zvonařce v Brně—Stavebně Historický Průzkum," 1997; "Minutes from the Brno City Parliament session–number Z3/O19," 20 June 2000.

5. AQUA PROTEC s.r.o., *Environmental Pollution Risk Analysis—ZETOR s.p. VANKOVKA*

Complex (Brno: May 1997).

6. "Program 1994–1999," at www.vankovka.cz/index.php?lang=cz&page=9&program=1.

7. Milena Flodrová and Libor Teplý, *Proměny Vaňkovky—The Changing Faces of Vaňkovka* (Brno: FOTEP, 2005).

8. Jiřina Bergatt Jackson and Collective, *Brownfields Snadno a Lehce* (Prague: Institut pro Udržitelný Rozvoj Sídel, 2004).

Chapter 9. Fighting Poverty and Environmental Injustice in Cities

1. Italo Calvino, *Invisible Cities*, trans. William Weaver (Orlando, FL: Harcourt, 1974); different types of inequality between countries in World Bank, *World Development Report 2006* (Washington, DC: Oxford University Press and World Bank, 2006).

2. Urban population in slums in developing countries from UN-HABITAT, *State of the World's Cities 2006/7* (London: Earthscan, 2006), pp. 16, 111; Mumbai and Nairobi from Gora Mboup, senior demographic and health expert, UN-HABITAT, Nairobi, e-mail to Molly Sheehan, 5 October 2006; Zuenir Ventura, *Cidade Partida* (São Paulo, Brazil: Companhia das Letras, 1994).

3. Janice E. Perlman, "Marginality: From Myth to Reality in the Favelas in Rio de Janeiro 1969–2002," in Ananya Roy and Nezar AlSayyad, eds., *Urban Informality: Transnational Perspectives from the Middle East, Latin America, and South Asia* (Lanham, MD: Lexington Books, 2004).

4. Glenn H. Beyer, ed., *The Urban Explosion in Latin America: A Continent in Process of Modernization* (Ithaca, NY: Cornell University Press, 1967); Peter Wilsher and Rosemary Righter, *The Exploding Cities* (London: A Deutsch, 1975); Franz Fanon, *The Wretched of the Earth* (London: MacGibbon and Kee, 1965); John F. C. Turner, *Uncontrolled Urban Settlement: Problems and Policies* (Pittsburgh, PA: University of Pittsburgh Press, 1966); squatters' interest in better opportunities for children from Janice Perlman, *The Myth of Marginality: Urban Poverty and Politics in Rio de Janeiro* (Berkeley: University of California Press, 1976), and from Joan M. Nelson, *Access to Power: Politics and The Urban Poor in Developing Nations* (Princeton, NJ: Princeton University Press, 1979).

5.   For Millennium Development Goals, see www.un.org/millenniumgoals; United Nations Population Division, *World Urbanization Prospects: The 2005 Revision* (New York: 2005).

6.   Janice Perlman, "Re-Democratization in Brazil: A View from Below, The Experience of Rio de Janeiro's Favelados 1968–2005," in Peter Kingstone and Timothy Power, eds., *Democratic Brazil Revisited* (Pittsburgh, PA: University of Pittsburgh Press, forthcoming); Kenya's tax on bicycles from VNG uitgeverij, *The Economic Significance of Cycling: A Study to Illustrate the Costs and Benefits of Cycling Policy* (The Hague: 2000), and from Jeffrey Maganya, Intermediate Technology Development Group, Nairobi, Kenya, discussion with Molly Sheehan, 8 May 2001.

7.   Daniel Kaufmann, Frannie Léautier, and Massimo Mastruzzi, "Globalization and Urban Performance," in Frannie Léautier, ed., *Cities in a Globalizing World: Governance, Performance & Sustainability* (Washington, DC: World Bank Institute, 2006), pp. 38–49; Robert Klitgaard, Ronald MacLean-Abaroa, and H. Lindsey Parris, *Corrupt Cities: A Practical Guide to Cure and Prevention* (Oakland, CA: Institute for Contemporary Studies, 2000), p. 32.

8.   Perlman, op. cit. note 6; T. Abed and Sanjeev Gupta, eds., *Governance, Corruption, and Economic Performance* (Washington, DC: International Monetary Fund (IMF), 2002); Vito Tanzi and Hamid Davoodi, *Corruption, Public Investment and Growth*, Working Paper 97/139 (Washington, DC: IMF, 1997); Sanjeev Gupta, Hamid Davoodi, and Rosa Alonso-Terme, *Does Corruption Affect Income Inequality and Poverty?* Working Paper 98/76 (Washington, DC: IMF, 1998); Ratih Hardjono and Stefanie Teggeman, eds., *The Poor Speak Up: 17 Stories of Corruption* (Jakarta: Partnership for Governance Reform, 2002).

9.   UN-HABITAT, *The State of the World's Cities 2004/2005* (London: Earthscan, 2004), pp. 134–57.

10.   Box 9–1 from the following: results of the original study from Perlman, op. cit. note 4; recent study from Janice Perlman, "The Chronic Poor in Rio de Janeiro: What has Changed in 30 Years?" in Marco Keiner et al., eds., *Manging Urban Futures: Sustainability and Urban Growth in Developing Countries* (Burlington, VT: Ashgate, 2005), pp. 165–85 (41 percent of the original study participants, or 307 of the original 750, were found; the 307 original interviewees were interviewed, along with a random sample of their children (367) and grandchildren (208)); Ignacio Cano et al., *O Impacto da Violência no Rio de Janeiro*, Working Paper (Rio de Janeiro: Universidade do Estado do Rio de Janeiro, 2004); levels of violence from Luke Dowdney, *Children of the Drug Trade* (Rio de Janeiro: Viveiros de Castro Editoria, 2003); police provoking violence from "Law-Enforcers on the Rampage; Brazil's Trigger-Happy Police," *The Economist*, 9 April 2005.

11.   UN-HABITAT, op. cit. note 9, pp. 134–57; François Bourguignon, "Crime, Violence, and Inequitable Development," paper prepared for the Annual World Bank Conference on Development Economics, Washington, DC, 28–30 April 1999.

12.   For traditional views of environmentalists and development specialists, see Eugene P. Odum, *Fundamentals of Ecology*, 3rd ed. (Philadelphia: Saunders, 1971), and Michael Lipton, *Why Poor People Stay Poor: Urban Bias in World Development* (London: Temple Smith, 1977); Janice Perlman and Bruce Schearer, "Migration and Population Distribution Trends and Policies and the Urban Future," International Conference on Population and the Urban Future, U.N. Fund for Population Activities, Barcelona, Spain, May 1986.

13.   Perlman and Schearer, op. cit. note 12.

14.   Ibid.

15.   United Nations Population Division, *World*

Population Policies 2005, at www.un.org/esa/popu lation/publications/WPP2005/Publication_index .htm.

16. United Nations High Commissioner for Human Rights, "Statement by Mr. Miloon Kothari, Special Rapporteur on adequate housing as a component of the right to an adequate standard of living, to the World Urban Forum III," Vancouver, 20 June 2006.

17. Martin Ravallion, *On the Urbanization of Poverty*, Development Research Group Working Paper (Washington, DC: World Bank, 2001); urban estimate comes from Michael Cohen, "Reframing Urban Assistance: Scale, Ambition, and Possibility," *Urban Update*, Comparative Urban Studies Brief, Woodrow Wilson International Center for Scholars, No. 5, February 2004, p. 1; total assistance from OECD/DAC figures in Worldwatch Institute, *Worldwatch Global Trends*, CD-ROM, July 2005; lack of urban housing programs from Daniel S. Coleman and Michael F. Shea, "Assessment of Bilateral and Multilateral Development Assistance and Housing Assistance in Latin America, Asia, Africa and the Middle East," Interim Working Draft for the International Housing Coalition, 3 May 2006.

18. Aid from OECD/DAC, and private capital from UNCTAD, both in Worldwatch Institute, op. cit. note 17.

19. David Satterthwaite, "Reducing Urban Poverty: Constraints on the Effectiveness of Aid Agencies and Development Banks and Some Suggestions for Change," *Environment and Urbanization*, April 2001, pp. 137–57.

20. Urban area expenditures from Frannie Léautier, World Bank Institute, e-mail to Molly Sheehan, July 2006; urban dimension missing from Diana Mitlin, *Understanding Urban Poverty: What the Poverty Reduction Strategy Papers Tell Us* (London: International Institute for Environment and Development (IIED), 2004).

21. William Alonso, "The Economics of Urban Size," *Papers in Regional Science*, December 1971, pp. 66–83; Rémy Prud'homme, "Anti-Urban Biases in the LDCs,'" Megacities International Conference, New York University, 1988; Rémy Prud'homme, "Managing Megacities," *Le courrier du CNRS*, No. 82, 1996, pp. 174–76.

22. Alfredo Sirkis, Director of Urbanism, Rio de Janeiro, discussion with Janice Perlman, 25 August 2005.

23. Satterthwaite, op. cit. note 19, p. 140.

24. Alan Altshuler and Marc Zegans, "Innovation and Creativity: Comparisons between Public Management and Private Enterprise," *Cities*, February 1990, pp. 16–24.

25. UN-HABITAT, Global Urban Observatory, at hq/unhabitat.org/programmes/guo.

26. Information in this section on new federations of the urban poor provided by David Satterthwaite of IIED, 9 September 2006. See also David Satterthwaite, "Meeting the MDGs in Urban Areas: The Forgotten Role of Local Organizations," *Journal of International Affairs*, March 2005, pp. 87–113; Sheela Patel, Sundar Burra, and Celine D'Cruz, "Slum/Shack Dwellers International (SDI)–Foundations to Treetops," *Environment and Urbanization*, October 2001, pp. 45–59; and www.sdinet.org.

27. Sheela Patel, "Partnerships with the Urban Poor: The Indian Experience," *UN Chronicle*, March-May 2001, pp. 47–49.

28. See Robert Neuwirth, "Bricks, Mortar and Mobilization," *Ford Foundation Report*, springsummer 2005, pp. 13–18.

29. Patel, Burra, and D'Cruz, op. cit. note 26.

30. Ibid.

31. Solly Angel and Somsook Boonyabancha, "Land Sharing as an Alternative to Eviction," *Third World Planning Review*, vol. 10, no. 2 (1988).

32. Teena Amrit Gill, "Slum Communities Claim a Stake in Their Community's Future," *Ashoka*

Changemakers Journal, January 2002; Somsook Boonyabancha, *A Decade of Change: From the Urban Community Development Office (UCDO) to the Community Organizations Development Institute (CODI) in Thailand: Increasing Community Options through a National Government Development Programme*, Working Paper 12 on Poverty Reduction in Urban Areas (London: IIED, 2003).

33. Somsook Boonyabancha, "Baan Mankong: Going to Scale with Slum and Squatter Upgrading," *Environment and Urbanization*, April 2005; Satterthwaite, "Meeting the MDGs in Urban Areas," op. cit. note 26.

34. UN-HABITAT, op. cit. note 2.

35. Yves Cabannes, "Participatory Budgeting: A Significant Contribution to Participatory Democracy," *Environment and Urbanization*, April 2004, pp. 27–46.

36. Celina Souza, "Participatory Budgeting in Brazilian Cities: Limits and Possibilities in Building Democratic Institutions," *Environment and Urbanization*, April 2001, pp. 159–84; see Rebecca Abers, *Inventing Local Democracy: Grassroots Politics in Brazil* (Boulder, CO: Lynne Rienner Publishers, 2000) for a description of the "negotiated solidarity" that emerged from this process. Box 9–2 is based on Yves Cabannes, "Les Budgets Participatifs en Amérique Latine. De Porto Alegre à l'Amérique Centrale, en Passant par la Zone Andine: Tendances, Défis et Limites," *Mouvements*, September-December 2006 (with thanks to Alexandra Celestin for editing the English version).

37. Mona Serageldin et al., "Assessment of Participatory Budgeting in Brazil," prepared for the InterAmerican Development Bank, 2002.

38. Yves Cabannes, University College London, e-mail to Molly Sheehan, 11 September 2006.

39. Chase Bank and Roper Starch, *Global Leaders Survey* (New York: 1997).

40. Mega-Cities Project, at www.megacitiesproject.org.

41. Rachel Leven, "The Pharaoh's Garbage: Growth and Change in Egypt's Waste Management System," Tufts University NIMEP Insights Volume II, spring 2006; Wendy Walker, *The Torah Zabbaleen: From Tin Shacks to High Rises* (Cairo: Association for the Protection of the Environment, 2005).

42. Eugenio M. Gonzales, *From Wastes to Assets: The Scavengers of Payatas*, International Conference on Natural Assets Conference Paper Series No. 7, December 2003; Asian Development Bank, *The Garbage Book: Solid Waste Management* (Manila: 2004); Ronnie E. Calumpita, "Corruption Hinders Waste Management," *Manila Times*, 24 August 2004.

43. Wael Salah Fahmi, "The Impact of Privatization of Solid Waste Management on the Zabbaleen Garbage Collectors of Cairo," *Environment and Urbanization*, October 2005, pp. 155–70; Jack Epstein, "From Cairo's Trash, A Model of Recycling/Old Door-to-door Method Boasts 85 Percent Reuse Rate," *San Francisco Chronicle*, 3 June 2006; Mona Serageldin, Harvard University, discussion with Molly Sheehan, July 2006.

44. Marlene Fernandes, *Reforestation in Rio's Favelas*, Environmental Justice, Mega-Cities Project, 1998.

45. Inter-American Development Bank, *The Socio-Economic Impact of Favela-Bairro: What do the Data Say?* Working Paper (Washington, DC: 2005); Jorge Fiori, Liz Riley, and Ronald Ramirez, "Urban Poverty Alleviation Through Environmental Upgrading in Rio de Janeiro: Favela Bairro," draft research report, Development Planning Unit, University College London, March 2005; Metropolis, *Metropolis 2005 Standing Commission Report* (Barcelona: 2005); numbers in 2006 from John Fiori, director, housing and urbanism program, Architectural Association Graduate School, London, e-mail to Kenro Kawarazaki, Worldwatch Institute, 6 August 2006.

46. Sonia Rocha, *Workfare Programmes in Brazil: An Evaluation of Their Performance* (Geneva:

International Labour Office, 2001); Krista Lillemets, *Exploring Participation: Waste Management Cases in Two Favelas of Rio de Janeiro*, thesis submitted in partial fulfillment of the requirements of the Degree of Master of Science at Lund University, Lund, Sweden, November 2003.

47. Hugh Schwartz, *Urban Renewal, Municipal Revitalization: The Case of Curitiba, Brazil* (Falls Church, VA: Higher Education Publications, Inc., 2006); Robert Cervero, *The Transit Metropolis: A Global Inquiry* (Washington, DC: Island Press, 1998), pp. 265–96.

48. Michael Specter, "Environmental Rules, How They Dictate Region's Agenda," *New York Times*, 25 November 1991.

49. Dennis Hevesi, "Test Runs for Futuristic Bus-Tube System," *New York Times*, 21 April 1992.

50. For good evaluations, see Michael Eng, "A Great New Ride," *Newsday*, 10 June 1992; constraints from Molly Sheehan's interviews with Al Appleton, former Commissioner of Environmental Protection, New York City, with Gene Russianoff, New York Public Interest Research Group, and with Robert Newhouser, New York City Transit, all in October 2006.

51. Thomas J. Lueck, "A Plan That Means to Put More Rapid in the City's Transit," *New York Times*, 8 June 2006; Robert Paaswell, Albert Appleton, and Todd Goldman, *Next Stop, Bus Rapid Transit: Accelerating New York's Bus System into A New Century* (New York: Institute for Urban Systems, City University of New York, 2004).

52. World Bank, *Assessing Aid* (Washington, DC: 1998); World Bank, *The Role and Effectiveness of Development Assistance* (Washington, DC: 2002); Paul Wolfowitz, President, World Bank Group, Address to Board of Governors of the World Bank Group, Singapore, 19 September 2006.

53. Klitgaard, MacLean-Abaroa, and Parris, op. cit. note 7; Winthrop Carty, Ash Institute for Democratic Governance and Innovation, Kennedy School of Government, Harvard University, "Citizen's Charters: A Comparative Global Survey," translation from Spanish of *Cartas Compromiso: Experiencias Internacionales*, presented at the launch of the Mexican Citizen's Charter Initiative, June 2004.

54. Philip Amis, "Municipal Government, Urban Economic Growth, and Poverty Reduction—Identifying the Transmission Mechanisms Between Growth and Poverty," in Carole Rakodi with Tony Lloyd-Jones, eds., *Urban Livelihoods: A People-Centered Approach to Reducing Poverty* (London: Earthscan, 2002), pp. 97–111; Perlman, op. cit. note 10; Janice Perlman, "Violence as a Major Source of Vulnerability in Rio de Janeiro's Favelas," *Journal of Contingencies and Crisis Management*, winter 2005.

55. A. Zaidi, "Assessing the Impact of a Microfinance Programme: Orangi Pilot Project, Karachi, Pakistan," in S. Coupe, L. Stevens, and D. Mitlin, eds., *Confronting the Crisis in Urban Poverty: Making Integrated Approaches Work* (Rugby, U.K.: Intermediate Technology Publications Ltd., 2006), pp. 171–88; Franck Daphnis and Bruce Fergus, eds., *Housing Microfinance: A Guide to Practice* (Bloomfield, CT: Kumarian Press, Inc, 2004).

56. C. K. Prahalad, *The Fortune at the Bottom of the Pyramid* (Upper Saddle River, NJ: Pearson, 2006); David L. Painter, TCG International, in collaboration with Regina Campa Sole and Lauren Moser, ShoreBank International, "Scaling Up Slum Improvement: Engaging Slum Dwellers and the Private Sector to Finance a Better Future," paper presented at the World Urban Forum, Vancouver, June 2006.

57. UN Millennium Project Task Force on Improving the Lives of Slum Dwellers, *A Home in the City* (London: Earthscan, 2005), pp. 55–56; Jane Tournée and Wilma van Esch, *Community Contracts in Urban Infrastructure Works* (Geneva: International Labour Organization, 2001).

58. Bénédicte de la Brière and Laura B. Rawlings, *Examining Conditional Cash Transfer Programs: A Role for Increased Social Inclusion*, Social Protection Discussion Paper No. 06083 (Washington, DC: World Bank, 2006); SYDGM and World Bank, the 3rd Internatioanl Conference on Conditional Cash Transfer, Istanbul, June 2006, at info.worldbank.org/etools/icct06/welcome.asp; James Traub, "Pay for Good Behavior?" *New York Times Magazine*, 8 October 2006, pp. 15–16.

59. Janice Perlman, "Megacities and Innovative Technologies," *Cities*, May 1987, pp. 128–36; William McDonough, "China as a Green Lab," in Howard Gardner et al., "The HBR List: Breakthrough Ideas for 2006," *Harvard Business Review*, February 2006, p. 35; Mara Hvistendahl, "Green Dawn: In China, Sustainable Cities Rise by Fiat," *Harpers*, February 2006, pp. 52–54; Box 9–3 from Annie Sugrue, EcoCity Trust, Johannesburg, August 2006.

60. Box 9–4 from Jorge Wilheim, São Paulo, Brazil, September 2006.

61. Leonie Sandercock, *Cosmopolis II : Mongrel Cities in the 21st Century* (London: Continuum, 2003).

62. Small Arm Survey, at www.smallarmssurvey.org; International Action Network on Small Arms, *Reviewing Action on Small Arms* (London: 2006), pp. 17–22.

63. Satterthwaite, op. cit. note 19, pp. 146, 148; Göran Tannerfeldt and Per Ljung, SIDA, *More Urban, Less Poor: An Introduction to Urban Development and Management* (London: Earthscan, 2006); Esquel Foundation, at www.synergos.org/latinamerica/ecuador.htm.

64. William Easterly, *The White Man's Burden: Why the West's Efforts to Aid the Rest Have Done So Much Ill and So Little Good* (New York: Penguin Press, 2006), p. 378.

65. Urban Sustainability Initiative, at bie.berke ley.edu/usi.

66. UN Millennium Project Task Force, op. cit. note 57, p. 101; Yves Cabannes, "Children and Young People Build Participatory Democracy in Latin American Cities," *Environment and Urban-*ization, April 2006, pp. 195–218; Sheridan Bartlett, "Integrating Children's Rights into Municipal Action: A Review of Progress and Lessons Learned," *Children, Youth, and Environments*, vol. 15, no. 2 (2005), pp. 18–40.

67. "YA! Youth Activism," *NACLA Report on the Americas*, May/June 2004, p. 48.

68. Rose Molokoane, presentation at Future of the Cities panel, World Urban Forum, Vancouver, 23 June 2006.

69. Watson quoted in Ernie Stringer, *Action Research* (Thousand Oaks, CA: Sage Publications, 1999).

Index

Worldwatch Papers

On Climate Change, Energy, and Materials

169: Mainstreaming Renewable Energy in the 21st Century, 2004
160: Reading the Weathervane: Climate Policy From Rio to Johannesburg, 2002
157: Hydrogen Futures: Toward a Sustainable Energy System, 2001
151: Micropower: The Next Electrical Era, 2000
149: Paper Cuts: Recovering the Paper Landscape, 1999
144: Mind Over Matter: Recasting the Role of Materials in Our Lives, 1998
138: Rising Sun, Gathering Winds: Policies To Stabilize the Climate and Strengthen Economies, 1997

On Ecological and Human Health

165: Winged Messengers: The Decline of Birds, 2003
153: Why Poison Ourselves: A Precautionary Approach to Synthetic Chemicals, 2000
148: Nature's Cornucopia: Our Stakes in Plant Diversity, 1999
145: Safeguarding the Health of Oceans, 1999
142: Rocking the Boat: Conserving Fisheries and Protecting Jobs, 1998
141: Losing Strands in the Web of Life: Vertebrate Declines and the Conservation of Biological Diversity, 1998
140: Taking a Stand: Cultivating a New Relationship With the World's Forests, 1998

On Economics, Institutions, and Security

168: Venture Capitalism for a Tropical Forest: Cocoa in the Mata Atlântica, 2003
167: Sustainable Development for the Second World: Ukraine and the Nations in Transition, 2003
166: Purchasing Power: Harnessing Institutional Procurement for People and the Planet, 2003
164: Invoking the Spirit: Religion and Spirituality in the Quest for a Sustainable World, 2002
162: The Anatomy of Resource Wars, 2002
159: Traveling Light: New Paths for International Tourism, 2001
158: Unnatural Disasters, 2001

On Food, Water, Population, and Urbanization

172: Catch of the Day: Choosing Seafood for Healthier Oceans, 2006
171: Happier Meals: Rethinking the Global Meat Industry, 2005
170: Liquid Assets: The Critical Need to Safeguard Freshwater Ecosytems, 2005
163: Home Grown: The Case for Local Food in a Global Market, 2002
161: Correcting Gender Myopia: Gender Equity, Women's Welfare, and the Environment, 2002
156: City Limits: Putting the Brakes on Sprawl, 2001
154: Deep Trouble: The Hidden Threat of Groundwater Pollution, 2000
150: Underfed and Overfed: The Global Epidemic of Malnutrition, 2000
147: Reinventing Cities for People and the Planet, 1999

To see our complete list of Papers, visit www.worldwatch.org/taxonomy/term/40

Price of each Paper is $9.95 plus S&H

AMERICAN ENERGY
The Renewable Path to Energy Security

The Most Comprehensive Overview
of Renewable Energy Available!

CONTENTS:

▶▶ 21st Century Energy

▶▶ Vision for a More Secure and Prosperous America

▶▶ Building a New Energy Economy

▶▶ A Cleaner, Healthier America

▶▶ Resources and Technologies

▶▶ American Energy Policy Agenda

Get Your Copy of this Landmark Report Today!

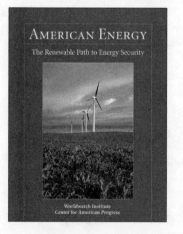

The report is downloadable for free at www.americanenergynow.org

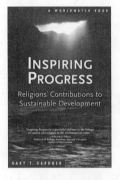

Vital Signs Online

With Vital Signs Online, Worldwatch presents its newest compilation of significant global trends in 10 categories—with written analysis, Excel charts and worksheets, and PowerPoint slides to illustrate each trend.

Categories include: Food, Agricultural Resources, Communications and Transportation, Energy and Climate, Environment, Global Economy, Resource Economics, Health and Disease, Population and Society, War and Conflict.

You can now get written analysis in addition to Excel worksheets and PowerPoint slides—and each trend is downloadable for just $1.00.

▶▶ 114 Trends with Analysis in PDF format. Each trend includes analysis to provide context and interconnections, plus charts, graphs, and tables to provide visual comparisons of the trend over time.

▶▶ 172 Trends with Excel worksheets and PowerPoint slides—easy to customize for your research and presentation needs.

▶▶ Bundled options for purchasing all trends in a topic area, or for purchasing specific trend analysis with all included data files.

Download your choice of indicators at
www.worldwatch.org/vsonline

▶▶ Only $1.00 per download!

From reviews of earlier editions of *State of the World*:

"The most comprehensive, up-to-date, and accessible summaries...on the global environment."
   —*E.O. Wilson, Pulitzer Prize winner*

"The environmental movement as we know it today could not exist without the extraordinary researchers at Worldwatch."
   —*Bill McKibben,*
      best-selling author of **The End of Nature**

"Worldwatch Institute's *State of the World* report helps point the way to a more sustainable future."
   —*E-The Environmental Magazine*

"Top-rated annual publication on sustainable development."
   —*GlobeScan survey of Sustainability Experts*

"An authoritative publication..."
   —*International Herald Tribune*